WORLD INDUSTRY STUDIES 3

Blue Gold:

The Political Economy of Natural Gas

WORLD INDUSTRY STUDIES

Edited by Professor Ingo Walter,
Graduate School of Business Administration,
New York University

1 COLLAPSE AND SURVIVAL: INDUSTRY STRATEGIES IN A CHANGING WORLD
Robert H. Ballance and Stuart W. Sinclair

2 THE GLOBAL TEXTILE INDUSTRY
Brian Toyne, Jeffrey S. Arpan, Andy H. Barnett, David A. Ricks and Terence A. Shimp

3 BLUE GOLD: THE POLITICAL ECONOMY OF NATURAL GAS
J. D. Davis

Forthcoming

THE CHEMICAL INDUSTRY IN THE WORLD ECONOMY
Thomas J. Ilgen

POLITICS VS ECONOMICS IN THE WORLD STEEL TRADE
Kent Jones

MICROELECTRONICS: BUSINESS STRATEGY, GOVERNMENT POLICY AND INTERNATIONAL COMPETITION
Thomas A. Pugel

THE WORLD MINING INDUSTRY: INVESTMENT, STRATEGY AND PUBLIC POLICY
Raymond Mikesell

Blue Gold:
The Political Economy of Natural Gas

J. D. Davis

Professor, Institute of Economics and Planning,
Roskilde University, Denmark

Springer Science+Business Media, B.V.

First published in 1984

British Library Cataloguing in Publication Data

Davis, Jerome D.
Blue gold.—(World industry studies; 3)
1. Gas industry
I. Title II. Series
338.4'76657 HD9581.A2
ISBN 978-94-015-6835-7 ISBN 978-94-015-6833-3 (eBook)
DOI 10.1007/978-94-015-6833-3

Library of Congress Cataloging in Publication Data

Davis, Jerome D.
Blue gold.
(World industry studies; 3)
Bibliography: p.
Includes index.
1. Gas industry. 2. Gas industry—Political aspects.
I. Title. II. Series.
HD9581.A2D38 1984 338.2'7285 84-6350

Set in 10 on 11 Times by Phoenix Photosetting, Chatham
and printed in Great Britain by Billing and Sons Ltd
London, Worcester

Contents

List of Abbreviations

AGA	American Gas Association
AGIP	Azienda Nazionale Generale Italiani
ANGTS	Alaskan Natural Gas Transportation System
BGC	British Gas Corporation
BNOC	British National Oil Corporation
BP	British Petroleum
CeFeM	Compagnie Française du Méthane
CFP	Compagnie Française des Pétroles
DETG	Deutsche Erdgastransport GmbH
DONG	Danish Oil and Natural Gas Company
DSM	Dutch State Mines
ELSAM	Jysk-fynske elsamarbejde
ENI	Ente Nazionale Idrocarburi
ERA	Economic Regulatory Commission
ERAP	Entreprise Recherches des Pétroles
EWE	Energieversorgung Weser Ems
FERC	Federal Energy Regulatory Commission
FPC	Federal Power Commission
FTC	Federal Trade Commission
GC	Gas Council (UK)
GdF	Gaz de France
IEA	International Energy Agency
IGU	International Gas Union
INTECS	International Economics
IRR	internal rate of return
LNG	liquefied natural gas
LPG	liquid petroleum gas
MEGAL	Mittel-Europäische Gasleitungsgesellschaft
MFN	most favoured nation
NAM	Nederlandse Aardolie Maatschappij
NGPA	Natural Gas Policy Act (US)
NPV	net present value
OAPEC	Organization of Arab Petroleum Exporting Countries
OECD	Organization for Economic Co-operation and Development
OGEC	Organization of Gas Exporting Countries
ÖMV	Österreichische Mineralölverwaltung Aktiengesellschaft

OPEC	Organization of Petroleum Exporting Countries
RAP	Régie Autonome des Pétroles
RWE	Rheinisch-Westfälisches Elektrizitätswerk
SAGAPE	Société d'Achat de Gaz Algérien pour l'Europe
SEGEO	Société Européenne de Gazoduc Est-Ouest
SNAM	Società Nazionale Metanodotti
SNEA	Société Nationale Elf-Aquitaine
SNEA(P)	Société Nationale Elf-Aquitaine (Production)
SNGSO	Société Nationale des Gaz du Sud-Ouest
SNPA	Société Nationale des Pétroles d'Aquitaine
TAG	Trans-Austrian Gasline
TENP	Trans-European Natural Gas Pipeline
UNECE	United Nations Economic Commission for Europe
VALHYD	Valorization Hydrocarbon Development Plan (Algeria)
VEGIN	national association of gas boards (Netherlands)
VEW	Vereinigtes Elektrizitätswerk Westfalen
WAG	Western Austrian Gasline

Foreword

Few industries have gone through as dislocating a series of economic and political pressures over the past decade as has the international gas business. The 1973 oil crisis, followed by an even more severe shock in 1979, fundamentally altered the economics of energy. Scarcity seemed here to stay; even if future market adjustments in supply and demand were to bring substantial relief to energy users, one could never again be sure of avoiding yet another repeat performance. As in all painful experiences, an important lesson was learned.

With a wide variety of energy sources having significant substitutability, the oil shock triggered an equally wide array of responses. Research on breeder reactors accelerated, and the construction of nuclear power reactors moved ahead rapidly. Wind, solar, geothermal and other 'clean' sources of power became the vogue. Oil shale and tar sands were opened up at enormous cost. National strategic petroleum reserves were established. Wood-burning stoves became *de rigeur* in many an affluent community. Remarkable progress was made in reducing energy use in residential, industrial and transportation applications. So successful was the sum total of these efforts around the world – though not without substantial costs – that the shortage gave way within less than a decade to a substantial oil glut, and many a well-intentioned scheme once again became uneconomic. The lesson had none the less been learned, and the link between economic activity and energy use had certainly been made far more elastic.

Gas as an energy source has been fully a part of these developments over the years. For political reasons, governments in many countries initially kept gas prices low, discouraging exploration activity and new production and encouraging conversions to gas; only later did they yield to market pressures via decontrol. Countries scrambled for assured gas supplies, ranging from the North Sea and the Alaska North Slope to the Soviet Union, North Africa and Australia. Flare gas and reinjection in the Middle East gradually gave way to downstream uses, particularly in the form of large-scale petrochemical facilities. Major technological as well as financing breakthroughs made it possible to transport LNG and pipeline gas over long distances at competitive rates.

Blue Gold is the story of the international gas industry: the external economic and political pressures it has had to endure over the years; the competitive structure, conduct and performance of firms in the industry; the role of government management and mismanagement; international competition and trade. Much has been written along these lines in relation to the world's petroleum industry – little attention has been given to gas. The present volume does much to remedy this gap and, like other volumes in this series, offers a balanced and definitive assessment of an industry that has taken on truly global scope and significance.

INGO WALTER
New York University

Acknowledgements

I should like to acknowledge the financial support of the Danish Institute of International Affairs. International Economics (INTECS) and the Danish Oil and Natural Gas Company provided support indirectly through consulting assignments that enabled me to learn some of the fundamentals of industrial contractual and negotiating relationships. I am similarly indebted to the Institute of Political Science, Aarhus University, Denmark, and its staff, particularly Anne Marie Christensen and Birgit Sommer who took time off from other obligations to type this manuscript. Lee Niedringhaus Davis, my wife (herself the author of a book on liquefied natural gas), provided me with encouragement, editorial insight and advice without which this book would never have been undertaken or finished. Edith Penrose, Melvin Conant and Peter Odell have each in their own way provided me with valuable counsel. The London libraries of the Institute of Gas Engineers and the British Gas Corporation have provided useful services, as has the natural gas industry in general. Finally, I am indebted to my editor, Nicholas Brealey, for his patient forbearance in the months when the process of committing words to paper seemed next to impossible.

Acknowledgements are due E. N. Tiratsoo, for permission to reprint his tables in this text (Tables 2.1 and 2.3); Lauri Karvonen and *Cooperation and Conflict* for excerpts from 'The Political Economy of Natural Gas Markets'; Professor P. R. Odell for permission to adapt his material into Figure 2.7; Jensen Associates for permission to reprint their material as Figures 4.1 and 11.1, Shell International Gas and Malcolm Peebles for Figure 7.1 and portions of the glossary; Philippe Kahn and the Centre de Recherche sur le Droit des Marchés et les Investissements Internationaux (CREDIMI) for the two gas contract texts in Appendix IV, the Controller of Her Britannic Majesty's Stationery Office for Table 5.1, Ruhrgas for permission to reproduce their most recent European map as Figure 8.1, and Nederlandes Gasunie for reproduction rights to Figure 7.2, *Oil and Gas Journal* for rights to Figure 6.1. Special thanks are due Jonathan Stern for the material reproduced in Table 8.12 with apologies for early morning telephone conversations.

Note on Units of Measurement and Abbreviations

The international gas industry uses a wide variety of unitary measures in its statistical tables, and it is important to make clear from the first how they are to be used in this book. Energy measures – megajoules, millions of tons of coal or oil equivalent, Teracals, thermies – can be confusing to the general reader. In so far as possible, I have utilized million British thermal units (10 therms) when discussing natural gas pricing, but have utilized coal and oil equivalent measures when price accuracy is less important.

Volumetric measures can be even more confusing. Here, I have opted for cubic meters as the prevalent measure, although cubic feet are also utilized. Conversion between the two volumes varies according to the temperature, pressure, amount of vapour and other conditions. Unfortunately, units where these factors are explicit – 'normal' cubic meters, standard cubic meters, and standard cubic feet – are often used inconsistently or are not specified in the gas literature on which this volume is based. Thus, unless otherwise specified, I have cited cubic meter figures and used the strict volumetric relationship between cubic feet and cubic meters (35.315 cubic feet to the cubic meter, 0.0283 cubic meters to the cubic foot). The more precise relationships between these volumetric measures and between these measures and energy measures are expressed in Appendix I.

A final note should be made of the use of the term 'billion'. Throughout this volume, 'billion' is 1,000 million, 'trillion' is 1,000,000 million.

Below is a brief list of the more commonly used abbreviations in this book:

bbl	barrel
Btu	British thermal unit
ft^3	cubic foot
inHg	inches of mercury
kcal	kilocalorie
kWh	kilowatt hour
lb/in^2	pound per square inch
lbf/in^2	pound-foot per square inch
m^3	cubic meter

m^3(st)	standard (SI) cubic meter
MMBtu	million British thermal units
mmHg	millimeters of mercury
mtoe	million tons oil equivalent
MW	megawatt
Nm^3	'normal' cubic meter
tce	ton coal equivalent

1

Introduction

The 'bird's-eye view'

The enormous natural gas flares that mark the Middle Eastern and North African oil fields can be clearly observed at night by astronauts hundreds of miles above the earth's surface. Given their altitude, these same astronauts might also see the bright flares of the giant Siberian oil field Samotlar. Next to this field, beneath the permafrost, lie the richest and most extensive natural gas reserves discovered to date. The gas in these fields is worth so much that the Russians have taken to calling it 'blue gold' (referring to the colour of the flame), the title of this book.

Probably unnoted by our astronauts are the means of transport for the natural gas. The first of these are the enormous pipelines, which utilize the finest of steel technology, running for thousands of miles between natural gas resources on the one hand and natural gas markets on the other. The cost of such pipelines is astronomical. For the currently discussed Soviet–European natural gas deal, the estimated cost of the pipeline is put at $10 billion – and this does not include the cost of installing the line, expropriating property, and the like. The second means of transport is liquefied natural gas (LNG) – natural gas cooled to an incredible −162° C., liquefied, and reduced in volume by a factor of 600 – which is carried in huge supertankers that criss-cross the world's oceans. These carriers, costing upwards of one-quarter billion dollars apiece, are technologically as complicated (although in cryogenic terms) as the space capsule containing our astronauts. (The costs of an LNG project are exorbitant as well; a project envisaging the delivery of Algerian LNG to the US could amount to about $5 billion including interest.)

As visible to the astronauts' 'bird's-eye view' as the night-time flares in the Middle East are the locations of the world's metropolitan centres: the Boston–New York–Philadelphia–Baltimore–Washington corridor, London and Southeastern England, the Ruhr valley, the

Tokyo–Yokohama–Nagoya–Osaka corridor – all pulses of energy easily registered from outer space, and, in addition, all major markets for natural gas. Not only must natural gas be delivered to these markets from distant sources, it must also be delivered at the appropriate quantities and prices. Seasonal and daily variations must be accounted for, as well as variations in demand stemming from economic prosperity or recession. To these ends, investments in distribution networks, storage facilities and gas appliances must be sufficiently low to guarantee a market rate of return to the distributor after he has paid for the natural gas itself, and yet at the same time provide the consumer with a guaranteed and reliable source of energy. The distribution and marketing of natural gas are expensive and exacting – a business in which small miscalculations can run to hundreds of millions of dollars.

Curiously, whereas the politics of the Organization of Petroleum Exporting Countries (OPEC) and the oil multinationals catch headlines worldwide, the no less heady mixture of money, politics and power characteristic of natural gas deals is broadly ignored. This is remarkable not only in that natural gas is second only to oil in the energy consumption of most advanced industrialized nations (and its market share relative to oil products is increasing), but also in that a prominent role as the fuel of the future is increasingly forecast for it.

Popular unawareness of the political economic issues behind natural gas is none the less explicable. Unlike the history of oil production, the natural gas industry has no one enormous trust (such as the Standard Oil trust) to offer; no colourful personalities like a Gulbenkian, a Rockefeller, a Deterding; and few if any Horatio Alger rags-to-riches stories. The triumph of the natural gas industry, even more than that of the oil industry, is one of organization. Yet to characterize the natural gas industry in this way is not to characterize those in it as grey bureaucrats. Senior gas executives tend instead to resemble tough battle-scarred veterans, one result of an industry 'rough and tumble' that rarely, if ever, makes the front page. The pressures of negotiating and signing a twenty-five-year contract covering billions of dollars worth of natural gas deliveries are not for the faint-hearted and are a certain means to premature old age.

The gap in public awareness also extends to the main issues surrounding this fuel: how it is produced, transported and marketed, and how the industry is organized in political economic terms. This book aims to fill this gap. Of necessity it is fairly general in its orientation. It will not dwell on every finesse in natural gas negotiation strategy; likewise it may ignore the expanding market for LNG in a country like Korea or the actual organization of the

National Energy Board in Canada. Its purpose is to explain 'who gets what, how' in the world of natural gas.

The growth of natural gas markets

Why is it that one culture seizes an opportunity to improve its lot while other cultures, presented with the same opportunity, either misuse it or ignore it altogether? With regard to natural gas, the first historical mention of its existence is possibly that of Moses encountering the 'mysterious burning bush' in the Sinai. Natural venting of gas deposits close to the surface elsewhere in the Middle East, in Baku and in Persia has led to similar religious cults. Thousands of miles distant, however, in China, at a time roughly contemporaneous with the Athenian Empire, natural gas was commercially applied. The Chinese, we are informed, not only used natural gas as a fuel, but they also developed gas distribution lines utilizing bamboo pipes. (We are not informed about pipeline pressures and other hazards of this first-recorded practical use of natural gas.)

Much as different cultures in the year 500 BC utilized natural gas for different purposes, so modern states, the embodiments of widely varying business cultures and industrial policies, have attempted to solve the problems of natural gas production, transmission and distribution in widely diverse and distinctively national manners. It is interesting that, given the same industrial parameters, nation states can have such varied patterns of behaviour.

One telling common experience, nevertheless, is that worldwide use of natural gas as a fuel is of relatively recent date. In the United States, Great Britain, Germany, France, Holland, Japan, Tsarist Russia (and the early Soviet Union), the first gas industry was based not on natural gas but on 'town gas', a fuel made by the carbonization of coal. This town gas, which often had an energy content of just 40 per cent or so of that of natural gas, was either a joint product of the coke works or, more often, an intended product of a specific carbonization process. In the nineteenth century, town gas was far more convenient than coal or oil, relieving the consumer of the unpleasantness of tussling with brown coal or using whale oil. Town gas (or, more accurately, manufactured gas) was introduced to the world capitals in the period 1812–25, and enjoyed considerable growth, particularly in street and domestic lighting, until the rise of the electrical power industry threatened its decline. Manufactured gas survived this threat as it continued to be economic in heating and cooking, and its byproduct coal tar had become the prime organic chemical feedstock. It remained a more or less prominent source of energy until the post-World War II period when oil and oil products threatened to bury the gas industry entirely (oil additionally replacing coal tar as 'petrochemical' feedstock). Town gas is still utilized today, as in the German Ruhr district.

That the gas industry did not die is due entirely to the introduction of gas that occurs naturally in geological reservoirs beneath the earth's surface. Although natural gas had been used commercially in the United States, Russia and Japan as early as the nineteenth century, this use was intermittent and transient. Typically, an entrepreneur in a small town would discover a natural gas reservoir, attach low-pressure distribution pipes to wells drilled in this field (if indeed drilling was necessary), and commence a very profitable, if short-run, business. The reservoir (typically tiny in comparison with the enormous finds made later in Texas and Oklahoma) would become depleted, and the 'boom' would be followed by a no less impressive 'bust' (Peebles, 1980, pp. 53–4). The solution to this problem was soon obvious: to discover additional gas reserves to tap when the nearest field became depleted. Unfortunately, this required the long-distance transmission of this gas, and the technology of the nineteenth century was not up to the task of designing pipelines that could withstand the pressures involved. It was not until the 1920s that American engineers developed the all-welded steel pipe capable of withstanding such pressures. The first long-distance all-welded pipeline was laid from northern Louisiana to Beaumont, Texas, in 1925. This line, some 14–18 in. in diameter and running for 217 miles, was as significant for the beginnings of the natural gas industry as was the first oil well drilled some seventy years previously for the oil industry. Even so, despite the existence of adequate pipeline technology, natural gas did not really 'take off' outside the United States until the 1960s.

Table 1.1 summarizes the introduction of town gas and natural gas to the major world economies. As can be seen, if one excludes purely local use of natural gas (in Niigata, Japan, and Russian Baku), well over 100 years separate the initiation of manufactured gas and the start of natural gas. The story of the fuel that is the subject of the book really commences in 1925 in the United States and twenty-five years later elsewhere in the world.

The political economy of natural gas

Critical to the viability of natural gas markets is their organization, an organization sought both by the parties directly involved and by the

Table 1.1 *Introduction of town gas and natural gas into major national economies*

Nation	*First recorded major use of town gas*	*First recorded major use of natural gas*
United States	1816	1821[a] 1925[b]
United Kingdom	1812	1964[c] 1967[d]
Germany	1825	1959
Netherlands	1820	1951
Japan	1857	1907[e] 1969[f]
Russia/USSR	ca. 1850	1871[g]
France	1819	1939

Source: Peebles, 1980, *passim*; Tiratsoo, 1979.

Notes: a First local use in Fredonia, New York.

b Development of all-welded high-tensile pipeline initiates long-distance transmission.

c Imports of LNG into Canvey Island.

d Landing of first North Sea gas in UK.

e Local distribution in Niigata.

f Commencement of LNG imports to Japan.

g It is assumed that the first practical use of natural gas in Russia occurred shortly after the discovery of oil at Baku.

nations in which they operate. Without long-term contractual relations covering the various phases of a natural gas deal (typically production, transmission and distribution) and without some sort of 'constructive' government interference in natural gas markets, trade in natural gas would be notoriously unstable, even volatile – if indeed it existed at all. Why is this?

Mention has been made of the billions of dollars involved in Siberian transmission lines and in natural gas contracts. These billions are intimately linked: in the absence of a contractually guaranteed return on the money invested, would the billions for this investment actually be forthcoming? Let us take two examples. The Alaskan Natural Gas Transportation System (ANGTS) is perhaps the most expensive undertaking ever in the history of the natural gas industry. Currently estimated to cost some $45 billion (including interest), the project is, as of this writing, stalled. One of the major reasons for the lack of progress is that the companies sponsoring the project want governmental assurance that they can dispose of the natural gas at a sufficiently high price to cover the costs of the mammoth system. Our second example is from the past: the development of the British offshore industry some fifteen years ago.

This effort required a commitment by the oil industry of hundreds of millions of pounds in offshore transmission lines and platforms. As is known, the British offshore gas resources are now highly developed. But would the oil companies have pushed so hard if they had not been virtually guaranteed a reasonable return on their investment today?

The interesting thing about these two examples is that they underline the need for future market stability if the large investments necessary to produce and transmit natural gas are to be undertaken. A free market, with natural gas prices rising and falling steeply from one year to the next, was anathema to the oil firms in 1968; it is anathema today. The establishment of a long-term contractual relationship insures them against the 'instability' of the free market.

Yet contractual relationships do more than this. Natural gas is a specialized business and is therefore generally subdivided into three specialized phases – production, transmission, and distribution – each normally undertaken by a separate corporation. Contractual relationships, often with a healthy or unhealthy dose of governmental regulation, serve to ensure that the future partners necessary to the success of the respective projects do not abdicate their responsibility. This is doubly important because the industrial dynamics of the three respective stages differ significantly from each other, a point to which we shall return in Chapter 2.

Finally, the contractual relationships set out the criteria for terms of gas delivery. To take a not uncommon example, on what criteria should the price of the natural gas be based: the average cost of the natural gas produced? the marginal cost of the natural gas produced? the price of an alternative fuel – and, if so, which alternative fuel? At what point will price criteria obtain? Clearly, the price in which the distributor is interested is the one that will ensure both profits and customers. The price in which the transmission company is interested is the price from producers that enables it to transmit the natural gas to consumer markets, cover its costs and sell to distributors at a profit. The relevant price to the producer is the one at which he can recover his costs (defined here as a reasonable return to invested capital) with a premium to cover the further risks of exploration and production. To add to the suspense, a considerable rent element is also often involved in natural gas marketing – a rent that each of the parties to a deal will be eager to appropriate for himself.

Curiously, with regard to both of the two examples of British and Alaskan natural gas, the prices can be criticized from a free market viewpoint. The price of Alaskan gas delivered to major American trunklines is estimated to be in the region of $9.00/million British thermal units (MMBtu – roughly 1,000 ft^3, or 35.315 m^3). Crude oil, at about $35.00/barrel, runs about $6.00/MMBtu.[1] By contrast, natural gas from the British fields contracted for in 1968 costs less than $1.00/MMBtu today – one-sixth the cost of crude, one-ninth the cost of Alaskan natural gas.

Understandably, price discrepancies such as these might make readers accustomed to thinking in terms of the freely competitive market place pull their hair in dismay. Yet both ANGTS and the British Southern Basin are consistent. The contractual prices specified in both are and were the ones essential to start highly capital-intensive projects in terms of guaranteed future returns on investment. That British gas seems underpriced today and Alaskan gas overpriced is the inevitable result of the political economy of natural gas. As surplus profits or deficits crop up that were unexpected when contractual relationships commenced (a form of unexpected resource or oligopoly rent), the parties can become discontented with the contractual relationship and demand its revision.

In this focus on contractual relationships between firms, the role of government should not pass unnoticed. One of the reasons for the low British prices was the existence of a government-owned monopsony transmission and distribution company, the British Gas Corporation (BGC), and a government policy that effectively prevented any gas producer from making a contractual relationship with a third party. One of the reasons for the lack of progress on the Alaskan pipeline is the unwillingness of the American government to write a 'blank cheque' covering all possible economic losses that the project might incur in the future. For government (to paraphrase Clemenceau's dictum on war and generals), natural gas is too important to be left to the natural gas companies. Many American consumers in the winter of 1982–3 would have agreed with this dictum upon looking at their gas bills, which have soared since the deregulation of gas in the USA began several years earlier. Surprisingly enough, if they were totally honest, most natural gas companies would agree.

This theme of the drive for market stability is developed in the next two chapters. Chapter 2 looks specifically at the industrial dynamics of natural gas: how the commodity characteristics of the fuel shape the strategies of producing, transporting and selling it. Chapter 3 shows generally how both government and industry have evolved solutions to the problems of market stability that inevitably arise from the industrial dynamics of natural gas.

From this more theoretical discussion, we move to an investigation of the particular ways in which the leading gas nations have approached these problems. National solutions, as we shall see, differ in a number of respects, depending on how the natural gas industry has historically evolved in these countries and on how various governments have chosen to approach it. In the United States, 3,000–4,000 producers of natural gas sell to dozens of pipeline transmission companies, which in turn sell to hundreds of utilities. The American system has been closely regulated – thereby achieving a form of 'regulated' stability. The British example is in strong contrast to the American. Here a national gas company, the British Gas Corporation, has enjoyed a monopsony position regarding the purchase of natural gas produced in the British North Sea and integrates within itself the transmission and distribution functions of the industry. (Both the American regulated stability and the British monopsony stability are in the process of change.) Continental European solutions to instability are a veritable pot-pourri, whereas inter-European trade is rather highly coordinated. The nation with the greatest reserves of natural gas, the Soviet Union, has chosen to develop its gas according to a centralized bureaucratic plan; for the Soviets, 'instability' is not represented by market instability, it is rather an inability to fulfil the 'plan'. The Japanese are perhaps uniquely dependent among consuming nations in their reliance on imported liquefied natural gas. By bringing this gas directly to their centres of population, the Japanese avoid the expense of long transmission lines. Through their premium pricing policy (particularly as regards the use of natural gas in electrical power generation), the Japanese are capable of paying record high prices for this LNG upon its delivery. Chapters 4–8 deal with the specific 'ins and outs' of the various national solutions to instability (with the Japanese approach considered in the context of the more general discussion of LNG in Chapter 9).

International trade: a new form of instability?

The Soviet–European gas pipelines and the LNG carriers, unnoticed by our hypothetical astronauts at the commencement of this chapter, constitute the element of international trade in natural gas. Most national consumption follows a certain pattern. As pointed out earlier, the natural gas closest to the major markets is developed first, and depleted first. Only when proximate reservoirs begin to decline are natural gas fields further away brought into consideration. These new sources may be either domestic or foreign. Thus in the United States, the period 1925–50 marked the peak of deliveries from Pennsylvanian, West Virginian and Californian gas fields and the commencement of deliveries from Louisiana, Texas,

Oklahoma – thousands of miles away from the consuming centres in California and on the Atlantic Seaboard. Similarly in the Soviet Union, natural gas from the Carpathians and from Azerbaijan declined relatively early. Necessary increments to gas supplies for the Central Russian area were found first in Central Asia (Uzbekistan, Turkmenia) and later in Western Siberia. Once again these new sources necessitated huge capital outlays and (in the case of Western Siberia) considerable engineering difficulties. In Western Europe, natural gas from the Groningen field is increasingly being supplemented by more expensive natural gas from the North Sea, gas that has incurred a high development cost and has to be transported a relatively long distance. Additional supplemental sources of natural gas are now being sought from the Soviet Union and Algeria. Of more long-term potential interest is natural gas imported as LNG from Nigeria, from Qatar, from the Camerouns, and even from the Canadian Arctic Islands.

International trade has been as subject to the 'law of increasing distances' as domestic trade. The first imports of natural gas into the United States from Canada at Niagara, New York, occurred early in the twentieth century – but these imports were initially destined for distribution in the cities along the border. The distances involved were not significant and the amounts supplied were small. Similarly, British imports of LNG to Canvey Island in the mid-1960s were largely supplemental to the sources of town gas and refinery gases then being marketed in the United Kingdom.

With the need to exploit resources at greater and greater distances from major consumption centres, an increasing reliance on international trade becomes a virtual certainty. The future of international trade in natural gas in the next decades will depend on two critical issues: the competing modes of international transport, and the willingness of nations with an exportable surplus of the fuel actually to export it.

The means of international transport may well signal some changes in the stability of world trade in natural gas. Chapter 9 introduces the problems involved in LNG transport; Chapter 10, the relative virtues of LNG versus gas pipelines. It has been hypothesized that LNG trade in ships is by definition more flexible than natural gas piped across frontiers: first, because the LNG carrier can be rerouted to alternative customers; second, because buyer and seller are more definitively linked by a pipeline, which does not allow for third-party opportunities for either producer or consumer. Evidence on this point, nevertheless, is mixed, as is exporter/importer experience with LNG.

Producer willingness to export natural gas is also questionable. It does not necessarily follow that a country with an enormous exportable surplus of gas will be willing to export that surplus. Various factors enter the picture here: subjective attitudes towards depletion policies and national identity, the existence of other more valuable ways to utilize the natural gas domestically, the relation of natural gas prices to oil prices on a Btu basis (this last is important for OPEC countries), and previous experience with export technologies (or, more specifically, LNG liquefaction plants and ships). Chapter 11 investigates the future of the international gas trade as a function of the incentive to export.

Natural gas: fuel of the future?

There is little doubt that supplies of natural gas are immense and that its particular combustion qualities are highly desirable. Its market growth worldwide has been remarkable, and the factors for future growth all seem to be present. Will natural gas in fact live up to its potential? The answer to this question probably lies in the rapidly changing practices in trading natural gas. It is one thing for natural gas trade to expand within national borders. Here, national law reigns unchallenged and the state can act as mediator in case of conflict. Given market instability – and the consequent risk-minimization through contractual relationships and state intervention – producers, transmission companies and utilities can cooperate in the secure knowledge that legal or political recourse exists should 'things go wrong'.

Internationally, this form of risk-minimization is at best inchoate. Not only is the risk factor greater, but the sums of money required for transmission lines or LNG projects are immense. The technological risk can be considerable as well. Algeria, for example, regarded by the major natural gas importers as perhaps the epitome of contract breakers, feels somewhat justifiably aggrieved owing to the sub-par performance of much of the LNG equipment installed on Algerian soil. A good deal of this equipment when installed was essentially 'state of the art', and it is not too surprising that some of it has since had to be scrapped. However, what is logical in the eyes of the Western firms installing the equipment is easily misperceived as thievery by the Algerians. The resulting troubles lead to even further strains.

In sum, what is needed internationally is the sort of regime that is now available only nationally, whereby parties to international natural gas deals can cover their risks and avoid the undesirable consequences of market instability. The trick is how to achieve this regime.

Note

1 Unless otherwise explicitly stated, energy measurements will be made in millions of British thermal units (MMBtu), volumetric measurements in cubic meters (m^3) or cubic feet (ft^3). See Appendix I for the relationships between these units of measure.

2

The Dynamics of an Industry

The nature of a fuel

Philip Chantler, in his analysis of the British town gas industry (1938), speculated how very different the structure of the gas industry might have been had it been possible for customers to arrive at the gas works and take delivery in containers. Clearly this was impossible. The gas could only be delivered through a network of underground pipes, which meant the creation of a central utility industry with local monopoly power and with public regulation. Thirty-five years later, with the virtual disappearance of town gas as a fuel, one can similarly speculate how different the structure of the natural gas industry might have been if the production of natural gas had occurred locally at the point of consumption. It could be argued from this point of view that the natural gas industry would have become a localized form of industry. (There were over 1,049 different town gas firms in the UK alone upon their nationalization in 1949.) Yet natural gas, as is commonly known, is not a local fuel; in most cases it must be transported hundreds or thousands of miles. Concomitantly, the natural gas industry is (with some exceptions) characterized by a high degree of concentration, a concentration necessitated by the economies of scale of transportation.

In any industry, there are certain economic imperatives that at the same time circumscribe its activities and define its dynamics. Thus the automobile industry struggles with the problems of 'static' economies of scale. Depending on whether automobile manufacturers produce and sell the somewhat mythical million cars apiece annually, they can be said to have bad or good years. The degree of success enjoyed by the individual companies in achieving this goal not only explains the nature of industrial competition in the branch, but also indicates where activities must be limited. Similarly, the characteristics of 'dynamic' economies of scale (in which the costs of production of the additional '*n*th' unit are constantly falling) help to explain the workings of the computer industry, where a technological advance by one firm can lead to its total dominance of the market (e.g. IBM).

The commodity characteristics of natural gas, too, serve to define the economic imperatives of the natural gas industry.

The nature of natural gas

The contrast between the Middle Eastern gas flares and the hidden Siberian 'blue gold' underlines what is perhaps the most puzzling characteristic of the natural gas industry: Why should natural gas be burned off in the one place and developed in the inaccessible wastes of the other? The answer to this question lies in part in gas resource location, in part in its energy content, and in part in the relationship between natural gas and oil and the economics of the joint production of the two fuels.

As Table 2.1 shows, most of the world's non-associated 'giant' gas fields are inconveniently tucked away in such remote corners as Northwestern Australia, New Zealand, Arctic Canada and Sarawak, far from the major existing natural gas markets. Gas produced from these fields is worth only what it can bring in these markets minus transportation costs. The price that a nation such as Algeria can get for its gas may be as low as 10 per cent of the eventual price at which the gas is sold in the US market (Tiratsoo, 1979, p. 293). However, Table 2.1, compiled in 1978, is already somewhat dated; it excludes some recent finds, certain of which are really quite significant.

The second of the factors determining flaring or production concerns the energy content of the gas. In this respect, Table 2.1 is somewhat misleading in using volumetric (trillion ft^3) measures. To give one example, both the Kapuni and the Maui gas reserves are located in New Zealand. The Kapuni field contains an estimated 630 billion ft^3 of reserves (about 22 billion m^3); the Maui field contains an estimated ultimate 5.6 trillion ft^3 (some 200 billion m^3). If we assume that the unit costs of exploring, developing and marketing the gas from the two fields were identical, one would roughly estimate the cash value of the Maui field to be nine times that of the Kapuni field. Yet 44.2 per cent of the Kapuni field gas is incombustible carbon dioxide, a similar amount is methane, and the balance is composed of the higher hydrocarbons. In the Maui field, in contrast, there are only minute amounts of carbon dioxide, and the field yields a prime 81 per cent methane plus the higher hydrocarbons. Thus, the difference between the cash values of the reserves of the two fields is probably – under the assumptions mentioned – on the order of twenty. It is for

Table 2.1 *World 'giant' non-associated gas fields*

	Field	*Country*	*Age of reservoir*	*Date discovered*	*Ultimate gas reserves (trillion ft^3)*
1	Urengoy	USSR (W. Siberia)	Cretaceous	1966	176.5
2	Kangan	Iran	Permian	1973	170.0
3	Yamburgskoye	USSR (W. Siberia)	Cretaceous	1969	155.3
4	N-W Dome	Qatar	Permian	1976	100.6+
5	Zapolyarnoye	USSR (W. Siberia)	Cretaceous	1965	94.0
6	Krasniy Kholm	USSR (Orenburg)	Permian	1966	74.0
7	Hassi R'Mel	Algeria	Triassic	1956	70.0
8	Hugoton-Panhandle	USA (Kansas–Texas)	Permian	1926	70.0
9	Kangan	Iran	Permian	1973	70.0
10	Groningen	Netherlands	Permian	1959	60.9
11	Medvezhye	USSR (W. Siberia)	Cretaceous	1967	55.0
12	Bovanenko	USSR (W. Siberia)	Cretaceous	1971	53.0
13	Pars	Iran	Permian	1973	50.0
14	Pazanan	Iran	Oligo-Miocene	1938	50.0
15	Kharsavey	USSR (W. Siberia)	Cretaceous	1974	42.4
16	Taz	USSR (W. Siberia)	Cretaceous	1962	40.4
17	Dorra	Neutral Zone	Cretaceous	1974	35.0
18	Bahrain	Bahrain Island	Jurassic/Permian	1931	20.0
19	Kangiran	Iran	Permian	1968	20.0
20	Semakovskoye	USSR (W. Siberia)	Cretaceous	1971	19.0
21	Vyuktyl	USSR (W. Siberia)	Permian	1964	17.7
22	Layavozh	USSR (Pechora)	Carboniferous	1965	17.5
23	E J Bermudez	Mexico	Cretaceous	1976	17.5
24	Gazli	USSR (C. Asia)	Cretaceous	1956	17.0
25	Shebelinka	USSR (Ukraine)	Permian	1956	16.4
26	Severo-Urengoy	USSR (W. Siberia)	Cretaceous	1970	16.0

27	Komsomolskoye	USSR (W. Siberia)	Cretaceous	1966	16.0
28	Sredne-Vilyuy	USSR (E. Siberia)	Triassic-Jurassic	1963	15.9
29	Yamsovey	USSR (W. Siberia)	Cretaceous	1970	15.0
30	Nar	Iran	Permian	1974	14.0
31	Messoyakha	USSR (W. Siberia)	Cretaceous-Jurassic	1967	14.0
32	Arun	Indonesia	Miocene	1971	13.0
33	Kirpichli	USSR (C. Asia)	Cretaceous	1972	12.5
34	Gubkin	USSR (W. Siberia)	Cretaceous	1965	12.3
35	Leman Bank	UK (North Sea)	Permian	1965	12.3
36	Frigg group	Norway (North Sea)	Eocene	1971	12.1
37	Hateiba	Libya	Cret., Camb.-Ord.	1963	12.0
38	Blanco Mesaverde – Basin Dakota	USA (N. Mexico)	Cretaceous	1927	11.0
39	Rhourde Nouss group	Algeria	Triassic-Ord.-Cambrian	1962	11.0
40	Vyngapurovskoye	USSR (W. Siberia)	Cretaceous	1968	10.6
41	Russkoye	USSR (W. Siberia)	Cretaceous	1968	10.6
42	Gomez	USA (Texas)	Ord.-Cambrian	1963	10.0
43	Bagazhda	USSR (C. Asia)	Cretaceous	1971	9.6
44	Sui	Pakistan	Eocene	1953	8.6
45	Noviy Port	USSR (W. Siberia)	Cretaceous-Jurassic	1964	8.5
46	Krestichenskoye	USSR (Ukraine)	Permian-Penn.	1968	8.2
47	Jalmat-Monument-Eunice	USA (California–N. Mexico)	Permian	1929	8.1
48	Naip	USSR (C. Asia)	Cretaceous	1972	8.0
49	Indefatigable	UK (North Sea)	Permian	1966	8.0
50	N. Rankin	Australia (NW Shelf)	Triassic	1972	7.9
51	Shih-you-kou group	China	Triassic-Jurassic	1955	7.8
52	Severo-Stavropol-Pelagiada	USSR (Caucasus)	Oligo-Miocene	1950	7.3

Table 2.1 *World 'giant' non-associated gas fields — continued*

	Field	*Country*	*Age of reservoir*	*Date discovered*	*Ultimate gas reserves (trillion ft^3)*
53	Monroe	USA (Louisiana)	Paleocene-Cretaceous	1916	7.0
54	Lacq	France	Cretaceous-Jurassic	1951	7.0
55	Severo-Komsomolskoye	USSR (W. Siberia)	Cretaceous	1969	7.0
56	Peyatinskoye	USSR (W. Siberia)	Cretaceous	1969	6.6
57	Puckett	USA (Texas)	Ordovician	1952	6.5
58	Maastakh	USSR (E. Siberia)	Jurassic	1967	6.4
59	Badak	Indonesia	Miocene	1975	6.2
60	Drake Point	Canada (Arctic Islands)	Triassic-Jurassic	1975	6.1
61	Carthage	USA (Texas)	Cretaceous	1936	6.0
62	Bintulu	East Malaysia (Sarawak)	Miocene	1975	6.0
63	Katy	USA (Texas)	Eocene	1964	6.0
64	Rabbit Island group	USA (Louisiana)	Miocene	1940	6.0
65	Soleninskoye	USSR (W. Siberia)	Cretaceous	1969	5.7
66	Maui	New Zealand	Eocene	1969	5.6
67	Urtabulak	USSR (C. Asia)	Jurassic	1963	5.4
68	L-10	Netherlands	Permian	1975	5.3
69	Achak	USSR (C. Asia)	Cretaceous	1966	5.0
70	Kenai	USA (Alaska)	Tertiary	1959	5.0
71	Old Ocean	USA (Texas)	Oligocene	1934	5.0
72	Süd-Oldenburg	W. Germany	Permian	1968	5.0
73	Yefremovka	USSR (W. Siberia)	Permian	1965	4.6
74	Dhodak	Pakistan	Eocene	1976	4.5
75	Hecla	Canada (Arctic Islands)	Triassic-Jurassic	1975	4.2
76	Serecny-Yamal	USSR (W. Siberia)	Cretaceous	1970	4.1
77	Hewett group	UK (North Sea)	Triassic-Permian	1966	4.0

78	Gidgealpa	Australia	Permian	1964	4.0
79	Moomba	Australia	Permian	1964	4.0
80	Mari	Pakistan	Eocene	1957	4.0
81	Mocane-Laverne	USA (Oklahoma)	Pennsylvanian-Miss.	1952	3.8
82	Samantepe	USSR (C. Asia)	Jurassic	1964	3.7
83	Marlin	Australia	Eocene-Paleocene	1966	3.6
84	Rio Vista	USA (California)	Eocene-Paleocene	1936	3.5
85	Bierum	W. Germany–Netherlands	Permian	1963	3.5
86	Meillon	France	Cretaceous-Jurassic	1965	3.5
87	Khadzhiy-Kandym	USSR (C. Asia)	Jurassic	1967	3.5
88	Bayou Sale	USA (Louisiana)	Miocene	1940	3.5
89	Gugurtli	USSR (C. Asia)	Cretaceous-Jurassic	1965	3.5
90	Tang-e-Bijar	Iran	Oligo-Miocene	1975	3.5
91	Kettleman Hills North	USA (California)	Miocene-Eocene	1928	3.5
92	Gassi Touil	Algeria	Triassic	1971	3.5
93	Yetypur	USSR (W. Siberia)	Cretaceous	1971	3.5
94	Zapadno-Tarkosolinskoye	USSR (W. Siberia)	Cretaceous	1972	3.5
95	Bakhrabad	Bangladesh	Miocene	1963	3.5
96	Barqan	Saudi Arabia (Red Sea)	Miocene	1969	3.5

Scurce: Tiratsoo, 1979, p. 293.

Table 2.3 *World 'giant' oil fields with at least 3.5 trillion ft^3 of ultimately recoverable associated gas*

	Field	*Country*	*Year discovered*	*Age of reservoir*	*Ultimate oil recovery (billion bbls)*[a]
1	Ghawar	Saudi Arabia	1948	Jurassic	83
2	Burgan	Kuwait	1938	Cretaceous	72
3	Bolivar Coastal	Venezuela	1917	Mio-Eoc.	32
4	Safaniya Khafji	Saudi Arabia Neutral Zone	1951	Cretaceous	30
5	Rumaila	Iraq	1953	Cretaceous	20
6	Ahwaz	Iran	1958	Oligo.-Mio., Cret.	17.5
7	Kirkuk	Iraq	1927	Oligo.-Eoc., Cret	16
8	Marun	Iran	1964	Oligo.-Mio	16
9	Gach Saran	Iran	1928	Oligo.-Mio., Cret.	15.5
10	Agha Jari	Iran	1938	Oligo.-Mio., Cret.	14
11	Samotlor	USSR	1966	Cretaceous	13
12	Abqaiq	Saudi Arabia	1940	Jurassic	12.5
13	Romashkino	USSR	1948	Cretaceous	12.4
14	Berri	Saudi Arabia	1964	Jurassic	12
15	Zakum	Abu Dhabi	1964	Cretaceous	12
16	Manifa	Saudi Arabia	1957	Cret.-Jurassic	11
17	Fereidoon-Marjan	Iran/Saudi Arabia	1966	Cret.-Jurassic	10
18	Prudhoe Bay	USA	1968	Cret., Trias., Miss.	9.6
19	Bu Hasa	Abu Dhabi	1962	Cretaceous	9
20	Qatif	Saudi Arabia	1945	Jurassic	9
21	Khurais	Saudi Arabia	1957	Jurassic	8.5
22	Zuluf	Saudi Arabia	1965	Cretaceous	8.5
23	Raudhatain	Kuwait	1955	Cretaceous	7.7
24	Sarir	Libya	1961	Cretaceous	7.2
25	Hassi Messaoud	Algeria	1956	Camb.-Ord.	7
26	Shaybah	Saudi Arabia	1968	Cretaceous	7
27	Abu Sa'fah	Saudi Arabia/Bahrain	1963	Jurassic	6.6
28	Asab	Abu Dhabi	1965	Cretaceous	6
29	Bab	Abu Dhabi	1954	Cretaceous	6
30	Ta-ch'ing	China	1959	Cretaceous	6
31	East Texas	USA	1930	Cretaceous	5.6
32	Umm Shaif	Abu Dhabi	1958	Jurassic	5
33	Wafra	Neutral Zone	1953	Eoc.-Cret.	5

Source: Tiratsoo, 1979, p. 69.
Note: a Primary recovery only.

OPEC producers is no less true of natural gas producers in the United States. On the other hand, utilities can be unconcerned about whether or not the price of natural gas to the producer is high enough to cover the costs of replacing the natural gas sold – i.e. a sufficient rate of return to induce the producers to undertake the risk of exploring for, drilling up, and producing the increments of natural gas needed to replace the increments sold. (That a 'proper' economic criterion such as replacement costs exists does not lessen the problem; determination of this criterion under differing circumstances is complicated by the negotiation relationships between the stages of the industry and has, for example, plagued the American natural gas industry since its inception.)

The question of 'waste' is as confusing as the price issue. That oil and gas are economically joint products (much as wool and mutton are the inevitable products of sheep farming) has led to more misconceptions. Chief of these is that it is wasteful to flare natural gas. In terms of non-economic criteria, this is indisputably true. It is nevertheless the case that natural gas is utilized in the most efficient economic manner consistent with the interests of the oil (or gas) producer concerned.

Because natural gas is flared does not mean it could be made freely available to a pipeline. To clear up this misconception, one must understand how oil or gas fields are developed. Here, the primary concern is to maximize one's economic return through the most efficient manner of developing a field, so that the production of oil, condensates and gas together is maximized at a minimum cost. A simplified case illustrates the principles involved. Suppose a company discovers an oil field with sufficient gas to justify the costs of delivery to a pipeline. Yet at the same time, it is necessary to repressurize the reservoir to maximize recovery of oil. Reinjecting half the natural gas produced will repressurize the reservoir, but at the same time render commercial sale of the balance uneconomic. Therefore half of the natural gas must be flared. Clearly, in such a case the company will seek to optimize its revenue, weighing the possible gas sales revenue against the revenue from increased oil production. Such considerations also place a floor under the price at which the producer is willing to sell his gas. To carry the example further, a price of 10¢/MMBtu (1,000 ft^3)[1] may be too low – the producer will earn more through reinjecting half his gas and flaring the rest; 11¢/MMBtu may be a different matter; and so forth. The flaring that results from such considerations can hardly be called 'wasteful' in the economic sense. (The same principles apply to other fuels. The author has seen production plans that call for the flaring of condensate as the most economic solution to development of a natural gas field in the Danish sector, for example.) In virtually all cases, natural gas, oil and condensate will be sold for what the market will bring.

A slightly different matter is whether, in the judgement of other authorities, reinjection of natural gas will in fact yield the anticipated revenues. Therefore it is not uncommon, especially in nations where production plans are subject to governmental approval, to hear talk of 'wasteful' flaring as opposed to reinjection into an oil field. What is ignored in such conversations is that the two parties to the argument may have entirely different technical evaluations of the feasibility of reinjection, entirely different cost perspectives, and so forth. The concept of 'waste' always refers to an economic alternative use that is forgone. This does not mean that all will be agreed on the economic alternative. Here there are, and will always be, differences of opinion.

A final note is in order here. Decisions about the use of dissolved or associated gas are often irreversible. Thus to take our example of reinjection versus sale to a local pipeline, a rise in prices offered at the pipeline may not lead to abandonment of reinjection as an alternative. Once all the dimensions of planning an oil/gas field development are implemented – the placing of wells, the location of gas compressors, the design of treatment plants and of storage facilities – the oil field will be an organic whole. To restructure these facilities with a view to marketing the natural gas may involve prohibitive costs.

Transportation and the economies of scale

If exploration and development risk, joint maximization of returns from natural gas and associated fuels, and engineering complexity (well-drilling, separation and treatment plants, and gas-gathering pipelines) characterize the production of natural gas, what contrasts does the next step – that of transportation – provide? Simply stated, natural gas transportation – both the rationally planned pipeline routes and large LNG carriers – enjoys enormous economies of scale and decreasing unit costs. (These are the two main forms of transport that we shall consider in this context; others exist, but they are largely hypothetical[2].) While both pipeline and LNG forms of transport reward scale economies (and those who possess them), this similarity is somewhat superficial when one examines the two forms of transport in sequence. Let us begin with pipelines.

Figures 2.1, 2.2 and 2.3 illustrate the major scale problems of pipeline transport. Figure 2.1 shows a hypothetical starting position: the letters A, B, C, and so forth to the left represent the points at which natural gas from differing fields is collected for delivery to the various markets; the numbers to the right represent these markets. The solution to this problem will depend on the distance of the pipelines, the dimensions of the lines concerned, and the required 'load factor' (all to be explained later in this context).

A casual reader might link the points in Figure 2.1 as they are

Figure 2.1 *Natural gas collection points (A–F) and markets (1–5)*

linked in Figure 2.2. Yet, after giving the matter more thought and attempting to minimize the distance for which pipeline must be laid, a somewhat different solution might emerge (see Figure 2.3). While perhaps not the optimal solution, the natural gas lines illustrated in alternative B (Figure 2.3) are some 12–15 per cent shorter than those in alternative A (Figure 2.2). Oversimplifying, if we assume that the total distance covered is some 2,000 miles and that the cost of laying these lines averages $750,000 per mile, irrespective of pipeline diameter, then the difference between Figures 2.2 and 2.3 amounts to roughly $200 million. Two things emerge from this exercise: first, one does *not* construct pipelines on whim; secondly, the geographical extent of a pipeline company can make for considerable efficiencies.

Yet scale economies of laying pipelines are not a function of distance alone. In Figures 2.2 and 2.3 the diameters of the pipelines are given as well, and this detail is not incidental. The throughput of natural gas pipeline is a function of $\sqrt{d^5}$ in which d is the diameter. In other words, all else being equal, doubling the diameter of the

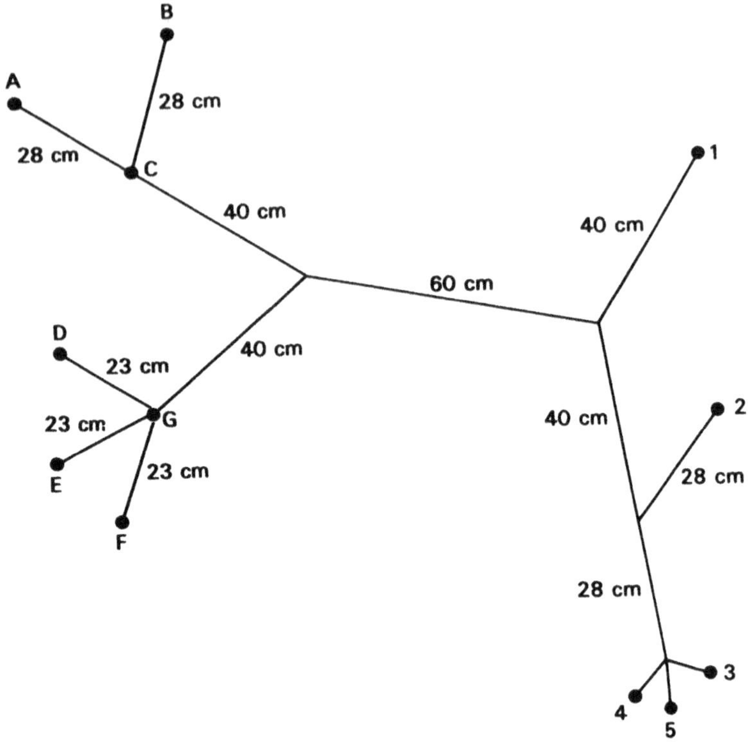

Figure 2.2 *Pipeline network: alternative A*

pipeline allows for an increase in the volumes delivered by a factor of 5.7.[3] Thus the economies achieved by a 60 cm line over a 40 cm line are quite considerable. The contrasting unit costs are shown in Figure 2.4 over varying distances.

If we assume that transport costs of natural gas are 'zero' at the point of delivery, then the unit costs of delivery (per MMBtu, for example) increase on the vertical index scale far more rapidly for the 40 cm line than for a 60 cm line. This difference becomes progressively greater the further the distance the gas is to be delivered (the horizontal scale). Over a 1,500 km distance, for example, the unit costs for gas in the 60 cm pipe are 70 per cent of what they would be for a 40 cm pipeline. Generally speaking, therefore, large-volume contracts mean large-volume deliveries and significant economies of scale in wide-diameter pipelines. (There are pipelines that dwarf our examples here. The Soviet Union, for example, has 7,000 miles (11,340 km) of 56 in. (142–3 cm.) lines.) Not only is the large-

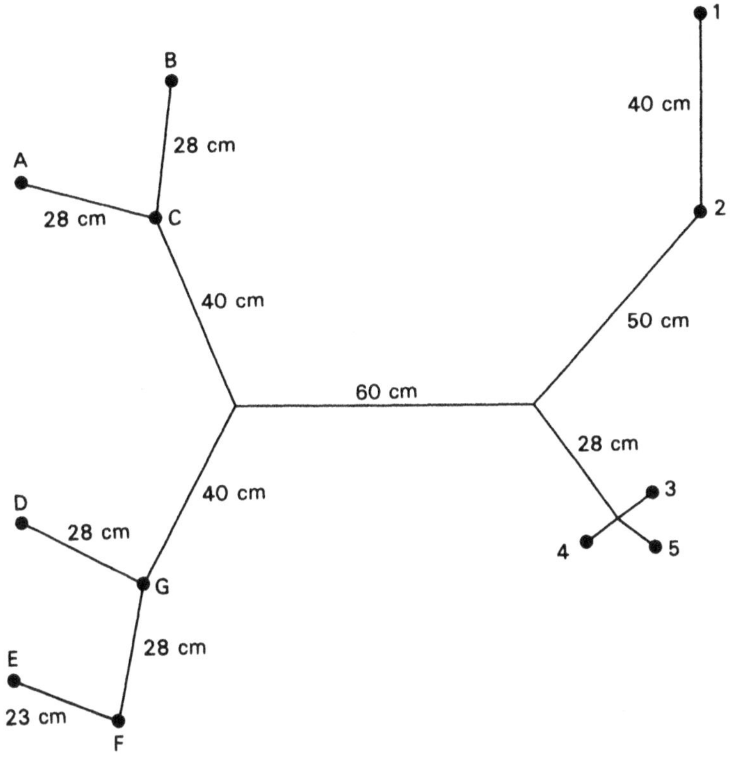

Figure 2.3 *Pipeline network: alternative B*

pipelines company favoured with the planning of the most geographically economic route, it can supplement this advantage through utilizing the largest and most economical pipelines.

Yet, returning to Figures 2.2 and 2.3, it is clear as well that the additional costs of increased pipeline dimensions must be accounted for in designing pipeline routing. Figure 2.3 has 35–40 per cent fewer kilometers of 40 cm pipeline, but it has 66 per cent more 60 cm line plus some 50 cm line. Contrary to what might be expected, it does not cost half as much again to lay 60 cm line as it does 40 cm line. While the material cost of 60 cm line would be fractionally more than that of 40 cm line, most of the other costs involved are fixed and therefore identical for the two lines. (These include 'right of way', laying the line, trenching and welding.)

An additional factor favouring companies possessing economies of scale is the 'load factor': variations in natural gas deliveries. Up to

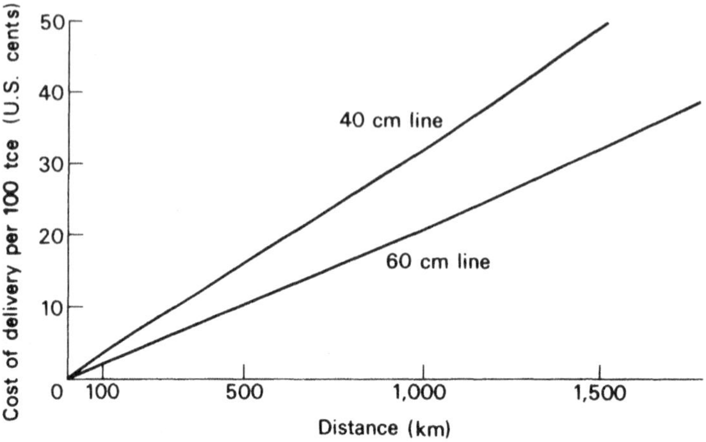

Figure 2.4 *Unit costs of delivery of natural gas through 40 cm and 60 cm pipelines as a function of distance and pipeline diameter (costs at delivery = 0 for both cases)*
Source: Manners, 1968, p. 72

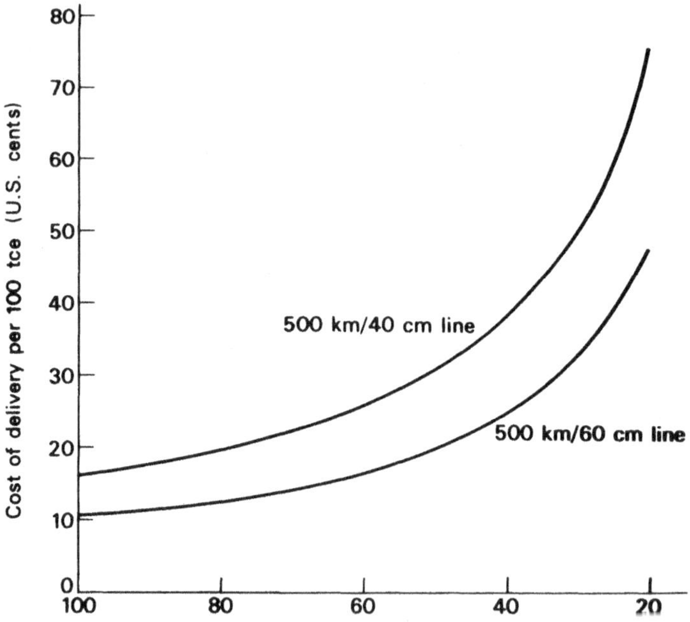

Figure 2.5 *Impact of load factor on costs of delivery over 500 km*
Source: Manners, 1968, p. 72

this point it has been assumed that the examples in Figures 2.2 and 2.3 handle deliveries with no seasonal or daily variations – a 100 per cent load factor. Given increased heating in winter and daily variations in both winter and summer, of course, this is an unreasonable assumption. How would the upper and lower limits of the variations of an 80 per cent load factor look? This requires solving for x in the following:

(1.1) $0.80x = (0.8:1) \times$ contracted amounts = lower limit or 0.8.

(1.2) $0.80x = (1:0.8) \times$ contracted amounts = upper limit or 1.25.

In other words, if the normal contracted amount is to be 100, an 80 per cent load factor allows for deliveries to be as low as 80 per cent or as high as 125 per cent of the daily or seasonal contracted amount. Such flexibility is much desired in the marketing of natural gas. It is less desirable, but necessary, in long-distance pipelines. This is evidenced by Figure 2.5.

Costs of delivery rise dramatically with lower load factors. This is due to the need for higher-tensile pipeline strength, greater compressor capacity and other capital expenditures. It should not surprise, therefore, that trunklines are generally designed for load factors of 85–95. The task of meeting seasonal and daily variations is generally left to the recipients of natural gas at the marketing end.

The structure imposed by the economies of scale – here largely that of a single monopsony purchaser forcing the field prices of natural gas down – has been a feature of much concern in the gas industry. To what extent, for example, do the owners of the hypothetical trunklines in our illustrations have an unfair advantage over the owner of field complex A, should that owner decide to sell his gas for energy consumption? This problem and others like it will be dealt with more extensively in the body of this book.

The economics of LNG transportation are also essentially those of scale, but with some significant differences from pipeline economics. LNG transportation is characterized by different stages. First, there is normally pipeline transportation for natural gas from the gas fields to a liquefaction plant, usually located in a major port. Here the natural gas is cooled to −162° C. through one of several cryogenic cooling processes.[4] Common to all of these methods is that, through utilizing the energy content of the natural gas, they consume up to 25 per cent of the original amounts of gas shipped to the plant in the liquefaction process itself. From the liquefaction plant, the LNG, now reduced in volume by a factor of approximately 600, is transferred to an LNG carrier. Such ships can now hold up to 165,000 m^3 of LNG (equivalent to 99 million m^3 or 3.5 billion ft^3 of gaseous natural gas). These carriers then transport the LNG to a terminal in the receiving land where it is re-gasified and delivered to utilities or trunklines for marketing purposes.

Each of these stages involves enormous capital investment and space-age technology. In place of the combination of routing, pipeline dimension and load factor characterizing pipeline transportation, LNG transportation involves the optimal use of technology at the successive stages of liquefaction, transport, and terminal receiving and re-gasification stages. Each of these stages must 'fit' with the others to achieve maximum profit. The carriers must be the proper size for the project, perform an agreed-upon number of voyages per year, and not be subject to malfunctions. The liquefaction plant must also be appropriate to the task at hand. The LNG receiving terminal must have appropriate capacity for re-gasification. (Often sea water is utilized to this purpose.) The chain of advanced technology is only as solid as its weakest link; if the carriers run it below capacity, if the terminal is damaged, if the liquefaction plant is not completed on schedule, millions of dollars will be lost.

The dimensions of the economic problems involved can be seen in a proposed El Paso project to ship LNG from Algeria to the United

Table 2.4 *El Paso II project: capital costs, operation costs per annum and resulting gas prices, 1977*

Stage	*Estimated capital cost ($m.)*	*Estimated operations cost p.a. ($m)*	*Cost per MMBtu/ delivered gas ($)*
Liquefaction and purchase from Sonatrach (Algeria)	2,300[c]	n.a.	1.39
Shipping[a]	1,791	137.8	1.13
Terminal/re-gasification	456.9	n.a.	0.32
Pipeline transportation to existing facilities	254.7	n.a.	0.19
Total cost/price	4,805	137.8	3.03[b]

Source: US Federal Energy Regulatory Commission, 1977a, Hearings Docket Nos. CP77–330 *et al.*, 25 October 1977.

Notes: a Six vessels owned by Sonatrach (Algeria) and six by El Paso/United Consortium.

b Or, in terms of volume, \$2.61/thousand ft^3 or 7.48¢/m^3. The cost per MMBtu includes returns to investment at each stage as approved by the Federal Energy Regulatory Commission.

c Includes amounts necessary to produce and gather natural gas from the Hassi R'Mel field, construct about 300 miles of pipeline to port of Arzew, and build liquefaction, storage and terminal facilities.

States to the amount of 410,625,000 MMBtu (357 billion ft^3, or roughly 10 billion m^3 re-gasified). Table 2.4 illustrates the capital costs of each stage and the resulting MMBtu prices to cover these costs, including a 'fair return' on invested capital.

Several things should be noted here. First, the costs are astronomical even in terms of 1977 dollars. Allowing for 10 per cent per annum real increase in costs – a not uncommon 'rule of the thumb' where modern technology is involved in such a project – the 1981 costs would be closer to 7 billion 1977 dollars. Second, the costs involved are reflected in the very high unit costs of $3.03/MMBtu (or $2.61/thousand ft^3) and these are costs *before* the gas is sent through the El Paso Trunkline to consumers in Southern California and elsewhere. Finally, the Algerians are the biggest investors in the project. Over and above the $2.3 billion invested in Hassi R'Mel, liquefaction, etc., Sonatrach in this proposal invests some $895 million in six LNG carriers. This last point has understandably caused some resentment in Algeria. Sonatrach must build all the facilities, ship the gas, liquefy it (losing some 15–25 per cent of the gas thereby) and yet receive only $1.39/MMBtu for this investment. Although half the ships, and the profits made through shipping, will be Sonatrach's, this firm incurs an investment of 67 per cent of the total. These high investment sums, together with various contractual terms, can do much to explain producer-nation dissatisfaction with LNG deals. Beyond this, little should be noted at this point other than that these issues will be further elaborated in the next chapter and in Chapter 9.

The quandaries of marketing

Consuming markets, the blobs of night-time energy noticed from outer space, present considerably different problems from those of natural gas production and transportation. Here the problems involve the supplying of natural gas for a series of end-uses, varying seasonally and diurnally, and in such a manner as to avoid waste and misallocation of resources.

The problems at the marketing end can be illustrated graphically and through use of an example. Figure 2.6 illustrates the problems of variation of consumption throughout a typical year for a North American distribution company. As can be noted, some 14+ per cent of annual sales take place in the month of December and only around 4.4 per cent in the month of July. If we assume an average year-round quantity of 8.33 per cent per month, this means a load factor (average load expressed as per cent of peak load) of some 59–60 per cent in this particular market. A high load factor (greater than 50 per cent)

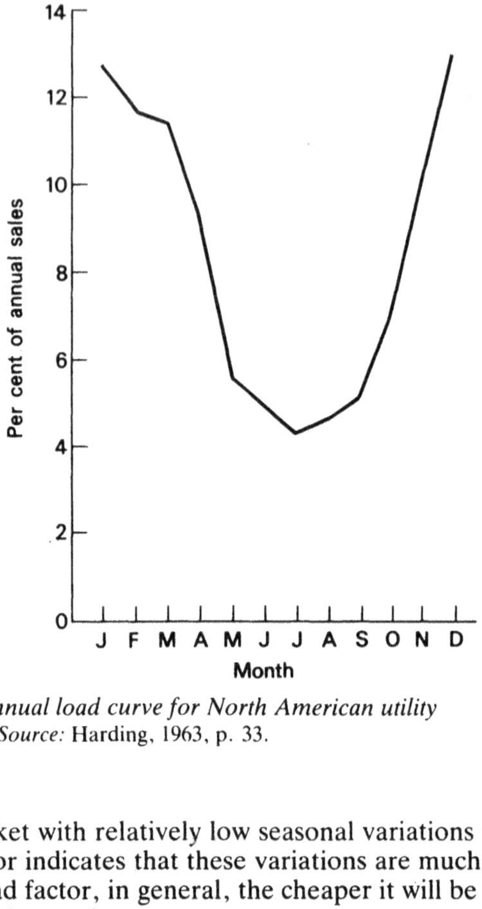

Figure 2.6 *Annual load curve for North American utility*
Source: Harding, 1963, p. 33.

signals a market with relatively low seasonal variations in demand; a low load factor indicates that these variations are much greater. The higher the load factor, in general, the cheaper it will be to supply gas to that particular market.

To use a hypothetical example, let us assume that we, a local utility, are considering purchasing natural gas from a trunkline at a quantity of 10 million ft^3 per day, at an 80 per cent load factor. Before doing so, we would first need to know what end markets we might hope to serve? What kinds of profits could we then hope to obtain? Finally, should the natural gas be purchased at all?

There are two classifications of markets we would want to consider: first, the division between the heating market and the non-heating market (the latter includes the use of gas for industrial, petrochemical and other purposes); and, second, the division into domestic residential, commercial, and industrial markets. With regard to the first classification, it is often reasonable for gas distribution firms to plan their distribution on a combination of both heating and non-heating markets. This is because of the great seasonal variations in peak demand in the heating market, a market directly affected by seasonal temperature swings. The non-heating market utilizes gas because of its 'form value': it might be desirable from the point of view of operational efficiency, as in a glass or ceramics works, or for alternative non-heating purposes such as a petrochemical feedstock, for example. Variations in consumption here are only marginally tied to seasonal swings of temperature and may have more to do with the general level of business activity and a firm's contraction or expansion plans for the future.

Perhaps more instructive from our point of view is the division of the market into residential, commercial and industrial markets. To begin with, a not untypical residential market might take around 40 per cent of the natural gas delivered. While immensely profitable, the residential market is also enormously costly in terms of investment and in terms of the very low load factor that characterizes this market (some 45 per cent in our example). The residential market is further subdivided into heating and non-heating uses, the latter comprising the market for cooking purposes and the like.

The commercial market comprises a wide variety of uses: schools, churches, hospitals and restaurants being added to what one normally considers 'commercial activities'. Here, the load factor is normally somewhat higher, and the 'price' somewhat lower than for residential uses. Not only are the distribution costs per unit of natural gas sold considerably lower, but the commercial category can be divided into various subgroups (as above). Experience has shown that there is a high diversity of natural gas usage among such subgroups (between restaurants, hospitals and churches for example), and through serving as many subgroups as possible the load factor becomes higher. Here the load factor might be of the order of 60 per cent, for a market comprising some 30 per cent of the total.

The industrial market, disregarding daily swings in usage, is the least seasonally determined of the various markets. Here natural gas is utilized primarily for processing purposes, and load factor (about 85 per cent) and distribution costs are less problematic than in the other markets. In their place, however, are other problems. First, the industrial market is generally dependent on the level of business activity for the firms concerned: a crisis in the steel industry or in the petrochemical industry could hit a gas utility hard if that utility were overdependent on firms in these branches. Second, more attention has to be paid to the price relationship between natural gas and alternative fuels from the industry's point of view. Owing to high levels of energy consumption and to a general desire to minimize energy costs, most industrial consumers of natural gas are very much aware of the energy alternatives. A final problem is that of growth and the projection of growth. As most natural gas utilities are forced to plan into the future, assessment of industrial growth and location possibilities are essential to overall grid planning.

In general, the trick in planning the sale and distribution of natural gas is to time demand needs of various customers so that the individual's maximum demand does not occur at the same time as the maximum demand of the others in the group; this further reduces group demand and smooths out the seasonal curve. The greater the 'diversity' factor, the higher the load factor (Harding, 1963, pp. 36–7). Simultaneously, the local gas utility wants to obtain the best possible 'investment to revenue ratio' for the natural gas sold (Harding, 1963, p. 50). This ratio is usually thought of in incremental terms when the gas utility is planning further expansion.

To return to our example, involving the acceptance of a contract of 10 million ft^3 per day at a seasonal 80 per cent load factor, it is clear that selling this gas at a profit in a system with a 65 per cent load factor can be difficult. There must be an element of flexibility built into the system in order to accomplish this goal. Several alternatives can be considered. For one thing, all the gas obtained could be sold to industrial users at a higher load factor. One could also sign contracts from alternative suppliers at lower load factors (say 10 million ft^3 per day at a 50 per cent load factor), enabling the sale of incremental amounts of additional natural gas to all existing markets. Less expensive, perhaps, depending on the additional markets to which this natural gas is to be sold, could be the sale of certain amounts of it on an 'interruptible supply' basis to industrial users. In such contracts, natural gas is sold cheaply to firms or utilities that can use it part time and switch over to other fuels if necessary. The low cost of the natural gas offsets the additional investment in double boiler/burner systems and the like. Selling 1.5 million ft^3 per day on an interruptible basis would enable our 10 million ft^3 per day to fit into the overall 65 per cent system load factor – if the load factor is to be maintained. A final alternative would be to store the natural gas during slack periods and market it at times of peak consumption. There are a wide variety of systems to choose from, varying from LNG storage tanks through 'line packing' (storage in the pipeline itself), storage in underground caverns (not unusual in Europe), and in depleted underground reservoirs, to more experimental methods (storage in underground acquifers, for example).

To express the problem less equivocally, one can assume that the cost of adjusting the additional amount of natural gas to the existing (or future) load factor is borne by a storage variable, which includes additional contracts and interruptible sales, plus increased investment in distribution lines and the like for the various markets. Similarly, one can assume net revenues from the investments. In our particular case, the question is one of marginal revenues and marginal costs. In other words, the value of our 10 million ft^3/day contract can be expressed as follows:

$$VC = MR - MC \tag{2.1}$$

where VC is the value of the contract (in net present value terms), MR is the marginal revenues (also in NPV terms) that the increased sales of natural gas should yield, and MC is the marginal cost (in NPV terms) of the additional sales (including the cost of natural gas concerned, added storage facilities and extended distribution lines) incurred through establishing the new markets.

Expression 2.1 is really far too straightforward, for there can be an n-tuple number of different manners of selling the natural gas, each yielding a marginal revenue and incurring a marginal cost. There also are limits to these solutions. If we assume economically rational behaviour on the part of the distribution company, the lower solution is bounded by the following expression:

$$\frac{\mathrm{MR}}{\mathrm{MC}} > 1; \tag{2.2}$$

in other words, installation of the new facilities should not be so expensive that the marginal cost of the new gas exceeds the marginal revenue that accrues from it. If the contract cannot satisfy the expression 2.2, then the distribution company should have not bought the contractual rights to the natural gas.

Let us assume that the universe of possible solutions is confined to three, then:

$$\frac{MR_1}{MC_1} > \frac{MR_2}{MC_2} \gtrless \frac{MR_3}{MC_3} > 1, \tag{2.3}$$

where MR_1 and MC_1 represent the most preferred distribution, storage and marketing plan, MR_2 and MC_2 represent the second choice, and so forth. Were that things were so simple! Clearly, the criteria by which marginal cost and marginal revenue are going to be measured will be modified. In part this will be due to differing firm goals regarding profit-maximization. The solution represented by MC_1 and MR_1

may maximize short-term profits, but it could well be that the other two solutions maximize longer-term profits or firm growth. Similarly, one of the alternative solutions to MC_1 and MR_1 may provide greater future security of natural gas supplies. Finally, these decisions are normally taken in a regulatory context. To what degree are increased prices necessary for a given solution, and how willing are the regulatory authorities to allow tariff rate increases? If the regulatory authorities are willing, how long will the necessary procedures take? These are but a few of the many decision parameters within which the public utility has to work.

In the distribution side of the industry, not unlike the transportation and production sides, the days of 'muddling through' are long over (although there are exceptions). The natural gas industry, argued F. B. Jones twenty years ago, was like 'Topsey'; it just 'growed' (1963, p. 32). While the 'Topsey factor' is perhaps not completely absent in today's industry, it certainly is in retreat. Natural gas planning at all levels has become a highly sophisticated exercise.

Do industrial dynamics justify state intervention?

To what degree is government involvement in the natural gas industry justified on the grounds of the industrial dynamics described in this chapter? This is not a rhetorical question. As will be noted in the balance of this book, state intervention and control of all three stages of the natural gas industry – particularly of the transmission and distribution stages – is an industry characteristic. The common economic rationale for such intervention is that the industry is not self-regulating: in other words, it is a natural monopoly. To what degree is this rationale justified in fact?

Natural monopolies are characterized by decreasing costs of production. Typically, capital overheads are high with a resulting high ratio of fixed capital costs to variable capital and operating costs. An industry with such a cost structure will always be able to provide additional capacity at a progressively lower unit cost. Adelman uses an electrical generating plant in his discussion of natural monpoly:

> For example, within very wide limits the larger an electrical generating plant, the lower its unit cost fixed and variable. Therefore additional capacity can always be had at a lower unit cost. (Adelman, 1972, p. 15)

This characteristic precludes local competition and makes the 'shopper's technique' (the ability of a consumer to shop around) impossible because no other supply is readily available to the shopper. He is forced to take the services offered, to go without or to move (Pegrum, 1965, pp. 567–8).

Such an industry is clearly not self-regulating and some degree of state regulation or control may be necessary to prevent monopoly abuse. To what degree does the natural gas industry qualify as a natural monopoly? To what degree is it self-regulating – with rising, not falling, incremental unit costs? The answers, unequivocal in the production and distribution stage, are perhaps more complex in dealing with the question of economies of scale in transmission.

As with oil, natural gas is subject to increasing costs of recovery. The cost of replacing natural gas consumed in the United States, for example, has within a few decades risen from pennies per MMBtu to up to $9.00/MMBtu. Algerian natural gas, once produced for around $0.10/MMBtu now costs some $2.50/MMBtu (including transportation to liquefaction plants in both cases). In Europe, replacement costs for consumed natural gas are running into several dollars per MMBtu. Even in the Soviet Union, the MMBtu costs of the huge West Siberian fields are a multiple of the MMBtu costs of developing the Central Asian fields, which in turn are several times more expensive than the costs of developing the Baku, Caucasus and Carpathian fields, the earliest fields exploited. It is hardly likely, therefore, that enormous surpluses of cheap natural gas will cause chaos in any of these markets. Lack of self-adjustment at the well-head does not provide a justification for arguing that natural gas constitutes a 'natural monopoly'. The reason must be found elsewhere.

The criteria of constantly decreasing unit costs and lack of 'shopper's technique', on the other hand, fit the distribution stage of the industry. Here, the fixed costs of existing mains and distribution lines, storage capacity and other plant make decreasing unit costs at the margin possible. Similarly, local purchasers of natural gas really cannot resort to the 'shopper's technique'. Not unsurprisingly therefore, distribution companies are widely recognized as local monopolies and subject to public regulation or ownership.

It is tempting to utilize the economies to scale argument of the natural gas transmission industry to argue that it constitutes a natural monpoly as well. What are the merits of this argument?

Where numbers of natural gas companies are few, it is possible to organize markets along spatial lines that simultaneously maximize profits, imposing a monopoly/monopsony solution to natural gas transmission. In the following example we shall dwell on a monopsony use of geographic space. (A simple reversal of the logic in Figure 2.7 makes it applicable to a monopoly.)

The horizontal axis in Figure 2.7 represents the geographic distance, the vertical axis the price of natural gas sold for distribution (Odell, 1969, pp. 9–15).[5] G_1 and G_2 represent not only the two main transmission companies involved in the example but also the geographic source of their natural gas. Assuming G_2 is selling natural gas at price P_2 (about $3.50/MMBtu), the natural gas can be distributed and sold profitably only in the geographical area *c–d* on the horizontal axis, the dashed sloping lines away from P_2 indicating the increasing costs of distribution incurred by the increasing distance.

If G_1 receives natural gas at point G_1, what price is he going to charge the distribution companies – P_1 or P_3? Should the price be set at P_3 (some $4.00/MMBtu), it will limit sales to the area delineated by *a* and *b*. Alternatively, G_1 could utilize a larger trunkline, higher load factor and other economies of scale, and market gas at the price

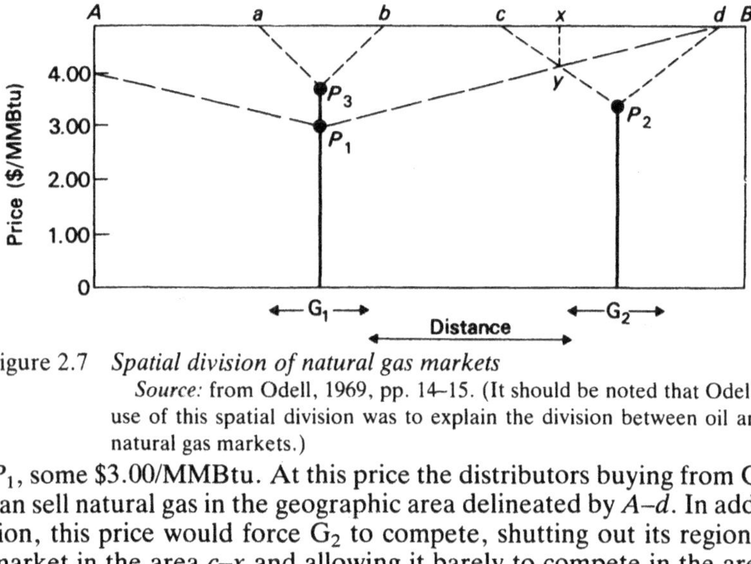

Figure 2.7 *Spatial division of natural gas markets*

Source: from Odell, 1969, pp. 14–15. (It should be noted that Odell's use of this spatial division was to explain the division between oil and natural gas markets.)

P_1, some $3.00/MMBtu. At this price the distributors buying from G_1 can sell natural gas in the geographic area delineated by *A–d*. In addition, this price would force G_2 to compete, shutting out its regional market in the area *c–x* and allowing it barely to compete in the area *x–d*.

G_1's preferences are important here. P_3 could optimize profits and enable market stability; P_1 could entail lower profits and would cut into G_2's bailiwick. It is important to note that G_1 is expected to maximize its profits, irrespective of the consequences to others. Such maximization may not include either the P_1 or the P_3 options, but profit-maximization can result in the exclusion of a large number of potential customers from the natural gas market. (In the P_3 option, they would be represented by the geographical spaces *A–a, b–c*.) Thus the spatial division solution to instability can entail either collusion or competition. In either case, the monopoly advantage implicit in pipelining gas is preserved as much as possible. The result is often high profits to the transmission companies, limited marketing of natural gas and a high degree of consumer/distribution company discontent.

This logic, combined with the recognition that pipelines are almost classic examples of decreasing costs with scale, has led economists to conclude that, when numbers of pipelines are small, a natural monopoly exists: 'A pipeline may be a decreasing-cost industry if and only if it is an industry entire of itself – if the market is so small as only to permit one or two to exist' (Adelman, 1972, p. 16).

To support the argument that, with increasing numbers, pipeline companies cease to be natural monopolies, economists generally refer to the purchasing of natural gas by American transmission companies. In areas such as the Permian Basin, the initial pipelines into the area at first had a monopsony position, but then lost this position as more reserves were found and more pipelines entered the region (MacAvoy, 1962). I am sceptical about the validity of generalizations made on the basis of the specific American organization of natural gas and petroleum markets. (See Chapter 4 for the impact of most favoured nation provisions on the pipeline monopsony positions in the USA.) Nevertheless, two things are clear: over time, both sellers to and purchasers from transmission lines may have recourse to the 'shopper's technique' in selling or acquiring natural gas supplies.

This possibility places constraints on transmission line policies. In Europe, for example, the prospect of cheap Russian gas being sold to Continental markets forced the downwards revision of natural gas prices charged by the companies in the late 1960s. Similarly, the perception that transmission lines in small markets in fact constitute natural monopolies supplies the necessary justification for most governments to establish their own transmission company monopolies, and this in spite of the fact that certain markets, such as the French or the British, were once large enough for transmission company competition. The existence of large national transmission companies in both countries today prevents this hypothesis from being more than a hypothesis. If international transmission companies were acceptable to European governments and if the European natural gas markets were integrated, one could very well argue that the natural gas transmission industry is self-adjusting. The point is that this is not the case.

The reason why this is not the case is that politics and political preferences intrude. National governments are often involved in their natural gas industries because of historical circumstances (such as the nationalization of manufactured gas industries and their consolidation into state enterprises in France and the United Kingdom). They may be involved in the natural gas industry in order to exert control over the manner in which their natural gas is produced, domestically have their own separate dynamic). It provides a useful tool to understand the bargaining relationships in natural gas markets.

It should be noted that the separate categorization of governmental and industrial activities in Table 3.1, used to highlight their respective roles, may also be somewhat misleading. Such activities are undertaken jointly and are seldom, if ever, mutually exclusive. Industry is, moreover, functionally dependent on government legislative power – and at times on government regulatory or arbitration powers – for its continued existence. The segmentation of industrial activities into vertical ownership and contractual activities is likewise a simplification. A variety of corporate behaviour patterns exist. Vertical integration may seem a poor substitution for the market structure that collective ownership of a common carrier pipeline can give the firms involved. Furthermore, the expertise vital for a firm's survival need not always come from a set of contractual relationships, common ownership of pipeline facilities or possession of producing fields and transmission lines; it could just as well be a characteristic of enlightened, well-informed management.

Governmental strategies

It is an oft-cited rule of thumb that firms cease being purely economic actors when, in pursuit of higher profit or a larger market share, they apply pressure on government to change the rules by which the market is ordered. Given this supposition, those industries only marginally subject to the discipline of the market will patently be more readily 'political' than others. Nowhere is this more true than of the natural monopoly nature of much of the natural gas industry. To be successful in this business, one must have the hardened qualities of a successful political tactician.

It is remarkable how little this is acknowledged within the industry itself. The recent establishment of the Danish natural gas net is an excellent case in point. Although small relative to other natural gas enterprises dealt with in this volume, the Danish case provides a wealth of illustrative material.

Danish natural gas was initially procured from a North Sea producer group, the Danish Underground Consortium, under a negotiation procedure set by a preliminary agreement between the Consortium and the Minister for Trade and Industry. The fuel was to be marketed through the national industry. When the Danish state-owned company, the Danish Oil and Natural Gas Company (DONG), encountered difficulties in its bilateral negotiations with the producers, the state entered into and enforced a compromise contract. When DONG needed loans to complete its share of the project, it was the state that guaranteed those loans raised on international capital markets. The overall structure of the Danish natural gas industry was a governmental creation as well. Distrustful of vesting too much power in DONG, government insisted on the creation of five regional distribution companies to which DONG was to sell its gas, but without fixing rules about how the negotiations were to be conducted (a fatal flaw as it turned out). Furthermore, the government set forth principles under which natural gas was to be sold to consumers; this led for a while to an awkward political compromise – the same gas was to be sold on a cost basis, on an oil product parity basis, and on a premium fuel basis. It was even specified in which areas natural gas was to be allowed to compete with other natural monopolies, e.g. the powerful electrical utilities. For 1982 and 1983, as uncertainty increased about the ability of the natural gas enterprise to sell the quantities of natural gas required by its contract with the North Sea producers, there was growing political discussion about compulsory measures that would ensure residential consumption of the planned quantities of the fuel.

Throughout this often vicious 'cut-and-thrust' battle, the managing director of DONG complained repeatedly of 'political interference' in the planning of the natural gas net. And yet the very existence of DONG depended on repeated government interventions on its behalf. What is interesting about the Danish example is that each of the parties was determined to conduct relationships on a voluntary, purely commercial basis. It seems to have come as a shock to all concerned – especially within DONG but also within the Social Democratic party – that the 'free play of market forces' meant constant government intervention to sort things out between conflicting parties and to define the issues in favour of the one or the other economic interest.

Although instructive about the degree of government activity that can take place in establishing a particular solution, it would be unwise to generalize too much from this one specific example. The wide variety of governmental roles in the natural gas industry defies short description. In this volume alone, the roles range from those of the Soviet Union, where the natural gas industry is the government, through strong planning systems (France and Japan), through nations that fully or partially own national gas companies (British Gas in the UK, Gasunie in Holland), to governments whose primary role is regulation of privately owned firms (the United States). The analysis of government policies is nothing less than the comparative analysis of national industrial policies.

Nevertheless, some general statements can be made about how governments influence the number of actors on the market, the terms of market exit, and the conduct of flexible transaction relationships (as outlined in Table 3.1). Very broadly speaking, the types of role government can play in this regard can be classified into three categories: the discrete support of inputs into the industry; the provision of an arbitration forum where differing interests can resolve their disputes; and direct participation in the industry. The first two functional roles are performed by virtually all governments, although in differing manners; the third is true of most governments, although perhaps not of all.

In common with other industrial endeavours, natural gas activities receive a wealth of public goods to ease their development: land and capital in particular, but also labour at times as well, are often provided on favourable terms. Discrete activities like the expropriation of rights of way, the provision of investment capital (or guarantees for loans), and guarantees against labour strikes (which could interrupt deliveries) have all been exercised by governments in support of the industry. Governments have been even more involved in aiding the procurement of natural gas supplies internationally. Government roles here have run the gamut from coordinating the policies of other nations against the Algerian demands for higher fob prices for their LNG exports, through interceding on behalf of their national firms, to outright subsidization of natural gas imports.

Another discrete government role has been that of arbitrator. Here the government may work both passively, seeing that national laws are enforced in terms of settling inter- and intra-industry conflicts, and more actively, in the establishment of elaborate regulatory and hearing paraphernalia. The latter is true of the United States; even though American gas regulation can hardly be called discrete, it has always been a highly public political issue in domestic politics.

Far more prominent has been direct government activity in national gas markets. In Great Britain, France, Italy and Canada, state-owned firms have played dominant roles in securing gas supplies for national markets and in fulfilling national energy policy goals. That the roles of British Gas, Gaz de France, ENI–SNAM, and TransCanada Pipeline/PetroCanada have from time to time created considerable political debate should not surprise, given the importance of these firms for the national economies of which they are a part.

Corporate strategies

Interlinked with governmental strategies are the strategies of the various corporations. The number of corporations in specific national markets varies in a manner that might not be foreseen from the simple examples in Tables 3.1 and 3.2. In the United States, for instance, there are literally thousands of producers, dozens of transmission lines and hundreds of distribution companies. In Japan, as yet a relatively small market, there are some 232 independent entities. In France, Great Britain and Italy, by contrast, there are one or two firms that almost totally dominate the natural gas industry. Common to all these markets is the use of legal contracts to enforce stability on the market.

Contractual solutions The natural gas market is characterized by a wealth of contracts of all kinds. There are contracts involving the delivery and installation of the capital equipment involved in natural gas trade. There are contracts ('operating agreements') between operators and other partners in the exploration and development of natural gas fields, and similar agreements for common carrier pipelines (prevalent in Europe, but not in the United States). There are joint venture contracts, contracts between affiliated parties, and contracts between non-affiliated parties ('arms' length contracts'). But the 'bread and butter' contract of the industry is that of natural gas purchase (or sale). This contract forms the basis on which producers, transmission firms and distribution entities raise and invest the billions of dollars needed prior to the actual commencement of deliveries. (Two examples of typical producer–transmission line contracts are included in Appendix IV of this book.) Although natural gas contracts of this sort vary, they do so only within the limits set for them by the nature of the trade, by historical precedent and by the perceived mutual needs of the parties involved. As a result, natural gas contracts can be characterized by a certain recognizable pattern of legal clauses that appear in contract after contract. These clauses are set out in Table 3.2 with the intended/incidental effects.

In addition to normal market barriers or political hindrances, private firms have their own methods to reduce the number of potential partners or to increase them. Chief among these is, of course, the

Table 3.2 *Natural gas contracts: provisions correlated with forms of instability*

Form of instability	*Contract clauses*	*Effects*
Number of actors	Limitations of parties to which a contract can be transferred/applied.	Restrict number of actors; ensures proper intended use is made of natural gas involved
	Most favoured nation	Weakens any monopsony/monopoly power of buyer/seller

continued

Table 3.2 *Natural gas contracts: provisions correlated with forms of instability — continued*

Form of instability	*Contract clauses*	*Effects*
Flexible basis for transactions	Pricing/escalation provision	Ensures prices of gas under contract evolve with normal economic developments
	Most favoured nation	Ensures equity between parties to a given contract and parties to a similar contract in the same region
	'Take or pay'	Ensures the volume of gas offered under contract will be purchased or that the seller will get an equivalent amount in cash
	'Deliver or pay'	Same as with 'Take or pay', except it is the buyer who benefits
	Load factor clauses	Ensure a given load factor to be fulfilled by the parties to a contract
	Renegotiation clauses	Allow the parties to negotiate changed circumstances not foreseen on entering the contract
Exiting	*Force majeure*	Specifies those conditions under which a party is not bound to perform according to contract
	Amounts to be delivered	Specifies the amounts to be delivered; seller/buyer obligation to see that the specified amounts are delivered inhibits exit
	Arbitration clauses	Gives parties recourse to enforcement/arbitration in event of non-performance/ exit

specification of partners in natural gas contracts and the inclusion of provisions making it difficult or impossible either to include other parties or to sell the contract further. But there are other indirect methods.

In the United States, the use of the most favoured nation (MFN) provisions has worked to undermine monopsony power. There are two kinds of MFN clauses in the United States: the two-party clause, in which the buyer is obligated to pay to the seller the equivalent of any higher price that the buyer may subsequently pay for a comparable quantity of gas from another seller within a defined geographical area; and the third-party clause, which obligates the buyer to pay the seller the equivalent of any higher price that a third party purchaser may subsequently pay for a comparable quantity within the defined geographical area (Neuner, 1960, p. 91). In the Permian Basin, in the San Juan natural gas province and in the Southeast Louisiana–Mississippi areas of the United States, these most favoured nation provisions have eroded the monopsony position of the early transmission company in each area. A pipeline enjoys economies of scale only until it is filled to capacity. If additional sources of natural gas are then still available, the monopsonist can build another line or increase the capacity of the old line. Because of the original monopsony prices that the transmission company would have achieved, it could outbid any other transmission company in any competition for additional supplies simply by paying more of the additional supplies and 'rolling in' the higher prices of the additional gas with the low monopsony prices for the original supplies. The MFN provision undermines this position. With the two-party provision, should an alternative producer in the same area offer natural gas, the monopsony buyer is in a difficult position. It must obtain the natural gas to preclude other transmission lines, but if it pays more for the additional supplies, that will automatically apply to the original monopsony price. If it does not bid for the natural gas, then another transmission company will deprive it of its monopsony advantage. If there is a third-party MFN provision, the situation is even more serious. For should the alternative producer sell to an alternative transmission line at a higher than monopsony price, then the monopsony price position is eroded and the transmission firm can do nothing about it. This feature of the most favoured nation provision shows how subtly contract provisions can work when it comes to controlling the numbers of buyers and sellers in a natural gas producing area.

The specification and indexation of prices is vital to ensure that the basis of transactions will evolve to the satisfaction of the contract parties. Pegging the price of the gas generally with an eye to the cost of other fuels normally sets the foundation for the marketing of the

Table 3.3 *Pricing provisions of Pac Indonesia and Algeria II import projects (US$/MMBtu, gross heating value, loaded fob)*

Pac Indonesia

Contract sales price:
Calculated quarterly.

$$P = P_o \times \left(0.5 \frac{A}{\$11.00} + 0.5 \frac{W}{135.0}\right)$$

P = calculated contract sales price.
P_o = $1.25.
A = applicable Indonesian crude oil price.
W = applicable value of the index of wholesale prices – all commodities.

Currency revaluation factor:
Applies to the contract sales price.

$$B = 1 + \frac{\Sigma \frac{c2}{c1} - 1}{11}$$

$c1$ = the commercial rate of exchange in effect on the date of initial deliveries for each of the currencies.
$c2$ = the arithmetic average of the commercial rates of exchange on the applicable dates in each quarter for each of the currencies.
B = 1 until its absolute value changes at least by 0.1%. Thereafter new value for B used only if it differs from old by 0.1% or more.

Algeria II

Contract ('invoice') price:
Calculated semi-annually.

$$P = P_o \left(0.5 \frac{F}{F_o} + 0.5 \frac{F^1}{F^1_o}\right)$$

P = invoice price.
P_o = base price equal to $1.30 as of 1 July 1975.
F = price of No. 2 fuel oil for New York harbor.
F_o = $12.642.
F^1 = price for No. 6 fuel oil, low pour, max. sulfur of 0.30%, delivered New York harbor.
F^1_o = $13.505.

Minimum price:
Calculated monthly.

$$MP = MP_o (E + 1).$$

MP = minimum price.
MP_o = base minimum price equal to $1.30/MMBtu as of 1 July 1975.
E = arithmetic average of the results obtained by applying the formula:

$$\frac{B}{A} - 1 \text{ to each of 6 currencies.}$$

Maximum $B = 1.25$.

Minimum contract sales price:

During its debt amortization period Pertamina will calculate a price sufficient to meet:

(1) repayment of principal amount (including interest during construction),
(2) payment of interest when due, and
(3) payment of projected costs of operation and maintenance.

A = average commercial exchange rate for each currency during July 1975.
B = average commercial exchange rate for each currency as measured by average purchase and sales rates for telegraphic transfer for each business day of preceeding month.
E = 0 until its value increases by at least 0.1% as compared to 0. Thereafter new value for E used only if it differs from old value by 0.1% or more.

Floor MP = US $1.30.

Recalculations of the minimum price will be made *once* according to the following formula:

$$MP^1 = \$.80 \frac{X}{2{,}300} + \$.15 \frac{Y}{60} + \$.35$$

MP^1 = recalculated minimum price.
X = actual capital costs incurred by Sonatrach (in millions of dollars).
Y = actual operating costs of Sonatrach during the first years of operations (in millions of dollars).

Source: Office of Technology Assessment, 1980, p. 78.

natural gas. Too high a contract price means that natural gas will be confined to premium uses alone; too low a contract price will enable the purchaser to sell natural gas further at cut-rate prices, thereby threatening other fuel markets. Indexation is also vital to flexibility, by ensuring that price evolution for natural gas sales in the future will not occur in isolation from developments in the rest of the economy. Table 3.3 illustrates the various price-stabilizing terms for two US LNG contracts. Although the clauses are from LNG contracts, they are similar to those for pipeline contracts. There are three aspects of pricing flexibility: the establishment of a contract or invoice price tied to some other economic measure, the use of currency provisions, and the inclusion of minimum price clauses.

Indexation of the contract price is of course of considerable importance. Not shown in the table is the importance of the size of the original contract price and the place at which it is charged. Here there is a clear conflict of interest between seller and buyer. The seller would like to have as high a price as possible, levied as close to the point of final sale as possible. (This ensures a high contract price and the maximum effect of an indexation clause.[1]) The buyer, on the other hand, wants the exact opposite.

Currency clauses are designed to ensure that untoward exchange fluctuations do not unduly affect the profitability of a joint enterprise. Surprisingly, they are also often a feature of domestic contracts in Europe and the Third World. They are seldom used in the United States. The currency clause, it should be noted, is expressed separately in the Indonesian contract but is tied to a minimum price clause in the Algerian case.

The minimum price clause in both contracts assures the investing partners, in this case Pertamina and Sonatrach, that the prices in the contract will always be sufficient to cover their capital and operating costs. This is the buffer effect of the minimum price – the price below which the contract price cannot fall. Its contribution is obvious. It assures the producers that they will under no circumstances lose money in the course of the contractual relations. Minimum prices similar to these are a common feature in European natural gas contracts with regard to both pipeline and LNG supplies.

Other clauses mentioned in Table 3.2 bear directly or indirectly on transaction flexibility. The MFN clause (already covered in another context above) here has the same intention as the indexation of contract prices. Another clause is the load factor clause – a clause not only specifying the nature of expected deliveries, but giving an indication of the necessary investments that both parties must undertake.

A slightly different set of clauses also ensures payment should a partner be unable to perform according to obligations. These are the 'take or pay' and (more rarely) 'deliver or pay' clauses. As indicated in the name, these clauses, respectively, require the purchaser to take what is provided by the seller under the terms of the contract or to pay the seller the equivalent amount ('take or pay'), or the reverse ('deliver or pay'). Generally, in both cases there is agreement that the undelivered but paid-for amounts will be 'made up' at some future time. This is often hard to accomplish, however.

It may seem untoward to include renegotiation clauses as a means of ensuring flexible transactions. Nevertheless, these clauses are an expression of a very real need on the part of all involved in a contractual relationship periodically to review the relationship and to determine what changes are in their mutual interest to implement. Even the most perceptive vision of the future has its cloudy aspects, and this is no less true of natural gas negotiators than it is of other mortals. Such clauses give all parties a chance to change contractual terms so that all are content. This is an extremely important condition for the success of any natural gas trade.

There are two sets of clauses to ensure that parties will not suddenly exit from the contractual relationship. The first of these, *force majeure*, specifies under what conditions the parties are released from their obligation to honour their contractual commitments. These clauses may be long and specific – describing situations such as nuclear war, the North Sea freezing over, and such – or they may be cursory and vague. Such specifications of exemption from contractual obligations serve as a means to determine whether one party or the other to a contract is in fact 'exiting'. After this is certified, the second of the two clauses, that specifying arbitration arrangements (the picking of national jurisdiction, or court, that will 'try' contract violations), enters the picture. Arbitration may not have to be resorted to; often the mere threat serves as a sufficient inducement to performance, and the case need go no further.[2]

The emphasis on mutual interests in the contractual relationship should not disguise the very real conflicts that occur during contract negotiations. The separate parties all come with their own distinct preferences, which must be reconciled over time. Such negotiations can be very bitter and prolonged, lasting up to two years and resulting in contracts running to hundreds of pages. Table 3.4 shows how interests can differ among the various parties to a single or several contract relationships.

These differences tend to be logical. Producers want a fast start-up and a quick build-up of production, both of which will maximize their cash flow rates of return. Transmission firms want a somewhat slower build-up to enable them to dispose of the natural gas contracted for. Both producer and transmission company are interested in high load

Table 3.4 *Conflicting interests and contract terms*

Contract term	*Producer interests*	*Transmission company interests*	*Local seller + consumer interests*
Load factor	High as possible; reduces development costs	High as possible; reduces transport costs	Low as possible; reduces marketing costs plus enables entry to flexible energy markets
Production profile	Quick start-up of production increases IRR on investment	Varies depending on pipeline capacity available plus cost of increased capacity	Slow start-up gives opportunity to establish markets for gas
	High annual rate of production enables stronger gas flows	High annual rate may lead to marketing difficulties	Lower production rate ensures longer period for selling gas
Price	High initial price ensures quick return to capital	Purchase as low as possible, selling price as high as possible without destroying final purchaser–consumer markets	Low initial price enables establishment of domestic gas markets[a]

	Price escalation clause as favourable as possible (tied to oil product plus crude prices)		Price escalation clause formulated to allow inroads on important oil product markets
	Base price increases at a constant compounded rate	Price escalation set to rent capture plus future final purchaser demand	
		Base price (purchased) increases irregular; if regular, orderly arithmetical increases	
Penalties	'Take or pay'	Re. producer: 'deliver or pay' Re. final purchaser: 'take or pay'	'Deliver or pay'
Amount of gas	Generally lower amounts	Amounts must be sufficient to cover pipeline costs	Generally higher amounts to enable better long-term planning of gas marketing

Source: J. D. Davis, 1983, p. 6.
Note: a May not apply to utilities etc. that want to keep parity between various fuels+electrical power, etc.

factors, which generally reduce costs. Distribution companies, on the other hand, faced with enormous seasonal variations in market demands, tend to prefer low load factors. Similarly, differences can exist to the types of penalty clauses which contracts contain, escalation clauses, and many other issues.

Integrative solutions Supplementing contracts as 'market stabilizers' are other corporate organizational forms – common ownership, vertical integration, interlocked directorates (see Table 3.1). While these forms supplement and underpin existing contractual relationships, they can also serve to minimize transaction costs. This minimization has both an internal and an external aspect. Internally, vertical integration between stages can enable better coordination of gas production and distribution load factors for the enterprises thus integrated. It can also enable more effective inter-firm pricing policies. (This can be beneficial when the trade is international in that it can allow for transfer pricing to avoid tax incidence from different tax regimes.) Vertical integration can force complementary management skills together, providing for better expertise and decision-making capability. Externally, a vertically integrated firm can be better placed to compete. It can also be better equipped to withstand challenges from its competitors. A transmission firm with a large producing subsidiary, for example, could be better placed to negotiate a large contract than a firm with only a transmission function.

In terms of our contract discussion, the internal aspects of vertical integration are the most striking. Contracts are no longer really 'arms' length' contracts, as parties to these agreements are in effect 'prisoners' within the same corporate group. This considerably eases contractual relationships and problems such as transactions flexibility and exiting. Given all these advantages, therefore, it is hardly surprising that contractual relationships are not infrequently supplemented by some other organizational form of coordination between firms. Indeed, today most firms outside the United States have obtained at least some degree of vertical integration either in a formal sense, with one firm owning subsidiaries engaged in the various industrial stages, or (particularly in Europe) informally through common ownership, interlocked contractual arrangements, and the like.

Natural gas: the nature of competition

To what degree is the buying and selling of natural gas subject to 'marketplace discipline' at all? In these pages, state intervention, vertical integration, common ownership of pipelines, and contractual relationships have all been stressed. The consequences of these phenomena are not to be mistaken. They constitute constraints on the 'free play' of market forces. These constraints, it has been emphasized, are necessary to the success of the industry. That this is so is frequently misunderstood by layman and specialist alike, giving rise to serious misconceptions about the nature of the industry.

The first of these misconceptions confuses constrained competition with the absence of competition. Stating that the market for natural gas is 'constrained' by non-market forces is not the same as contending that natural gas is somehow 'removed' from the discipline of the market. This is to mix an economic conception of perfect competition with the economics of oligopsony–oligopoly markets or bilateral monopoly relationships. These are all subject to the final test of success: success or failure on the market. An analogy from an industry characterized by difficulty of access and dominance of the few – the automobile industry – is perhaps illustrative. Few would contend that the automobile industry is 'unconstrained'. It is a market typified by oligopoly behaviour, one in which governments are notorious for intervention. This does not mean that Volkswagen or General Motors have set market forces in abeyance.[3] Rather, as competition from well-designed Japanese automobiles has made rather obvious, even mighty General Motors is subject to some of the same types of criteria for success or failure as smaller firms. This holds true for the natural gas industry as well. No amount of governmental intervention, and no degree of intra-industry organization, is going to make a 'bad' natural gas contract 'good'. Within limits, therefore, the natural gas industry is subject to the same rules of market success or failure as are other industries.

Nor is it pre-ordained that state firms or state regulation are necessarily less efficient than interrelationships between private firms. Goldberg, in fact, provocatively likens the problems of government regulation to those of 'administered contracts', the type of regulation endemic to the natural gas industry:

> . . . [I]f the regulatory relationship were replaced by a private contract, the problems faced by the private gent would differ mainly in degree than in kind from those that plague regulators and provide a field day for their critics. Indeed even in a rather simple private sector contract like a university food service contract one can observe on a lesser scale the whole panoply of regulatory horrors.
>
> Such contracts typically last for three years with the price for the last two years to be determined on the basis of expected costs plus a specified rate of return on sales. The university like the regulator

must determine which costs are allowable; it must further decide on how detailed its monitoring of the firm should be; and on how much discretion the firm's management should have in making decisions which will affect costs . . . (Goldberg, 1976, pp. 441–2)

The reverse of governmental regulation of the natural gas industry (directly through a governmental regulatory authority or indirectly through a state-owned national firm) is not necessarily a more efficient industry. Rather, it is an industry that will be plagued with many of the administrative contractual problems cited by Goldberg. Given the transaction costs thus incurred, the absence of governmental regulation may or may not lead to a more efficient natural gas industry.

A final misconception is that natural gas industry behaviour should conform with some economic criterion for efficient performance. In fact, the natural gas industry almost never does, and where it does it is more the result of accident than of plan. For this lapse, the industry is frequently taken to task by economists. Yet there can be some very good reasons for failure here. To begin with, what criterion should be chosen for the negotiation of a North Sea contract between producer and a transmission company? Should the price be based on the cost of gas production (the rent element accruing thereby to the transmission company)? Should the price be based on the cost of replacing that natural gas to be consumed under contract? Or should it be based on the prices of other fuels – and if so, what other fuels? Here the difficulties of finding a single criterion on which the principals can agree, and then defining that criterion to the satisfaction of both, are obvious. Even assuming the unlikely event that a criterion is chosen and agreed upon by the parties, what guarantee is there that the agreed-upon price that is right in the year 'zero' will be appropriate in year 15? Both parties are going to be locked into a contractual relationship for that long – and to adequately ensure that the price will follow 'lock-step' with the demands of the criterion, the contract would have to be made infinitely complex, thereby raising transaction costs, problems of interpretation, and so on.

In practice, the North Sea producer and the transmission company will generally enter negotiations with some sort of economic criteria, normally differing for each party. In this case, both sets of criteria serve as an intellectual justification for obtaining the negotiating goals of the two parties involved. Thus in my personal experience, a North Sea negotiation commenced recently in which the transmission company was determined to buy the natural gas concerned 'at cost' (defined loosely as a 20 per cent IRR on investment); the producers entered the negotiations determined to tie natural gas prices to the most expensive oil products on the market. Intellectual justification of these goals was shaky on both sides and the process of negotiation and the resulting compromise buried whatever small bit of economic theory that had existed at the outset.

The process of negotiation undermines economic criteria in other ways too. Economic efficiency is only one of the negotiating goals. There are many others that in the last instance can be of more importance. One of these is the initiating of a contractual business relationship. For the parties concerned, solving for the problems of instability as outlined in this chapter is of far more import than whether or not the contract reflects 'cost-plus' criteria. In addition, other goals than gas price and delivery terms may be the subject of negotiations: access to pipeline transportation, security against oil revenue taxes, questions of mutual training facilities, and many other issues. The minutiae of contractual negotiations also serve to divert attention away from theoretical economic criteria: it is difficult to think about the theory of pricing after a day negotiating 'toilet visitation' rights on a North Sea platform.

The result of these processes is a contract that may make little economic sense. This view of the economic processes underlying natural gas markets may be imperfect, but it is a fair explanation of the real world political economy behind the natural gas trade, a view that will be elaborated in the following chapters.

Notes

1 This point is worth developing a little as it is important. Assume that the price of the contract can be charged either at a platform in the middle of the North Sea, or landed at the beach. If the price is charged at the platform, it must be lower by a couple of cents than the landed price to pay the purchaser for the transportation charges involved. If we assume that indexation causes the price to double, the following situation occurs:

(1) $$P_N = P_0 \times 2$$

(2) $$P_N = (P_0 + 0.03) \times 2,$$

where the first equation refers to a price charged at the platform (P_0 being the original contract price, P_N the price at time N) and the second equation refers to a price paid 'landed'. Clearly the landed price – closer to the point of ultimate consumption – is the better of the two for the seller, the platform price for the buyer.

2 One such notable case occurred between Sonatrach and Trunkline (an American transmission company). Sonatrach attempted to renegotiate prices for LNG that it

had contracted to deliver to Trunkline. Trunkline refused to enter into these renegotiations. Sonatrach attempted to suspend deliveries, in effect cancelling the contract. Trunkline threatened to take the matter to arbitration. Sonatrach, surprising many, 'caved in'.

3 For such an argument, see Galbraith (1970), pp. 38–40.

4

United States: Regulated Stability

Introduction

Not only is the American gas industry the largest in the Western world, but in terms of organization it is also probably the most complex, and definitely the most unique. It has been estimated that there are at least 3,800 corporations producing natural gas in the US. Sixty-three companies are engaged in inter-state gas transmission. There are about 1,700 distribution companies. Together, these companies manage and run over 78,000 miles of gas-gathering lines, some 263,000 miles of transmission lines and some 688,000 miles of distribution mains, enough in total to wind around the globe forty-one times. There is also incredible diversity within these figures, and US gas literature bristles with terms not found elsewhere: inter-state and intra-state sales of natural gas, sales for resale, 'over the fence' sales, 'field prices', 'incremental' and 'rolled-in' prices, 'new natural gas', 'biennium gas', to mention just a few. As a general rule, a proliferation of specialized terms leads one to suspect the not so invisible hand of the legal profession. American natural gas is no exception to this rule. The US gas industry is first and foremost a regulated industry, and it is in the means by which it is regulated that its uniqueness lies. In the chapters to come, various national solutions to the tendency towards disorder in natural gas markets will be reviewed. In almost all of these cases, stability is sought through actions such as limiting numbers, monopoly and monopsony positions, shadow pricing, vertical integration, and extensive government ownership–direction of the natural gas industry. In the US, however, apart from approximately 500 municipally owned utilities, the entire US gas industry is privately owned.

Although the American natural gas industry is divided into three parts – production, transmission and distribution – regulatory levels do not match the three parts: distribution is regulated by both municipalities and the individual states; inter-state transmission is regulated by the federal government and intra-state transmission by the individual states; production is regulated by both state (prorationing of production) and federal authorities (inter-state and, since 1978, intra-state prices). There are different views of this regulation; it has caused economists to despair, consumers to leap to its defence, and producing and transmission companies to lobby intensively. Efforts to change regulation or, in some instances, to abolish it entirely have given authorities considerable frustration and difficulty. Undeniably, however, regulation has led to a particular set of American solutions to the general problems of instability. The manner in which new legislation alters or eliminates this regulation will lead to a slightly different solution set. In more senses than one, understanding natural gas in the United States is understanding the American regulatory system and its effects.

Early days: town gas, natural gas and utility regulation

Curiously, although the American natural gas industry was a leader in the introduction of gas to industrial, commercial and domestic markets, it got its start rather late. This was more a function of the relatively recent widespread use of natural gas than of any slowness on the part of the American gas market. Natural gas was initially exploited commercially in 1821 in Fredonia, New York, where use was made of natural gas seepages first to illuminate a local inn and then to light the town. This started a 'boom and bust' pattern of natural gas usage, a pattern to which we referred in Chapter 1. The distribution of town gas, in contrast, had a much more secure economic existence. It is not surprising, therefore, that in the period 1816–1925 town gas was much the preferred alternative.

The supply of both town and natural gas during this period was a bit rough and tumble. In addition to the 'fly-by-night', 'here today, gone tomorrow', natural gas distribution companies, there were the municipal town gas and light works. Regulation of these was initially the responsibility of the various cities they serviced, cities often known for their political machines and corrupt politicians. Therefore there was, early on, a drive to regulate gas utilities in the public interest. Many of the reformist politicians of the early 1900s got their start investigating corrupt practices in gas and electrical distributing companies. In New York, a young lawyer, Charles Evan Hughes, took on Tamany Hall, investigated the city's utilities and revealed a series of abuses. He recommended that the price of town gas be reduced from \$1.00 per 1,000 ft^3 to \$0.75, and that the price of electricity be reduced from 15¢ to 10¢ per kilowatt hour for prime customers. Riding on this plank, Hughes achieved national fame and the governorship of his state, introducing two comprehensive state public service commissions (Anderson, 1980, pp. 13–15). Some two-thirds of the states followed New York in the next decade. Two other governors, Robert M. La Follette of Wisconsin and Hiram Johnson of California, also achieved national reputations in this movement. Still another, Woodrow Wilson of New Jersey, eventually rose to become President of the United States.

Free enterprise buccaneering was if anything more pronounced in the exploration and development of oil and gas resources in the American Southwest. Following the first commercial producing well drilled at Titusville, Pennsylvania, in 1859, the oil industry blossomed. Reputations and fortunes were made and lost overnight. These were the years of the formation and break-up of the Standard Oil Trust and the great discoveries – Spindletop, East Texas, and other fields – that have since given the oil industry its more colourful characters. In the spirit of this freebooting period, any associated natural gas that was found was usually (wastefully) vented or flared. Non-associated gas was often disposed of in the hopes that accompanying gas liquids/oil would at some point begin to flow. However, the oil industry soon began to be regulated too: state legislation was introduced to allow for more orderly production and development of the fields that were found. Legislation was also introduced to curb venting and flaring of natural gas. Most of this legislation, nevertheless, was not on the books until the 1930s and 1940s.

Linking the production of natural gas to its consumption on a long-term basis first became feasible with the invention of the all-welded high-tensile steel pipeline in the mid-1920s. The first of many long-distance transmission lines was laid from Northern Louisiana to Beaumont, Texas, a distance of some 217 miles. Other lines followed rapidly. In 1931 Chicago was connected to the Panhandle fields with the first 1,000 mile pipeline in history. The natural gas era had begun.

Given the 'free-wheeling' within the gas utilities and the oil industry at this time, there could be little doubt that conflicts in natural gas pricing and distribution would occur – and occur they did. Far from being grateful, local utilities rapidly became suspicious of the new fuel. Town gas, then as now, was a relatively expensive fuel source. It had already survived one challenge, from the electrical utilities, which deprived it of the municipal lighting market. Because natural gas was purchased for pennies per thousand ft^3, then transported at an undetermined cost to the transmission companies and sold for considerably more than the purchase cost, suspicions were aroused that the transmission companies were making windfall profits from their operations.[1] This was only one of the problems. Existing town gas works fought the encroachment of natural gas into their areas. Elsewhere, the transmission companies refused to service communities that wanted gas but also sought some control over the rates charged. A lawsuit to gain insight into pipeline earnings and gas prices (*Missouri vs. Kansas National Gas Co.*) was brought before the Supreme Court in 1924. The court rejected the plaintiff's arguments on the grounds that state authorities had no jurisdiction over inter-state commerce.

By the terms of this decision, no one possessed any authority over the industry at either its producing or receiving ends in cases where the industry was involved in inter-state commerce. Most of the companies engaged in inter-state transmission delivered their gas directly to the local distribution companies, and these companies perforce had no means to exercise control over prices paid for their natural gas if it was obtained from outside their state (Pegrum, 1965, pp. 656–7).

Charges by the manufactured gas companies of price-fixing led to investigations by the Federal Trade Commission (FTC). Holding companies comprising hundreds of gas and electrical utilities figured prominently in the 'Crash' of 1929 and were highlighted in an FTC examination of the causes of the débâcle – an examination brought to an end in 1933 with the publication of an eighty-volume report. The case for some form of control seemed overwhelming. Regulation was soon to follow.

Early regulation: 1938–54

The hole in regulatory legislation indicated by *Missouri vs. Kansas Natural Gas* was filled by the Natural Gas Act of 1938, which was passed unanimously by Congress. It succeeded earlier legislation designed to improve public control over the holding companies, which by this time presided over considerable assets. Forty-four holding companies, it is estimated, controlled some 66.4 per cent of manufactured gas and some 29.3 per cent of natural gas produced in the United States (Pegrum, 1965, p. 660). While there were some advantages to holding company arrangements, there were also considerable abuses: charging of excessive fees to operating companies; use of objectionable, even fraudulent, means of accounting; the manipulation of securities markets; the retention of millions from operating companies under the excuse of payment of federal taxes; and other practices that were not in the public interest. The Public Utility Act of 1935 put these holding companies under the Securities and Exchange Commission, which was given the job of controlling their activities and eliminating unnecessary companies (Pegrum, 1965, p. 661).

Perhaps because of the success of the Securities and Exchange Commission in controlling utilities, focus in the literature on natural gas has been concentrated on the Federal Power Commission (FPC), the authority vested with the powers to control inter-state commerce in natural gas. This is something of a pity in that the FPC in its heyday really accounted for only a fraction of the regulation implemented in the field of natural gas production, transport and sales. Table 4.1 illustrates where the Federal Power Commission fits into the bigger picture. The FPC controlled the general traffic in inter-state gas from 1938 until its dissolution in 1977, but there were significant limitations to this regulation. For one thing, it regulated the rates of return for natural gas transmitted through inter-state lines for resale to local distribution companies (city gate prices) but not for natural gas sold to industry, even though this gas might share the same pipeline. Furthermore, and this is important given the nature of competition in pipelines, the FPC could not order interconnections of gas lines even under emergency conditions. The degree to which production of natural gas for eventual inter-state commerce actually fell under the powers of the FPC was a hotly debated topic for fifteen years. Finally, the US Supreme Court clarified the issue in a 1954 decision (*Phillips Petroleum Company vs. State of Wisconsin*), ruling that the FPC's jurisdiction extended to 'the rates of all wholesalers of natural gas in interstate commerce whether by a pipeline company or not and whether occurring before, during, or after transmission by an inter-state pipeline company'. This somewhat unfortunate decision led the FPC to attempt to set 'fair field prices' for natural gas – an incredibly complicated jurisdictional matter that was to lead to much controversy.

At the same time, the natural gas industry was regulated by the individual states through both public utility commissions, which generally control public utility prices to consumers alone, and conservation commissions, which are empowered to control rates of production from oil and natural gas fields.

As a result, by 1954 the US gas industry was subject to a patchwork quilt of regulatory authorities whose competences often overlapped and whose activities were widely regarded within the industry as generally iniquitous.

Regulatory consequences: 1938–54

The period 1938–54 witnessed rapid growth in the natural gas industry. In 1938, about 65.12 billion m^3 of natural gas were consumed in the continental US; by 1954, consumption was at the 238 billion m^3

Table 4.1 *Overlapping federal and state regulatory institutions and functions*

Stage	*Inter-state*		*Intra-state*	
	Institution	*Function*	*Institution*	*Function*
Production	FPC (1938–78) FERC (1978–)	1954 FPC controls price of natural gas sold to inter-state lines 1978 FERC supervises implementation of the Natural Gas Policy Act (both inter- and intra-state sales)	Conservation/pro-rationing committees (Texas Railroad Commission, etc.)	Production regulation/unitization. Attempt to control gathering lines, etc.
Transmission	FPC (1938–78) FERC (1978–)	1938 Natural Gas Act (regulation of transport or sale of natural gas in inter-state commerce) (sales for resale)	State public services commissions	Control of intra-state transmission and sale of natural gas (can include controls of industrial sales by transmission companies)
Distribution	Securities and Exchange Commission	1935 Public Utility Act (control of securities and financing of utilities and holding companies)	State public services commissions	Regulation of utilities at the consumer level

level, a 365 per cent increase in a period of sixteen years. By the same year around 140,000 miles of transmission lines had been laid. The 'Big Inch' and 'Little Big Inch' oil pipelines, built to supply the East coast with oil from the American Southwest, were converted to natural gas, aiding in the opening of the domestic, commercial and industrial markets of the Central Eastern Seaboard. During this period the well-head price of natural gas on a national average increased from 1.0–1.7¢/ft^3 to around 10.1¢/ft^3.

Most of this gas was located in Texas and Louisiana. (Even today, slightly over 73 per cent of all natural gas comes from these two states; this figure rises to more than 90 per cent if one includes the neighbouring states of Oklahoma, Kansas and New Mexico.) Its transport was something of a triumph of technology and organization. It was during this period that entrepreneurs laid the foundations for the large transmission lines that characterize the industry today. While men such as Rockefeller, Deterding and Gulbenkian have been immortalized in the oil literature, few readers may have even heard of Paul Kayser, an attorney who founded the El Paso Gas Company to carry gas by pipe from Lea County, New Mexico, to El Paso. Kayser and his company survived the depression by running lines to the copper mines in Arizona. During the war, El Paso extended its gas lines into California, opening up this valuable market. The company was on its way to becoming the pre-eminent transmission firm in the United States. Another personality, Gardiner Symonds, originally a Chicago banker, became the founder of the Tennessee Gas Transmission Company, part of a conglomerate later renamed Tenneco. Symonds' empire today consists of no fewer than three major US transmission companies: the Tennessee Gas Pipeline Company, which supplies the American Northeast with gas from Louisiana and Texas, and the two branches of the Midwestern Gas Transmission Company, one running from the NE line in Tennessee to the Chicago metropolitan area, the other connecting the Trans-Canada pipeline and the Midwest. In addition, Tenneco comprises a major oil company and has interests in chemicals, packing, and agriculture and land development. Texas Eastern, another of today's leading transmission companies, was orginally formed by Texans interested in bidding for ownership of the 'Big Inch' and 'Little Big Inch' oil lines to the American Northeast. Under the energetic leadership of George and Herman Brown, the firm acquired the lines, which proved capable of carrying some 140 million ft^3 a day to the Northeast. Texas Eastern then added some twenty-one compressor stations, increasing the capacity of the lines to some 435 million ft^3 per day. Not content with this, the Texas Eastern management acquired another transmission company, the Transwestern Pipeline

Company, to service California and the West from Texas and Oklahoma. It has, moreover, a 28 per cent holding in another inter-state transmission company, the Algonquin Gas Transmission Company, a major supplier to New England markets (Peebles, 1980, pp. 56–64; Ridgeway, 1973, pp. 337–58).

Despite this growth, in 1954 there was no truly national natural gas market in the United States. Instead there were two markets, separated by the Continental Divide. On one side of the Divide were California, Montana, Utah, Colorado and West Texas. On the other were the producing areas of the Gulf coast and mid-continent and the consuming areas on the East coast and in the Midwest. Imports, such as they existed, were split along the same lines (Adelman, 1962, p. 47). Prices paid in 1955 at the consuming end, together with production prices and transport tariffs, are given for New York City in Table 4.2.

Table 4.2 *Natural gas and competitive fuels: New York residential market, 1955 (US $/MMBtu)*

	No. 2 fuel oil[a]	*Residual fuel oil*	*Natural gas*
Price to producer	—	—	0.114
Transmission tariff	—	—	0.221
City gate price/price at New York	0.752	0.387	0.335
Distribution costs	0.326	—	1.755
Price to consumer	1.078	—	2.090
Allowing for efficiency[b]	1.8911	—	2.60

Source: Adelman, 1962, pp. 70–1.
Notes: a Otherwise known as gasoil.
b Estimated at 57 per cent for oil, 80 per cent for gas.

The $0.114/MMBtu field price for natural gas given in Table 4.2 deserves further comment. It is considerably above earlier field prices for natural gas, having more than doubled in the period 1949–54. The reason for this increase was two-fold. First, during the early years of natural gas production, there were few alternative buyers to the one inter-state pipeline company that serviced a particular gas-producing area. The inter-state companies used their monopsony buying power to force prices to rock bottom. In the words of Breyer and MacAvoy:

> In the 1950s, pipelines pushed field prices below competitive levels wherever possible, . . . [T]he field price for natural gas was often depressed below the competitive level because of the lack of effective competition among buyers. During the next few years, several

> pipelines sought new reserves in old field regions where previously there had been a single buyer, and this new entry raised the field prices to a competitive level. (Breyer and MacAvoy, 1974, p. 62)

Thus monopsony equilibrium price, a price that would just cover the costs of natural gas production, was being replaced by a more competitive equilibrium price requiring at least one alternative purchaser, such as is specified in the discussion in Chapter 3.

But this was not all. The escalation clauses in the oil contracts were increasingly being supplemented by contingency clauses, in particular most favoured nation (MFN) provisions. These clauses were themselves part of the negotiation process, as pointed out by Adelman:

> Clearly the contingent clause was an appreciable part of the bargain: the seller was getting something additional and hence accepted a lower initial price than he would otherwise have taken; the buyer would not pay as high a price for a reserve if he had also to sign the contingent clause. It would be interesting, if it were possible, to estimate the current values of the contingent clauses from year to year; i.e. what would a seller have required and a buyer have been willing to offer to pay, for a contract striking the clause? Probably the clauses were worth more to sellers than either buyers or sellers believed them to be worth – i.e. price increases after 1946 were greater than expected. (Adelman, 1962, p. 43)

Generally, therefore, the interaction between MFN clauses and increased competition for field gas provided a 'ratchet' effect on field gas prices. It must be emphasized that this is a generality. Small producers in Texas, especially those with remote gas fields, continued to feel the sting of the oligopsony position of the major pipelines; as recently as 1960, monopsony still existed for major producing regions, most notably the West Texas–New Mexico fields, that were essentially serviced by only one pipeline.[2]

As previously noted, the period 1938–54 witnessed a rapid growth in the inter-state/intra-state pipeline system. The companies reacted to the contingency clauses in a predictable manner, usually refusing to sign any more additional contracts in each gas-producing area in order to hold the prices of contracted gas down. Yet the pipeline companies were confronted with a 'prisoner's dilemma'. By boycotting an area to protect old prices, they opened a vacuum for other buyers to fill – a development that would lead them to pay higher prices anyway. The one authority that could have preserved monopsony price structures and the irrelevance of most favoured nation clauses was the Federal Power Commission. But rather than limiting extensions of inter-state lines into these regions, the FPC actively promoted pipeline competition for access.

> [W]hen some pipelines, or local gas distributors, or local public utility commissions protest the approval of new [interstate pipeline projects] on the logical and factually correct ground that this will feed excess demand and tend to put up the price of gas, they have little success before the Federal Power Commission. (Adelman, 1962, p. 47).

Control of producer prices would be forced directly on the FPC by the federal courts, but not until 1954.

What of the city gate price (the price of sale to local distribution lines) of natural gas versus the price of No. 2 fuel oil in New York, as presented in Table 4.2? Two things should be noted here. First, natural gas clearly has an advantage over No. 2 fuel oil and residual fuel oil in New York City with regard to large industrial markets. The additional cost of distributing gas to large industrial users is far lower than the cost to residential consumers. (This extra cost can run to some 10¢/MMBtu and, even where it is incurred, the greater efficiency of natural gas as a fuel would tip the balance to gas.) Even in residential markets, natural gas has some advantages: it has an extra application (cooking) not available to oil-heated homes, and it provides a certain degree of ease of consumption: one applies the energy by turning a switch, much as in the case of electricity – a far more expensive alternative. The second and more important factor to note is that the price to the residential consumer varied greatly, both within and without New York City. 'It should not be supposed that the very high distribution cost is typical for American cities. [This] shows variations from a low of 24.8 cents to a high of 193 cents for 1953–1954' (Adelman, 1962, p. 71). It is likely that the advantages of natural gas are more marked elsewhere.

Regulation: 1954–78

Nowhere is the bargaining/legislative regulatory aspect of US natural gas more apparent than in the question of field pricing of natural gas. The doubling of field prices within the five-year period 1949–54 discussed previously did indeed awake political concern. It was also recognized that the Natural Gas Act of 1938 lacked clarity about the degree to which the FPC could in fact control field prices directly. (There has been little disagreement over the FPC's authorization to control field prices indirectly in instances when it approves inter-state sales and determines that field prices are excessive relative to costs of service.) The first case in what was to be a tug-of-war was that of the *Colorado Interstate Gas Company vs. the FPC* (1945). Here the Supreme Court held that in cases where the inter-state natural gas company owned production and gathering facilities that provided gas for inter-state commerce these facilities were subject to the control of the FPC. Despite the attempted passage of the Kerr Bill in 1950, designed to reverse this decision, the Supreme Court extended FPC powers even further when, in *Phillips Petroleum Company vs. the State of Wisconsin* it unequivocally extended FPC price and rate controls over all stages in wholesale sales of gas for resale in inter-state commerce. An attempt to reverse this decision in 1956 was defeated by a veto from President Eisenhower, a veto justified by what the President saw as an objectionable amount of lobbying by gas industry representatives.

The impact of the *Phillips vs. Wisconsin* decision was not long in coming. As Breyer and MacAvoy state:

> The very numbers of producers of natural gas created overwhelming difficulties in rate of return regulation. In 1954 there were more than 5000 producers, and by 1960 more than 2900 applications for increased rates were awaiting FPC action . . . By 1960 the Federal Power Commission had completed only 10 of these cases. The remaining backlog led the Landis Commission, appointed by President Kennedy . . . to conclude that 'without question' the Federal Power Commission represents the outstanding example of 'breakdown of the administrative process'. (Breyer and MacAvoy, 1974, p. 68)

The only way around this backlog was for the FPC to issue 'interim prices' to be applied to produced gas pending the ultimate determination of price. As the backlog increased, the FPC (in Phillips II), rather than hearing and determining individual cases, divided the US into five major producing areas and set prices for each area. An additional problem was that, through setting prices based on historical data that applied to old fields (fields starting production or in decline), the FPC could not take account of the marginal costs of the discovery and production of 'new' gas. The result was the establishment of a two-tier price system: one price geared to 'old gas', the other to 'new' (Permian Basin case). The other significant legal precedents are highlighted in Table 4.3.

Nor did regulatory 'initiative' stop here. In the later 1960s, a change in FPC hearing procedures was implemented whereby the FPC instituted a bargaining process, rather than going through

Table 4.3 *FPC regulation of field gas prices, 1945–65*

Date	*Case*	*Principles*
1945	Colorado Interstate Gas Company vs. FPC	Supreme Court decides that, when a natural gas transmission company owns production and gathering facilities, these are subject to the control of the FPC
1950	Kerr Bill	Attempt to amend the Natural Gas Act of 1938 through exempting 'arms' length sales' of natural gas from FPC regulation; vetoed by President Truman
1954	Phillips Petroleum Company vs. State of Wisconsin	Supreme Court determines that FPC has 'jurisdiction over the rates of all wholesalers of natural gas in inter-state commerce whether by a pipeline company or not and whether occurring before, during, or after transmission'
1956	Harris–Fulbright Act	Exempt well-head prices from FPC regulation; vetoed by President Eisenhower
1963	Supreme Court affirmation of FPC Phillips II decision	Abandonment of case-by-case price regulation; country divided into five regions and prices set by area
1965	FPC Permian Basin decision	Introduction of two-tier price ceiling system, one for 'old' and the other for 'new' gas

lengthy hearings to determine the 'relevant cost/price criteria' applicable to various cases (the original Phillips case, for example, consumed eighty-two hearing days and some 10,626 pages of testimony). An example is the Southern Louisiana case, where the producers, the distribution companies and industrial customers negotiated a set of prices and then took them to the FPC for approval. The FPC considered the recommended prices, set them a 'shade' lower than those agreed upon between the parties, and gave its approval. This administrative precedent relieved the FPC of the burdens of conducting its own hearings and collecting its own evidence; this burden was passed to the purchasers and producers. Although the new FPC method of conducting hearings has been criticized by some as not approximating 'rent capture' – in that the prices ultimately decided on are the result of bilateral negotiations between two groups and not the product of objective economic analysis – this sort of bilateral negotiating situation is not unusual elsewhere in the world of natural gas. Theoretically, too, the producers could wait for an alternative bid from another set of consumers, much as has been outlined in the previous chapter.

The pattern of regulation of natural gas field prices had essentially been set by the late 1960s. There were two basic types of regulatory category: general categories, where all natural gas exploited in an area or at a specific time could be regulated, and discretionary regulations, which sought to set certain prices for certain gas developed under particular circumstances. There was a tension between these two sets of regulatory categories. For one thing, the FPC would for purposes of reducing its (and producer) transaction costs attempt to cover as much natural gas pricing as possible through general regulations. Then, as the inherent inequities of these general regulations became obvious, as well as the deleterious effects on production of higher-cost natural gas resources, the FPC would switch over to more discretionary forms of regulation. This might, for example, involve the granting of certain favours to specific categories of new gas to make their production more economic. This would be done at a cost to the FPC itself of increased regulatory burdens (increased transaction costs).

The 'pendulum-like' back and forth movements of the FPC for the period 1954 to early 1977 are illustrated in Table 4.4. It should be noted in this context that these policy changes only partially alleviated the increasing shortages of natural gas dedicated to inter-state transmission. The nationwide price increases of 1974 and 1976 in particular reflected an increasingly desperate situation, which culminated in severe curtailment of natural gas deliveries in the winter of 1976–7 and which led to the enactment of the Emergency Gas Act of 1977 and the payment of very high prices for 'emergency sales'.

Rising natural gas field prices were not the only occurrence during the period 1974–8. The period also witnessed the replacement of the FPC by the Economic Regulatory Administration (ERA) and the Federal Energy Regulatory Commission (FERC) as part of the formation of the Department of Energy in 1977. Of the two bodies, FERC took over most of the personnel and juridical competences of the FPC. The ERA has a predominant voice with regard to natural gas imports into the United States.

The FERC in the fledgeling Department of Energy had its job cut out for it, as became clear when Congress finally passed the Natural Gas Policy Act (NGPA) of 1978, an Act so complicated that its implications are still not fully understood. This Act, the result of

Table 4.4 *Policy types/costs: FPC regulation of natural gas field prices 1954–77*

	Policy category	
Year	*General/inclusive (low transaction costs)*	*Discretionary/exclusive (high transaction costs)*
1954–60	—	Case-by-case, FPC rulings re. field gas prices
1960–3	Field gas prices set by regions instead of individual cases	—
1965	—	Permian Basin case leads to different price ceilings for 'old' versus 'new' gas
1974 (June)	Introduction of single nationwide rate for new gas produced on or after 1 January 1973 ($0.42/thousand ft^3)	—
1974 (December)	Nationwide rates increased to $0.50/thousand ft^3 with annual 1¢ escalations	—
1976 (July)	Nationwide rate increased to $1.42/thousand ft^3 for gas from wells commenced on or after 1 January 1975 with 1¢ per quarter escalations	—
1977 (February)	—	Emergency Natural Gas Act of 1977 – federal allocation of natural gas until 30 April 1977; FPC authorizes emergency sales of less than 1 trillion ft^3 (2.84 billion m^3) at prices averaging $2.25/thousand ft^3

Note: all prices were subject to adjustments for energy content, state severance taxes and the like.

much lobbying/political pressure and bargaining between the producers and the producing states on the one hand and the consumers and the rest of the country on the other, specified some twenty-five different ex-field pricing maxima for natural gas. Included in these were the intra-state natural gas producers, who were brought under FERC auspices. Table 4.5 summarizes this Act, designating the various categories of natural gas by paragraph heading, correlating the legal categories with the actual conditions under which the gas is (will be) produced, and specifying the date of ultimate decontrol of the gas along with the percentage share of the entire US market that the gas had (a) at the time of passage in 1978, (b) in 1980, and (c) that it is anticipated to have in 1990. Two other pieces of legislation were passed simultaneously with the Natural Gas Policy Act: the Powerplant and Industrial Fuel Use Act of 1978 which attempts to restrict industrial boiler and powerplant usage of the fuel, and the Public Utility Regulatory Policies Act of 1978, which attempts to modify existing rate structures to a more equitable basis among large and small consumers.

A more than cursory look at Table 4.5 will reveal how revolutionary the NGPA really is. The amount of gas sold under existing contracts – contracts not subject to deregulation (Section 104) – is expected to decline from 49.2 per cent of total gas sales in the US to some 6.4 per cent by 1990. This gas plus Section 105 natural gas (deregulated in 1985) are together projected to decline from 86.4 per cent of the US market to 10.3 per cent in 1990. The slack is to be taken up by Section 102 natural gas (the so-called 'new natural gas') and by Section 107 natural gas ('high-cost natural gas'), which are together expected to increase from some 4.4 per cent to 63.1 per cent of the gas supplied to inter-state and intra-state consumers in 1990. Whether events will live up to the expectations behind the NGPA is something else together.

The problems of regulated stability

The regulators

It is not without reason that US natural gas occupies a special place in the literature on economic regulation. Depending on the economic method and viewpoint involved, regulators of natural gas have been either notoriously unsuccessful or conditional successes. The workings of the FPC in particular have come in for acidulous comment and analysis

The FPC (now FERC), as we have noted, was charged with the regulation of natural gas pipelines and with the regulation of field

Table 4.5 *Maximum gas ceiling prices under Natural Gas Policy Act with market shares thus affected, 1978, 1980, 1990 (est.)*

Section of Act	*Gas category*	*Ceiling description*	*October 1981 ceiling prices per MMBtu*	*Date of deregulation*	*Percentage of total production by section 1978*	*1980*	*1990 (est.)*
102	New natural gas						
	• new onshore wells at least 2.5 miles from nearest marker well or at least 1,000 feet deeper than any completion within 2.5 miles	$1.75 as of 20 April 1977 plus monthly inflation and escalation adjustments	$2.909	1 January 1985	3.9	17.3	41.9
	• New onshore reservoirs			1 January 1985			
	• new Outer Continental Shelf (offshore) leases effective on or after 20 April 1977			1 January 1985			
	• reservoirs discovered after 27 July 1976 on old offshore (OCS) leases			Not deregulated			
103	New onshore production wells						
	• wells with surface drilling starting after 19 February 1977, satisfying applicable federal or state well-spacing requirements and that are not within a proration unit	$1.75 as of 20 April 1977 plus monthly inflation adjustments	$2.514		3.9	14.1	5.9
	–gas from wells deeper than 5,000 ft			1 January 1985[c]			
	–gas from wells shallower than 5,000 ft			1 July 1987[a c]			

104	Gas dedicated to inter-state commerce before the NGPA enactment (9 November 1978)						
	• post-1974 gas		$2.080	Not deregulated	49.2	32.7	6.4
	• 1973–4 Biennium gas		$1.760[b] $1.348[c]	Not deregulated			
	• replacement contract gas or recompletion gas	The just and reasonable price as of 20 April 1977 plus monthly inflation adjustment	$0.988[b] $0.756[c]	Not deregulated			
	• flowing gas		$0.497[b] $0.424[c]	Not deregulated			
	• certain Permian Basin gas		$0.589[b] $0.517[c]	Not deregulated			
	• certain Rocky Mountain gas		$0.589[b] $0.497[c]	Not deregulated			
	• certain Appalachian Basin gas		$0.469[b] $0.442[c]	Not deregulated			
	• minimum rate gas		$0.259[l]	Not deregulated			
105	Gas sold under existing intra-state contracts						
	• if contract price was less than $2.078 on 9 November 1978	The lower of (a) the contract price under the contract terms as of 9 November 1978 (b) the Section 102 price		1 January 1985[f]	37.2	24.2	3.9
	• if contract price was more than $2.078 on 9 November 1978	The higher of (a) the contract price as of 9 November 1978 plus monthly inflation adjustment and (b) the Section 102 price		1 January 1985[f]			
106	Sales of gas made under 'rollover' contracts						
	• inter-state	The higher of (a) the just and reasonable price as of the rollover date plus monthly inflation adjustment and (b) $.54 as of April 1977 plus monthly inflation adjustment[d]	$0.770[b]	Not deregulated	0	4.6	11.7

Table 4.5 *Maximum gas ceiling prices under Natural Gas Policy Act with market shares thus affected, 1978, 1980, 1990(est.) — continued*

Section of Act	*Gas category*	*Ceiling description*	*October 1981 ceiling prices per MMBtu*	*Date of deregulation*	*Percentage of total production by section 1978*	*1980*	*1990 (est.)*
106	• ntra-state	The higher of (a) the price paid under the expired contract as of the rollover date plus monthly inflation adjustment or (b) $1.00 as of April 1977 plus monthly inflation adjustment[id]	$1.433[g]	1 January 1985[h]			
107	High-cost natural gas[j]						
	• gas produced from wells 15,000 ft or deeper drilled after 19 February 1977	Section 102 price or higher incentive price	market price	1 November 1979	0.5	3.3	21.2
	• gas produced from geopressured brine, coal seams and Devonian Shale	Otherwise applicable or higher incentive price	market price	1 November 1979			
	• gas produced from tight sands		200% of the Section 103 price	Not deregulated			
	• qualified production enhancement (only for 105 gas)		Section 109 price	Not deregulated			

108	Stripper well natural gas • non-associated natural gas produced at an average rate less than or equal to 60,000 ft^3 per day over a 90-day period	$2.09 as of May 1978 plus monthly inflation and escalation adjustments[k]	$3.116	Not deregulated	4.8	4.4	4.3
109	Other categories of natural gas • Prudhoe Bay gas • gas not otherwise covered	$1.45 as of April 1977 plus monthly inflation adjustment	$2.080	Not deregulated	0.0	0.0	4.7

Source: Office of Oil and Gas, 1981, Part One, pp. 14–15, 51, 43.

Notes:
a Beginning 1 January 1985, gas from wells shallower than 5,000 ft receive a price midway between the price specified by this formula and the 102 price.
b Small producers – independent producers not affiliated with a Class A natural gas pipeline company whose total jurisdictional sales on a national basis, including those by affiliated producers, do not exceed 10 Btu on a 14.73 pressure basis.
c Large producers – producers that are not small producers.
d Ceiling prices may be raised if just and reasonable.
e Inter-state production from 103 wells on dedicated acreage committed on 20 April 1977 is not deregulated.
f If contract price exceeds $1.00 by 31 December 1984, except a price established under an indefinite price escalator clause.
g Or expired contract price, whichever is higher.
h If the price is more than $1.00 on 31 December 1984.
i Natural gas production in which a state government or an Indian tribe has royalty or other interest is to receive the Section 102 price if it was not committed to inter-state commerce on 8 November 1978.
j High-cost gas provisions elective, i.e. do not apply if special tax provisions are utilized.
k These prices have been escalated monthly, in addition to the inflation adjustment factor, by 3.5 per cent annually. Starting April 1981, they will escalate by 4 per cent annually.
l Dollars per thousand ft^3.

prices of natural gas. Its activities in both areas have come under criticism.

What criteria did the FPC utilize in its regulation of inter-state pipelines, for instance? The standard practice had been to regulate the rates of return. These rates should not, therefore, have been above the rates that companies such as United Gas, East Tennessee and the like had received for other investments with as little risk as natural gas pipelines. In fact, the rates of return allowed by the FPC were considerably above this level (Table 4.6). As can be seen in the table, only one case, that of Consolidated Gas Supply, even approximates the rates of return on equity in other equivalent investments.

Table 4.6 *Gas pipelines: allowed rates of return compared with estimated cost of capital for alternative investments (%)*

Company (year of rate decision)	*Est. cost of equity capital (alternative investments)*	*Allowed rate of return to equity (FPC decision)*
United Gas Pipeline (1961)	4.8–5.0	10.3
Panhandle Eastern (1961)	4.5–5.0	12.0
Alabama–Tennessee (1962)	5.0–5.2	9.3
El Paso Natural Gas (1962)	5.0–5.2	12.0
Natural Gas Pipeline of America (1968)	7.0–7.5	9.5
Panhandle Eastern (1968)	7.0–7.5	11.2
El Paso Natural Gas (1970)	9.0–9.5	13.7
Consolidated Gas supply (1970)	9.0–9.5	9.4

Source: Breyer and MacAvoy, 1974, pp. 31–3, Tables 2–3 and 2–4.

The process of setting rates for pipelines thus appears largely to have been the result of political bargaining between the Commission and the pipeline companies concerned. Again, this reduced the FPC case load. Had it set allowed rates of return even close to the rates of return to equity for other investment, it might well have provoked retrials and new rate-setting decisions far exceeding its capacity. As Breyer and MacAvoy state: 'To sum up, the commission may have limited extreme profit taking, but allowed rates of return seem to have exceeded costs. Although this state of affairs reduced the commission's case load, it added more to pipeline profits than to consumer price savings (1974, p. 33)[3]. This statement may be a bit unfair. Allowed rates of return were generally tied to a specific load factor, often up to 90 per cent; should the pipeline company be unable to maintain this load factor for the project reviewed, its rate of return could fall dramatically. Given the gearing of equity to loan capital, such a shortfall could have dramatic effects on the overall rate of return of the project concerned.

Within the realms of political bargaining, FPC/FERC decisions nevertheless often seem haphazard. A good example of this can be found in the rate hearings for the El Paso II LNG project designed to import LNG from Algeria. Here, the FERC ignored its own staff in allowing an internal rate of return (IRR) to equity in the shipping segment of the project of 18.41 per cent. (Staff argued that, with the various credit and subsidy arrangements allowed on shipping markets, the actual rate of return would be closer to 30 per cent.) Furthermore, in using the yardstick of IRR rather than booked rate of return, FERC overlooked the fact that, for shipping, an IRR of 18.41 per cent in this case approximated a booked rate of return of around 48 per cent (L. N. Davis, 1979, pp. 203–4, 208). FERC failed, in addition, to examine the listed cost of the receiving terminal (which had also been figured into the rate base). This terminal's estimated cost was 2.65 times that of a similar terminal given in the *Oil and Gas Journal* at about the same time (L. N. Davis, 1979, p. 207, fn. 23).

If the FPC/FERC regulation of natural gas prices (as discussed in the previous section) and its regulation of transmission rates were questionable in terms of economic efficiency and good regulation, the interface between federal regulation and local or state regulation would complicate the process even further. The impact of the Phillips decision on state regulatory authority with regard to Texas has been characterized by one authority as follows: 'All the other circumstances inherent to the petroleum industry had made the creation of a rational system of gas regulation in Texas unlikely. The Phillips decision made it impossible' (Prindle, 1981, p. 101). The regulatory body in the greatest natural gas producing state in the 'lower 48', the Texas Railroad Commission, had long had authority to regulate pipelines. This was no mean task; there are so many oil and gas pipelines in one Texas producing area that it is known as the 'spaghetti bowl'. In addition, the Texas Railroad Commission had authority over gas flaring and over prorationing (and thus indirectly over sales and sales prices). The difficulties of the state regulatory agencies multiplied[4] with the Phillips ruling and, more particularly, with the 1963 Supreme Court decision whereby production regulation (prorationing) in Kansas (and thereby Texas), insofar as it affected inter-state commerce, was rendered illegal. The result was that for years before the passage of the NGPA a field developed by two different corporations – one selling to intra-state (Texas) markets, the other to inter-state pipelines – was subject to differing regulatory jurisdictions, one of which disallowed the other its major purpose. This anarchy was not diminished by the Natural Gas Production Act but remains to plague producing companies, inter-state and intra-state pipelines, and consumers alike.

The dual market problem

The Natural Gas Policy Act of 1978 substituted one set of dual markets for another. One of the problems plaguing US natural gas supply prior to 1978 had been the coexistence of a regulated inter-state market and a non-regulated intra-state market. The NGPA removed this distinction and replaced it with another: regulated (primarily 'old' inter-state gas) markets and unregulated markets, with all the shades in between. Post-1978 legal language became if anything even more arcane than before, now replete with terms like Section 102 gas, Section 103 gas, Section 107 gas (deep uncontrolled), and so forth. Let us examine the problems of these dual markets briefly before concluding our analysis.

The impact of the pre-1978 coexistence of a controlled and an uncontrolled market side by side is easily demonstrated in Table 4.7. It can be seen that the inter-state market in the United States fell from almost 62 per cent in the mid-1960s to little over 53 per cent in 1977. This happened in a period when natural gas was becoming increasingly utilized as a fuel in the Northeastern corridor, in the Great Lakes region, and in California and the Northwest. Table 4.7, if anything, understates the situation in 1977. Since natural gas contracts run anywhere from fifteen to twenty-five years, old natural gas contracts would clearly bias the proportion of inter-state sales upward in the period 1967–77. More impressive are the statistics on how much 'new gas' (recently discovered gas) was dedicated to inter-state

Table 4.7 *US natural gas consumption and inter-state sales 1963–77 (trillion ft^3 p.a.)*

Year	*Total consumption*	*Of which through inter-state sales*	*Percentage of total* %
1963	16.07	9.8	61.0
1965	17.46	10.8	61.9
1967	19.87	12.3	61.9
1969	22.8	14.1	61.8
1971	24.94	15.1	60.5
1973	25.21	14.7	58.3
1975	22.82	12.9	56.5
1977	22.79	12.1	53.1

Source: Subcommittee on Energy and Power, 1980, pp. 442, 447.

as opposed to intra-state lines. Oklahoma, the third-ranked natural gas producer in the US 'lower 48', is not untypical. Although Oklahoma was a major supplier to inter-state pipelines, in the period 1972–7 over 85 per cent of newly discovered gas here was dedicated to intra-state usage at higher prices.

The ludicrous aspect of this problem is that, during the severe natural gas shortage of 1977, the producing states suffered a surfeit of natural gas dedicated to intra-state usage. This in turn was the major factor in finally bringing about reforms, resulting first in the Emergency Act and later in the Natural Gas Policy Act.

With the NGPA, the rules made an about turn. Suddenly, it made comparatively little difference whether natural gas was dedicated to inter- or intra-state usage. More important was its classification in the Act, and whether the FERC could be lobbied to increase price ceilings. The natural gas lobbies began work almost immediately: Section 104, 106 and 109 gas, they contended, was priced too low and should be repriced to coincide with the more expensive categories of natural gas in Section 102 (deep and new gas) and Section 103 (gas at depths of 10,000–15,000 ft); Section 103 gas prices should be increased to 150 per cent of the amounts allowed in the NGPA; and so forth. Although the significance of these arguments should not be minimized – they do indeed involve billions of dollars – the details are not of particular interest in this context. What is important is the extent of the political lobbying that occurred and the insistence by the industry on further price increases for various categories of natural gas. Just as higher prices for intra-state gas could be utilized to argue against controlled producer prices for inter-state gas, so higher prices for new gas and deep gas could be utilized to argue against controls on the price of older, cheaper natural gas.

With the 1981 accession to the executive branch of the Republican party – a party pledged to the removal of all regulation of natural gas prices – debate began in earnest about when and how natural gas prices should be deregulated. According to our bargaining hypothesis presented in the previous chapter, this would not be expected to be an entirely frictionless affair. It hasn't been.

The large natural gas organizations were fragmented within. Of prime importance to all concerned was how the approximately 50–60 per cent of natural gas due for decontrol in 1985 should in fact be decontrolled. The American Gas Association, the American Petroleum Institute and the Interstate Natural Gas Association, to name three of the more prominent organizations, went through some rather divisive meetings, regarding both decontrol under the NGPA and the principles of decontrol generally. This lack of consensus was further exacerbated by declining natural gas sales, in part a function of higher prices paid by consumers and in part due to the general economic recession, which has intensified a demand slump.

Generally speaking, there were three groups with conflicting interests. Transmission companies were worried about the effect of higher prices and lower demand on their load factors. Producers were caught between their desire for general deregulation, with the resulting higher prices for natural gas at the well-head, and their desire to maintain or increase their sales. (These two objectives are in fact irreconcilable.) The natural gas utilities were worried about passing higher prices on to their customers. Rate increases must be approved by state and/or local authorities and this can take time. Furthermore, as prices increase, it becomes harder to retain industrial customers, and this aggravates load factor problems.

There were also more specific conflicts within these groups. Producers of 'old gas' were pushing for deregulation, while producers of Section 107 gas and other high-cost natural gas, sometimes selling for as much as $10.00/MMBtu in 1982 prices, were less sure about the virtues of deregulation. They knew that the reason they could demand and get the prices they did was owing to 'old gas', which, since it is 'underpriced', allows transmission companies to 'roll in' the higher-priced forms and maintain their old markets. Should prices for this 'cushion gas' increase, it would be harder to dispose of the higher-priced categories of natural gas. Transmission companies also varied in their interests in this regard. Many had become involved in the search for new gas. One result was often intra-corporate division, where the exploration department came out heavily in favour of instant deregulation while the transmission division opted for continued support of the NGPA and gradual deregulation.

These internal divisions led to colourful exchanges in various congresses and other industry meetings. An American Petroleum Institute meeting in 1982 is one case in point. As the various producers complained of regulation, representatives from the other gas industry sectors became more and more impatient. At last one of them stated the facts as he saw them: 'When you talk of the free market, you don't really mean free market. You really have in mind price parity with oil.' Oil prices weren't free market prices. OPEC had seen to that. 'The price of oil as a competitor against gas is artificially high' (*Oil and Gas Journal*, 12 April 1982, p. 44).

Towards unregulated stability? US natural gas in 1982

What happens to natural gas markets when one set of solutions is substituted for another? How can regulated prices be phased out and replaced by non-regulated prices without significant market dislocations, intra-industry infighting and political controversy? The answer to this question is still somewhat tentative, but it seems likely that the transition when it comes will be less abrupt and drastic than might have been expected. There are three reasons for this projection. First, it would appear that the role of the federal government/FERC will ease the transition. Second, a surprising amount of the natural gas that will be most affected – inter-state gas – is controlled by a few firms. These firms will probably set the pace for the rest of the industry and ease the problems of transition. Finally, new forms of marketing and contracting behaviour are emerging that limit the risk and uncertainty which would otherwise be intolerable to the individual firm.

The role of government, most particularly in 1982, was to refrain from unilateral initiatives to tinker with this or that price of natural gas covered by this or that section of the NGPA. Rather, government activity has been characterized by a 'wait and see' position while the natural gas industry argues its way to a form of internal consensus. FERC recognizes all too clearly that sudden changes are liable to disadvantage one or the other industrial sector. With the ultimate stakes in the billions of dollars, it has prudently adopted a gradualist position *vis-à-vis* industry arguments, a position also characteristic of the Republican administration of the early 1980s. It is therefore highly doubtful that there will be any sudden swings in policy that will upset the present equilibrium.

Industrial concentration, often regarded as an evil in itself, does have one redeeming aspect: it can simplify things enormously. The degree to which changes imposed externally on producing/transmission/distribution firms can be settled internally among the firms concerned reduces the degree to which thousands of firms might renegotiate thousands of contracts, with a high degree of external supervision.

If one disregards total figures, there is considerable industrial concentration in the US natural gas industry. Some twenty-three integrated producers supply about 70 per cent of all natural gas sold to inter-state pipelines. The ten largest producers account for some 45 per cent of all natural gas (both inter- and intra-state) produced in the United States and about 39 per cent of the natural gas sold. Together with the three largest importers of natural gas from Canada, they account for 45.2 per cent of all natural gas sold in the US.

Concentration is, if anything, somewhat higher in the transmission industry. The transmission companies mentioned earlier in this chapter – El Paso and Transcontinental Pipeline (which have joint ownership and closely coordinated operations), Tenneco (which owns

Tennessee Gas Transmission, Midwestern, East Tennessee), and Texas Eastern (which owns Transwestern and Algonquin) – are cases in point. There are thirty-two major inter-state transmission companies in the United States; the rest are extremely small companies or are distribution companies. Of the natural gas transported by these thirty-two 'majors', the proportion carried by our three corporate groupings amounts to some 31–32 per cent. If we expand the sample to five (including Pennzoil/United Gas, also a major natural gas producer, and Panhandle Eastern), this proportion increases to 51.3 per cent. Many of these companies rely on major utilities/distribution companies as their major outlets: over 50 per cent of El Paso's revenues have historically come from its sales to two California utilities; almost all of Transcontinental Pipeline's revenues come from fifteen major utilities along the Atlantic Seaboard; Texas Eastern is not only the major shareholder in Algonquin, but Algonquin is the second largest customer of Texas Eastern, and so forth.

There is considerable evidence for the hypothesis that concentration will ease the industrial costs of implementing deregulation. The large companies will probably seek their own equilibrium solution while taking their cues from their smaller competitors on the margin. Here, new forms of marketing and contractual behaviour are emerging, forms that insure the links in the natural gas chain against contractual and marketing risk and uncertainty.

Of what do these new forms of behaviour consist? If the theses in Chapter 3 are correct, one can expect various forms: a search by the transmission companies for alternative upstream sources of natural gas; a desire to pass increased costs of natural gas acquisition to the next link in the chain between producer and consumer; an increased readiness to sell contracted-for gas to non-traditional customers through non-traditional channels; and the emergence of new contractual forms to insure all parties involved in the industry. Company behaviour under deregulation confirms all these anticipated forms of behaviour.

With regard to upstream behaviour, the *Oil and Gas Journal* commented in early 1982:

> Gas pipelines in the US are rapidly stepping up their efforts to find and produce more oil and gas . . . Areas of activity for gas pipeline exploration/production programs have fanned out from some original concentration in the Gulf of Mexico to nearly all US oil and gas plays. (Cowan and Hagar, 1982, p. 55)

The reasons for moving upstream are mixed:

> Growth in transmission earnings for several gas pipeline companies

> have slowed, and a move into exploration/production is seen as a way to bolster overall company performance. In addition, several transmission companies have increased the amount of gas they provide their gas distribution networks through their upstream efforts.
>
> Meanwhile other pipelines are faced with dwindling reserves under purchase contracts and seek to augment these supplies with exploration/production programs of their own. (ibid.)

The sums of money are real enough, however, particularly for the smaller transmission companies: ONEOK has virtually doubled its exploration expenditures every year since 1979; Midcon's budgeted amount for 1982 was a 100 per cent increase over the annual average of the previous five years; ARKLA's 1982 exploration expenditures of $82 million were up from a low of $3 million in 1971; Transcontinental's budgeted increases were no less impressive, from the $10–20 million level in 1974 to a budgeted $382 million in 1982 (Cowan and Hagar, 1982).

In dealing with the higher costs of acquiring natural gas, each of the three links engages in a certain degree of 'buck-passing'. Public utilities tend to insist on maintaining old contract prices as much as possible. One of El Paso's principal markets for example, Pacific Gas and Electric, has insisted that El Paso utilize the 'marketing out' clauses that El Paso has in its contracts with producers selling it high-cost (Section 107) gas. El Paso has of course resisted this idea, preferring to pass on the higher costs in the form of higher prices to Pacific Gas and Electric. It has been estimated that the average price of US natural gas will 'fly up' by 43¢/MMBtu when decontrol of one-half of US production is implemented. This average says little about the extremes: some natural gas will rise $1.10/MMBtu; other natural gas will rise by only about 19¢ or so per MMBtu. The utilities naturally have a vested interest in assuring themselves of as much of the 19¢ increase per MMBtu gas as is humanly possible. The utilities are also plagued by the loss of industrial markets, as one after the other of these markets is taken over by other fuels. This has a negative effect on load factor that other customers must pay. Additionally, much of inter-state transmission/utility gas prices is connected to the fixed cost of already installed equipment (pipelines, compressors, storage reservoirs, etc.). With adverse load factors and diminished demand, a smaller quantity of natural gas will have to assume the defrayment of these fixed costs. This results in an even higher price to residential consumers and again in even lower demand in the residential market. Inability to sell the natural gas leads to gas surpluses further down the transmission/producer line. The result is that the producers want both higher prices and increased demand, the transmission companies want continued high load factors and an ability to pass cost increases of natural gas to the utilities, and the utilities want to avoid market parity between natural gas and oil, i.e. higher producer prices for which they have to pay.

Several contract provisions have become prominent in recent years. One of these is the 'marketing out' clause. This specifies that the pipeline company must pay a high price for the natural gas concerned (Section 107 gas mostly), but reserves the right that if the pipeline company cannot sell the natural gas purchased under the contract, it can return and offer a lower price for the gas. The producing company can either accept the new price or find another prospective purchaser at the old price. Clearly, this marketing out provision induces a degree of insecurity in the producers of high-cost gas. Another provision commonly included in natural gas contracts (particularly in 1982) is a maximum price provision, in which the seller pledges to the buyer that there will be a ceiling to the natural gas prices provided for in the contract. These provisions, even though not enforced, place the producers of natural gas in a quandary. They can either try to sell their high-cost gas at the highest price possible (and run the risk of having a 'marketing out' provision implemented), or sell their high-cost gas at a lower price and at lower risk to themselves. By mid-1982 two major US transmission firms had invoked 'marketing out' provisions for their high-cost gas purchases: Michigan Wisconsin announced that it would pay no more than $6.00/MMBtu for its more expensive varieties of natural gas; Transcontinental, once paying as much as $10.76/thousand ft^3 for Section 107 gas, would pay no more than $5.00/MMBtu (*Oil and Gas Journal*, 19 July 1982, p. 44).

The use of 'marketing out' clauses, particularly in Section 107 natural gas contracts, has risen considerably. It is estimated that over 70 per cent of the contracts signed in 1981–2 contained such clauses. About half of these contracts also included maximum price provisions.

Other contract provisions are on the way out. El Paso has served notice that it is abandoning most favoured nation clauses, clauses that in El Paso's case normally set future contract prices to the average of the three highest alternative contract prices. 'Take or pay' provisions are also being hotly contested. These provisions were originally inserted in contracts to assure the producer that some minimum amount of natural gas would be purchased annually. With increased prices and declining demand, however, they are creating major difficulties. The usual 'take or pay' provision averages 76–98 per cent of contract quantities. Pipeline companies argue that these high amounts are unreasonable:

> For example, if take or pay is 90 per cent of maximum volume, the pipeline is caught in a bind. The high take obligation doesn't afford flexibility in the current market. I think take or pay requirements will drop to 60 per cent of maximum volumes. (*Oil and Gas Journal*, 19 July 1982, p. 44)

Problems are particularly severe in California where, because of some large utility take or pay obligations for Rocky Mountain and Canadian gas, California producers have had to cut production to one-third of capacity. This has led to considerable losses.

Another industry practice on the rise is that of 'off-system sales' – a pipeline company faced with a surplus that it can no longer dispose of through its own contractual system sells this surplus to a party outside its traditional system. This of course can lead to selling wars when traditional customers of other pipelines switch their supply sources to the 'off-system' sales. Some of the largest transmission companies have either engaged in off-system sales (such as Consolidated), or have asked FERC for permission to do so (Michigan/Wisconsin, Transcontinental, and Natural Gas Pipeline Company of America).

Another byproduct of higher prices and decreased demand has been a lessening of investment in transmission and distribution systems. This not only enables the companies concerned to deal with the higher costs of gas acquisition, but it also is required owing to declining demand.

The result of these various practices has been to depress the price of gas from the producers. 'Buyers are more aggressive now than a year ago', a director of Transcontinental declared in 1982. His company had just notified producers with 'marketing out' clauses that Transco would pay no more than $5.00/MMBtu for deep gas (*Oil and Gas Journal*, 31 May 1982, p. 50).

If there was optimism among the pipelines, pessimism reigned, at least temporarily, among the producers. Lear, a medium-sized gas-producing company, based its drilling after deep gas in the period 1980–1 on anticipated contract prices of some $6.00/MMBtu. By mid-1982 this price had been revised downward to $4.50–$5.00/MMBtu, a price that tended to reduce Lear's incentive to 'go after deep gas'. Other producers were, if anything, bitter about the situation. One such company, Dyco Petroleum, stated that it would no longer sell deep gas to El Paso because of changes in El Paso's purchase contracts (*Oil and Gas Journal*, 31 May 1982, p. 50).[5]

Particularly interesting in this newly emerging contractual/equilibrium situation is the competitive position of oil products. In the late 1970s, deep gas prices were pegged at around 110 per cent of the prices of No. 2 fuel oil (gasoil). By 1983 they had slipped to about 75 per cent. The failure of the transmission companies to sell their natural gas successfully in competition with No. 6 (industrial) fuel oil on the Atlantic Seaboard had been a major reason for exercising their 'marketing out' rights and bidding less for natural gas today than they did a year or two previously. 'Deregulated gas prices are going to come down and continue to come down', stated an executive of Transcontinental Gas Pipeline Company to reporters from the *Oil and Gas Journal* (31 May 1982, p. 50). As of this writing there are few who would argue with this contention.

Conclusion: natural gas – the American solution

Of the various major consumers of natural gas, with the possible exception of Canada, none but the United States has attempted the regulatory solution to the market instability problem discussed in Chapter 3. In part, this is undoubtedly due to differing government management styles. There is a significant difference between French *planification*, for instance, and the US government style of managing industrial interrelationships. It is also due in part to the extremely large markets involved and to the great numbers of buyers and sellers throughout the entire American gas industry. (This is in noted contrast to the monopsony position of British Gas *vis-à-vis* British gas resources.) Whatever its origins, the US attempt to achieve market stability and equity through the regulatory process is of exceptional interest. So too might be the American contracting processes described in the previous sections, a theme to which we shall return.

Unnoticed, almost unmentioned, in this analysis thus far, has been the declining resources base of natural gas in the United States, a result to a certain extent of the tight regulations of the gas market in the past (MacAvoy and Pindyck, 1975). The nature of this shortage is perhaps most clearly seen in Table 4.8, a list of the 'giant gas fields' discovered to date in the United States. Of the fourteen fields assigned to this class in 1976, nine were discovered before 1940, three in the 1950s, two in the 1960s and none in the 1970s. The original reserves of these fields constitute about 20 per cent of all natural gas discovered in the United States, some 5,024 billion m^3. Today, the fields account for little more than 1,706.6 billion m^3. Three of the fields are no longer registered among the fifty largest fields in the US. The reserves of another are dedicated to an LNG trade with Japan. Together, these fields constituted about 14.5 per cent of economically recoverable reserves in 1980. Furthermore, some 736 billion m^3, or 43 per cent of the total, comes from the Prudhoe Bay field complex on the Alaskan North Slope. This gas can be delivered only upon the

Table 4.8 *Giant Gas Fields[a] in the United States/Alaska, 1976 (billion m^3)*

	Discovery date	*Original reserves*	*Remaining reserves*
Hugoton	1926	1,982.0	354.0
Prudhoe Bay	1968	736.0	736.0
Blanco Mesaverde	1927	311.0	133.0
Gomez	1963	283.2	70.8
Jalmat	1929	229.4	22.1
Monroe	1916	198.2	nr.[b]
Puckett	1952	184.1	33.3
Carthage	1936	169.9	32.5
Katy	1964	169.9	68.9
Rabbit Is. Complex	1940	169.9	130.3
Kenai[c]	1959	141.6	69.6[c]
Old Ocean	1934	141.6	34.0
Mocane-Laverne	1952	107.6	nr.[b]
Bayou Sale	1940	100.0	21.9
Kettleman Hills	1938	100.0	nr.[b]
Totals		5,024.4	1,706.6

Notes: a Here only fields of more than 100 billion m^3.
b No longer registered among the top fifty fields in the USA.
c Reserves dedicated to LNG exports to Japan.

building of a pipeline, the Alaskan Natural Gas Transportation System (ANGTS), currently estimated to cost up to $47.6 billion including inflation and interest charges. This cost is reflected in the probable price of the natural gas to the consumer, currently estimated at $9.20–9.30/MMBtu in constant dollars for the first year, declining to some $3.20/MMBtu in the twentieth year.[6] Seen in terms of the easily recoverable reserves from giant gas fields, it would appear *ceteris paribus* that additions to American gas supplies are liable to be expensive.

Another means of adding to United States' gas supplies is imports. There would appear to be room for expansion here. In 1980, some 22.65 billion m^3 of natural gas were imported from Canada, 2.83 billion m^3 from Mexico, and 2.26 billion m^3 from Algeria. Two problems exist in this regard: first, the exporters must be willing to export natural gas, and, second, the price has to be acceptable to FERC/ERA. As will be noted in a later chapter, LNG imports from Algeria have had an unfortunate past. The Mexicans, too, have been extremely sanguine about the terms on which they are willing to supply natural gas. In Canada (long the principal source of American in British society. Thanks to the Corporation, the man in the street for many years enjoyed a premium fuel at incredibly low prices. This fuel was made available either by the competence of the British Gas Corporation, to believe the Labour backbenchers, or by the misuse of monopoly power, to follow the reasoning of the Tory members of Parliament. Depending on party affiliation, British Gas comes in for fulsome praise or for round condemnation. Where gas issues in the United States are discussed in terms of regulation or lack thereof, natural gas in the British Isles is viewed in terms of the British Gas Corporation. The Corporation therefore shares fully in the controversies that most American transmission companies avoid.

A further indication of this contrast between the two countries is that, unlike the United States, Great Britain enjoyed no access to natural gas (outside of a rather small Algerian LNG trade) until 1965/6. The initial problems of gas in the United Kingdom were those of town gas, a form of fuel on the way out in the United States but still widely used in Great Britain. The origins of British Gas lay in an attempt by a Labour government, beginning in 1949, to reorganize and rationalize this industry, an attempt that was still underway when the news of the West Sole find off the coast of Norfolk signalled the natural gas revolution in Britain.

The challenge of a declining industry

The first challenge that confronted the British gas industry thus had nothing to do with the North Sea, but concerned the decline of the outmoded and increasingly expensive fuel, town gas.

The gas industry had had its origins in Britain. The London and Westminster Gas Light and Coke Company was given statutory authority to provide manufactured gas for lighting purposes in 1812. By 1882, some 500 statutory companies had been set up, two-thirds privately owned, one-third owned by British municipalities (Peebles, 1980, pp. 21–3). As in the United States, the gas companies' expansion was limited by the introduction of electricity. Yet by 1949, the year of the Labour nationalization of the British gas industry, there were no fewer than 1,049 'undertakings' in the UK. A good many of these were little more than small family firms. All were relatively inefficient, and there was considerable waste, chaos and mismanagement.[1]

The Gas Act of 1948 provided the first start, reducing the number of undertakings overnight to twelve. These twelve were called area boards; each was given responsibility for the operations in its own region. The boards were appointed by the Minister of Power and presented annual accounts to Parliament. A coordinating body, the Gas Council, was also created. Its membership comprised a chairman, a deputy chairman (both appointed by the Minister of Power), and the chairmen of the twelve area boards. This body, originally responsible for research, government advising and borrowing, was the first step on the long road to the consolidation of the industry into the British Gas Corporation in 1972–3.

The problems this amorphous group faced were immense. Not only was the industry itself inefficient, but the technology of making gas was fast becoming uneconomic. The price of coal used to make town gas doubled in the 1950s; the already high price of gas therefore rose by 50 per cent to even greater heights (Harlow, 1977, p. 140). To build a reliable and growing industry, therefore, the twelve gas boards and the Gas Council had their work cut out for them.

On the organizational side, rationalization began almost at once. Steps were taken to improve the marketing of coke and other gas manufacturing byproducts. Gas-making was centralized in the larger plants located nearer points of cheap coal supply. New distribution and transportation lines were built to connect districts previously supplied by separate undertakings. By 1962, the 1,050 plants had been reduced to 341. Of these, 151 were very small plants retained to supply outlying areas, while 74 supplied over 73 per cent of the manufactured gas (Harlow, 1977, p. 151).

Technological accomplishments were, if anything, even more striking. To gas produced through the carbonization of coal were added lean gas (produced from oil by the water gas process, or produced as a byproduct from coking) measuring some 300 Btu/ft^3, rich gases (largely ethane and methane purchased from refineries) measuring 900–3,000 Btu/ft^3, and an inert 'ballast' component consisting of nitrogen or carbon dioxide. This mixture, measuring some 400–500 Btu/ft^3, was called 'town gas', and was distributed through the mains from the gas works. Some of this gas was made at the works (made gas), but quite a considerable amount was bought from mines, coking plants and refineries (bought gas). The heavy predominance of coal gas in the 1950s (some 72–75 per cent of the total) caused concern, since coal, the basic feedstock for this gas, increased dramatically in price in this period and threatened to price town gas out of the market. The result was that the area boards and the Gas Council sought desperately to supplement coal gas with other newer types of gas.

During the mid-1960s, these efforts assumed three main forms. First, attempts were made to distil gas from oil. This tended to produce a 'lean' gas, much like that of the Lurgi gas-from-coal technology imported from Germany. The gas industry itself perfected two alternative oil-based processes – the gas recycle hydrogenator and the catalytic rich gas process – but both of these indigenous alternatives remained expensive.

Second, the Gas Council undertook to explore the British Isles for supplies of natural gas, spending about £1 million for this purpose between 1953 and 1958. The amounts found were small, however: by 1961, only two fields, one at Cousland in Scotland and the other at Eskdale in England, were being exploited by the area boards concerned (Select Committee on Nationalized Industries, 1961, p. 69).

Finally, under the leadership of the Gas Council and the initiative of the North Thames Gas Board (the largest and best-managed of the twelve area boards), several shiploads of liquefied methane (LNG) were delivered to facilities at Canvey Island from Lake Charles, Louisiana – the first international trade of LNG. This supplement worked out so well that the Gas Council received authority to purchase Algerian LNG, and subsequently signed the first long-term LNG contract worldwide.

Developments in the price of manufactured gas throughout this period served to spur the industry. As can be seen in Table 5.1, by 1961, it was clear that manufactured gas was becoming too expensive, ranging from 9.2d to 15d per therm on already installed plants. Bought gases were considerably cheaper, ranging from 6.5d to 8.9d per therm.

It was the importation of LNG to Canvey Island that proved a 'boon' to the industry, not only in facilitating its further consolidation, but also in preparing it technologically for the advent of North Sea gas. This was not immediately apparent. Both the import of Algerian methane and the construction of new hydrogenation plants (Lurgi) were cited as reasons for restructuring the industry by the House of Commons Select Committee on Nationalized Industries. This committee put the problem succinctly in its 1961 Report:

> It is clear that recent technological advance has brought the gas industry to a position which was not envisaged in the Gas Act, 1948. The emphasis of that Act was on the independence of Area Boards; technologically, the emphasis today is on their interdependence. (Select Committee on Nationalized Industries, 1961, p. 113)

By 1965, the picture had clarified. Whereas gas from oil processes produced a prohibitively high cost gas, the Gas Council could offer area boards imported methane from Canvey Island at a price of 7.5d/therm. Moreover, this purchase did not involve any heavy investment in capital plant or the incumbent risk. It is not surprising that eight of the twelve area boards availed themselves of the offer.

Furthermore, in order to deliver the methane, the North Thames Board constructed several hundred miles of high-pressure mains plus a conversion plant to town gas (Harlow, 1977, p. 147). This new set of mains had two far-reaching consequences. First, by centralizing the deliveries of variable quantities of natural gas to the eight area boards, it allowed increased Gas Council authority through the back door. It additionally showed the individual area boards the advantages of stronger, more centralized direction. Second, the existence of mains and conversion plant would put the industry in an excellent position to receive methane of roughly the same calorific value at roughly the same pressure from the North Sea once natural gas had been discovered there. Cooperation on this early LNG project thus materially aided the industry in preparing for what was to come. It led to the Gas Act of 1965, which in turn laid the basis for what would become the British Gas Corporation.

Ironically, the move away from coal gas was an unqualified success. By the mid-1960s, such gas had dropped to some 33 per cent of the marketed total. In that this move away from coal involved the expenditure of hundreds of millions of pounds on gas from oil works, the innovation of the Gas Council turned out to be based on the wrong assumptions. Natural gas from the North Sea was to make many works scrap iron within a few years after the first find. In that this move away from coal gas involved the importation of Algerian methane, promoted increased area board cooperation, and laid the basis for the first high-pressure natural gas trunkline (and the first conversion to natural gas on Canvey Island), the innovative drive of the Gas Council was to yield results. For the Council was able to meet the challenge of the North Sea with practical knowledge of the technological and organizational factors that were to play so important a role in the future.

The challenge of the North Sea

The 1958 Schloctern gas find in the Dutch Groningen province was followed by a rapid build-up of interest in the British offshore areas, culminating in the first licensing round in 1964. The first gas find was made in the next year and proved to be only one of many strikes. This rapid pace of events soon became something of a threat to the fragile nationalized gas industry.

True to form, the Gas Council and area boards rose to the new challenge. They confronted and solved two problems: preserving their sales monopoly over the British market itself, and taking control of the disposition of the total quantities of natural gas involved. Their solutions to these interlinked problems were to prove fateful for the future.

In the first instance, preserving their monopoly over the British gas market meant that the Gas Council established a monopsony position *vis-à-vis* the many corporate groups exploring for and producing natural gas from the North Sea. This monopsony position by definition would exclude competition for UK markets. By reducing the number of transmission companies to one (itself), the Gas Council forced the oil companies to sell to it and it alone as a *sine qua non*. This relationship could be only further improved by Gas Council integration 'upstream' into offshore exploration and production. In this manner the Gas Council stood to share in the profits of offshore production and, far more importantly, to gain insight into the economics of its future negotiating adversaries. This strategy, as will be explained, was a complete success.

Yet the very success of this monopsony strategy raised a severe problem: how was the Gas Council to dispose of the quantities of natural gas involved? This was not only an organizational and technical problem of the highest order, it was also a question of the fundamental role of the Gas Council in the future: failure or success here would spell failure or success in maintaining the North Sea monopsony (and the UK market monopoly). As stressed previously, a monopsonistic transmission line must successfully service all the natural gas fields found in its region; failure to do so leads other competitors into the area and to erosion of its monopsony advantage. American experiences have amply proved this point. Thus a monopsony strategy and vertical integration 'upstream' necessarily meant change in 'downstream' relationships.

The Gas Council and area boards responded to this challenge in two manners. One was vertical integration downstream, a trend that reached its ultimate conclusion with the founding of the British Gas Corporation in 1972. The second was the sale of natural gas at below-market rates. This underpricing was carried out with the blessings of several successive governments, which were not slow to realize the political nature of natural gas pricing to industry and households.

Once again, as with the negotiations with the oil companies and with the securing of cheap supplies of North Sea gas, the British Gas Corporation appeared to serve the national interest. This perception was not to be questioned until the advent of the Thatcher Conservatives in the early 1980s.

The strategy upstream: the Gas Council and the North Sea producers

Of the monopsonistic elements in the upstream strategy, the exclusion of potential alternative purchasers of North Sea gas was the most critical. This exclusion was accomplished early on, however, through government legislation. Two acts of Parliament were of critical importance here.

The first and perhaps the most important of these was the Continental Shelf Act of 1964, which gave extensive powers to the Gas Council and area boards, particularly as the two following provisions related to one another:

> The holder of the [production] licence shall not without the consent of the Ministry of Power use the gas in Great Britain and no persons shall without that consent supply the gas to any other person at premises in Great Britain.
>
> The Minister of Power shall not give his consent . . . to supply of gas any premises unless satisfied . . . that the supply is for industrial (i.e. non-fuel purposes in this context) purposes and that the Area Board in whose area the premises are situated has been given an opportunity of purchasing the gas at a reasonable price. (Peebles, 1980, pp. 30–1).

The first provision gave enormous discretion to the Minister of Power. The second limited possible oil company sales largely to petrochemicals – and only if the Gas Council and the area boards had been given the opportunity to buy the natural gas and refused it, a form of 'right of first refusal'. There remained the possibility of sales to the European continent. But such sales would have required political approval, and at the time were in fact out of the question as Dutch gas was gaining ground in those markets. Moreover, the costs of transportation would probably have been excessive (Dam, 1976, p. 75).

The second piece of legislation, the 1965 Gas Act, coincided almost exactly with the start of exploration on the British shelf. This Act imposed on the Gas Council the obligation to promote and assist the coordinated development of gas supplies in Great Britain as a whole and gave the Council the power to manufacture gas, to buy gas in Great Britain or elsewhere, and to supply gas in bulk to the area boards. The Act therefore authorized the Gas Council to 'enter into contracts for the purchase of gas and to construct a high pressure transmission system to distribute it to the twelve Area Boards' (Price Commission, 1979, p. 12). The significance of this Act cannot be overstated. Prior to its passage, it was theoretically possible for the oil companies to play the area boards off against each other in the hopes of obtaining a better contract. By authorizing the Gas Council to act on behalf of all the boards, the 1965 Gas Act completed the 'rule-making' phase of the negotiations that were to follow. It also established the Gas Council as the effective monopsonist.

The Council's exploratory activities, originally onshore, turned towards the offshore from 1962. To begin with, it joined the Amoco Group, one of the more active bidders for first round offshore licences in 1964. It was through this activity that the Gas Council obtained a 'window' into the nature of North Sea exploration and production.

Gas finds proliferated rapidly in the period 1965–8. British Petroleum (BP) found West Sole in 1965, followed by the Shell/Esso discovery of the enormous Leman field (1966). Two other major finds were also made in 1966: Indefatigable and Hewett. Of the larger fields known today from this period, only Viking was a late discovery (1968).[2] The Amoco Group was heavily involved in these finds and the Gas Council, through its interests in the Amoco Group, held a share in three of the five major fields: Leman, Indefatigable and Viking (see Table 5.2). If natural gas reserves were to be allocated according to share of field ownership, the Council would be one of the major owners of natural gas in the North Sea.

The advantage that this vertical integration into exploration and production might have given the Gas Council in the ensuing negotiations must remain hypothetical. It should be noted in this context, however, that natural gas negotiations elsewhere between an offshore group collectively owning a field and one or more prospective purchasers have led to considerable intra-group debate. These discussions have in some instances been more bitter than the negotiations between buyer and seller. The Gas Council could have been in the envious position of being the only prospective purchaser, while at the same time being one of the major sellers. This 'switching of hats'

Table 5.2 *Gas Council share in major British North Sea gas fields: Southern Basin*

Field name	*Estimated reserves (billion m^3)*	*Gas Council share (%)*
Indefatigable	127	19.3
Leman	197	14.8
Viking	82	0.6
Hewett	100	[a]
W. Sole	62	—
Reserve totals	568	9.5

Note: a No direct partnership, but Gas Council is 'interlocked' with the partners in this group in that both the GC and the major companies in the Hewett group are partners in the other fields – most notably Leman.

is not unique; similar cases have been known to occur on the European continent.[3]

Another feature of this vertical integration was that the Gas Council acquired a not inconsiderable expertise in the more elementary logic behind development costs and exploration risk. By the time of the major natural gas negotiations in 1968, the Council had been exploring for gas for nearly fifteen years. To charge the Gas Council with an ignorance of these factors, as many have, is not supported by the evidence. This could also explain the industry's failure to make a hard case in this regard (Dam, 1976, p. 82).[4]

Integration upstream, when combined with the legislative mandate of Parliament, might have enabled the Gas Council to play the various North Sea corporate groups off against each other. There is reason to believe that there were essentially four groups involved: the West Sole BP group, the two Indefatigable/Leman -Shell/Amoco groups, and the outsiders – the Phillips Group/Arpet Group involved on Hewett. The Phillips Group was the one that eventually broke ranks in the major round of negotiations in 1968.

Finally, the Gas Council could rely on the Ministry of Power to a high degree to follow a relatively pro Gas Council line in the negotiations by remaining on the sidelines. According to sources within the industry, during the fifteen-year period 1964–79, the minister empowered to act under the Continental Shelf Act intervened only twice in Gas Council (BGC) negotiations with the oil companies (Davis, 1981b, p. 127). Because of all the negotiating advantages enjoyed by the Gas Council, this impartiality has worked to the disadvantage of the oil companies, particularly those that felt that the prices offered by the Council should have been judged 'unreasonable' by the Minister of Power.

Given these factors, the negotiated outcome in 1968 was almost a foregone conclusion. The first round of negotiations with BP over gas from West Sole ended in February 1966 with a beach price of 5d/therm. This was a relatively high price compared with the prices on later contracts, but several mitigating factors should be noted. First, the conditions of delivery were particularly onerous; the Gas Council and the area boards had much planning to do regarding the marketing of West Sole gas. Therefore the quantities sold, and probably the load factor agreed to, were advantageous to the purchaser. Moreover, the contract would run for only three years. Second, it was alleged at the time that the 5d/therm price had been forced on the Gas Council by the Minister of Power. He certainly emphasized in announcing the contract that if quantities in excess of the 50–100 million ft^3 per day covered in the West Sole deal were to be negotiated in the future, prices paid to the companies would have to be lowered

(Dam, 1976, p. 76). Third, the first negotiating round was a 'learning process' for the Gas Council. Its representatives could be expected to apply the lessons learned from the BP contract to the contracts for Hewett, Indefatigable and Leman.

The second round of negotiations was prolonged and bitter, intersected by a devaluation of the pound. The Gas Council offered 1.8d/therm; the oil companies negotiating by groups demanded about 4.0–5.0d. The Phillips Group on Hewett broke oil company ranks in mid-March 1968 to settle for a price of 2.87d/therm. Arpet (also Hewett) was forced to follow suit; the other groups then fell into line. Prices were generally set at 2.75–2.90d/therm at a 69 per cent load factor. As can be seen in Table 5.3, it would not be until 1974 that the BGC would sign a gas contract for Southern Basin gas at prices above those offered to BP in 1966. Dam's observation that '[t]he prices

Table 5.3 *Initial prices for British North Sea gas: Southern Basin*

	Price per therm (10^5 Btu)		
Field	*d*	*p*	*US¢*
West Sole:			
1967–70	5.00	2.08	5.01
1971–		1.12	2.39
Hewett:			
Base price	2.87	1.195	2.83
Valley gas	2.025	0.844	1.997
Leman[a]			
First 600 million ft^3/day	2.87	1.196	2.83
Second 600 million ft^3/day	2.85	1.1875	2.81
Remainder	2.83	1.792	2.79
Indefatigable[b]			
All gas to 1983	2.90	1.208	2.89
Viking (1972)	3.60	1.50	3.50
Rough (1974)	—	3.4	7.87

Source: Davis, 1981b, p. 131.

Notes: a These prices are for the first fifteen years only. After that the prices are 2.87d, 2.80d and 2.75d per therm respectively for the specified amounts produced per day.

b These prices too are applicable for the first fifteen years of production. Afterwards the price will be 2.90d for the first 600 million ft^3/day, 2.83d for the next 600 million ft^3/day, and 2.78d per therm for the balance.

finally agreed upon were not based on any consistent theory but rather were compromises arrived at in hard bargaining' (Dam, 1976, p. 80) is very apt. This view of events also conforms closely to the bargaining-power approach adopted in this book (Chapter 3).

What of the contractual arrangements over time? One can argue that a major cause of bitterness over the outcome of the natural gas negotiations is related less to the deals themselves than to the manner in which energy prices have since changed. Particularly notable about the Gas Council price formulas was that they did not include any linkage with oil product prices. Such escalation factors as did exist were based on price indices, particularly those that would reflect oil industry costs. There is little to indicate that this linkage is the result of anything other than happenstance. At one point during the negotiations, the Gas Council is reported to have offered an index linkage of .333 tied to oil product prices. This offer was refused, undoubtedly causing oil company chagrin later in the day.[5] At the time, however, natural gas prices were not markedly out of line with oil product prices, particularly those for fuel oil; gas prices may in fact have exceeded Rotterdam spot market prices for fuel oil in 1969. Favourable delivery terms did aid in marketing the substantial amount of gas in the following years, yet it is difficult to see how this could have been otherwise.

Nor did oil company rates of return suffer. Much of the literature on the gas negotiations dwells on the 'cost' or 'cost-plus' theories of pricing, the argument being that the Gas Council adopted this point of view wholeheartedly. There is reason to believe that the initial offers by the Gas Council were made on this basis. Internal rates of return (IRRs) to the oil industry from the contractual sales of this gas were far in excess of 15 per cent and could have been as high as 30 per cent after royalties and taxes. In other words, the rates of return were extremely favourable to the companies in the North Sea – given the average return to capital in the industry in 1972, reported to be in the range of an 11.1 per cent return to net worth.[6] Thus the subsequent fall in exploration activity in the area probably had less to do with rent capture than with a perception that the Gas Council (and subsequently the BGC) would not need the gas until later in the day.[7]

The Gas Council too had its reasons to be unhappy with the contracts. Through buying the natural gas landed at the beach the Gas Council missed an opportunity to purchase the natural gas directly from the platform (*ab platform*) and itself transport it to land. *Ab platform* prices would have been lower, to account for the Gas Council's transportation expenses. More importantly, however, the escalation clauses would have applied only to this lower *ab platform* price, not to the higher price of the natural gas at 'the beach'. This oversight was to cost the British Gas Corporation hundreds of millions of pounds in the intervening ten–fifteen years. In the future, in negotiating for supplies elsewhere in the North Sea, the British Gas Corporation would attempt to buy natural gas *ab platform*. This has had fateful consequences when the BGC has been competing for supplies where the other (Continental) purchasers have agreed to a 'beach' price rather than insisting on paying for natural gas *ab platform*. (This is essentially what happened when the BGC was outbid by Continental buyers for Norwegian Statfjord natural gas in 1982.)

The strategy downstream

Concurrent with its integration upstream into the production of North Sea gas, the Gas Council integrated downstream as well.

It already possessed the high-pressure trunklines and distribution facilities, thanks to the importation of LNG and its marketing in eight different area board regions. This was a considerable advantage in the negotiations with the oil companies, which, in order to sell natural gas for industrial purposes (if allowed), would have had to invest in onshore trunklines and other facilities. There was little to indicate that the oil companies were particularly enthusiastic about this alternative – even if they had received ministerial assent to implement it.

The further incorporation of the area boards into the Gas Council was aided by a set of technical problems. First, North Sea gas has an MMBtu content of slightly over 1.00/thousand ft^3, whereas the MMBtu content of manufactured gas is around 0.4. Facilities therefore had to be converted. Second, the natural gas had to be sold. This led to pricing policies that were, in turn, much discussed. Finally, future gas supplies had to be secured in order to supplement what had been procured in the Southern Basin. All these problems were to demand the utmost of the gas industry.

Integration and conversion The conversion from town gas to high Btu content North Sea gas was not taken without considerable deliberation. It meant that much of the capital investment made in the years to 1967–8 would have to be written off, particularly the Lurgi plants and the oil gasification projects. The alternative would have been to retain these and use them to 'reform' natural gas, to bring its thermal value down to 400 MBtu. This latter course would have ensured that appliances, mains and distribution lines throughout the country would not have to be converted to contain and burn high-pressure, high-calorific North Sea gas. In effect, it would have meant the continuation or extension of the system already utilizing Algerian

LNG. On the other hand, such a solution would have maintained the high costs of running the plants concerned. Conversion to a pure North Sea gas system had the further advantage that the pipeline network would be delivering gas of a higher calorific content. It would therefore allow for a more efficient use of pipeline resources.

The decision in 1972 to opt for North Sea gas led to a ten-year conversion programme, costing some £577 million. To this cost must be added the £450 million write-off of obsolete plant, bringing the total to some £1,027 million. Nevertheless, allowing for inflation, the cost of the change-over was less than was estimated in 1966, and the introduction of North Sea gas was finished ahead of schedule (Peebles, 1980, pp. 37–8).

What if the decision to convert had not been taken? What would have been the impact on vertical integration if the reforming plants built by the individual area boards had remained, and the Gas Council had retained its role as seller of North Sea gas to the reforming plants? One can only speculate. Through eliminating the middle step of reforming (sales to reforming plants), the process of selling natural gas was unified. Later, such centralization would not only make political intervention in the natural gas transmission phase difficult, it would also further reduce the relative autonomy of the area boards.[8]

During the same year, 1972, the National Gas Act established the British Gas Corporation, a further concentration and centralization of the gas industry. The area boards were demoted to regions within the Corporation and their autonomous powers reduced.

Pricing Coincident with the drive towards both upstream and downstream integration, which reached fruition in 1972, was the need to sell the natural gas in the quantities contracted for. All the North Sea groups wished to sell their natural gas as quickly as possible. There was even a tendency to favour production profiles with a quick build-up in deliveries and a high load factor.[9] The only parties capable of fulfilling this demand from all the fields were the Gas Council and the area boards. Gas Council control over pipeline deliveries was strengthened by its ability simultaneously to take care of producer interests.

The resulting North Sea contractual terms made the phasing-in of North Sea gas of prime importance to the Gas Council. Table 5.4 illustrates Gas Council/British Gas Corporation progress in this regard. As can be seen, gas from the North Sea amounted to just 3 per cent of the total available gas in 1968; by 1974–5 it had risen to some 91 per cent – an enormous transition in just six years. At the same time, the total gas available increased by a factor of 2.6 during this period, an average of 15.7 per cent per annum compounded. The effect on the primary energy market in the UK was tremendous. In 1968, gas provided only 1.5 per cent of the country's primary energy supplies and ranked fourth after solid fuels, oil and nuclear/hydroelectric energy. Ten years later it provided some 19.4 per cent of the primary energy supplies, making market share inroads on solid fuels and holding other energy forms more or less constant in market share terms. To sell all this gas required a pricing policy that significantly underpriced natural gas with regard to alternative fuels.

To claim that the Gas Council consistently underpriced other fuels is perhaps to exaggerate. On the one hand, the Council has been

Table 5.4 *Gas available, 1955–75 (million therms)*

	1955–6	*1960–1*	*1965–6*	*1968–9*	*1970–1*	*1974–5*
Coal gas	2,085	1,706	1,269	665	238	
Water gas	367	416	386	92	2	
Lurgi	—	4	69	52	30	n.a.
Oil gas	31	85	692	2,134	1,208	
Total made	*2,483*	*2,212*	*2,416*	*2,942*	*1,479*	
Bought	416	678	1,424	2,063	3,106	1,266
Natural gas (North Sea)	—	—	—	160	2,157	12,424
Total available	*2,902*	*2,890*	*3,840*	*5,165*	*6,741*	*13,692*
Percentage of North Sea gas in the total	—	—	—	3	32	91

Source: Harlow, 1977, p. 170.

accused of selling in excess of fuel oil prices: Adelman, for example, criticizes proposed sales to British Steel in 1968 as being far in excess of what the market for fuel oil could offer (Adelman, 1972, p. 241).[10] Elsewhere (and more commonly) the Council is charged with selling at sacrifice prices. This seems particularly to have been the case with bulk industrial contracts in the period 1968–72. Here the Council charged an average price of some 4.5p/therm, which was considerably less than the price of domestic fuel oil during the same period (some 6.0–8.3p/therm) (Pryke, 1981, p. 13). More recently, the pattern has varied.

Furthermore, it is one thing to 'dump' natural gas on non-premium markets – markets that cannot utilize the unique advantages of the fuel; it is quite another thing to dump natural gas on premium markets, hitherto using fuels such as gas oil, that adequately price these unique advantages of natural gas. The British Gas Corporation, aided by advantageous North Sea load factors, targeted premium markets. This is in notable contrast to the situation on the Continent where transmission companies have traditionally delivered enormous quantities of natural gas to non-premium markets – particularly that of electrical generation in Holland and West Germany. So, if there was dumping, it was 'optimal' dumping.

Pryke is not in doubt about why pricing policy was formed as it was. With regard to industrial sales to 1972, he writes:

> Why did BGC sell North Sea gas at such bargain prices? As we have seen the industry had been able to buy large supplies of natural gas at a very low price and the Gas Council and the Government decided that it was economically advantageous to absorb it at a rapid rate. Once it had committed itself to this policy, BGC became worried that it would not be able to dispose of the large supply that was available. It was therefore led into selling it in a hurry at a knockdown price. (Pryke, 1981, pp. 13–14)

The obligation to supply potential customers within 25 yards of a BGC main has in fact worked to the benefit of the domestic household users of natural gas. As of 1980, the real cost of natural gas to residences had fallen below the price levels of 1976. According to the Price Commission, of the various fuel uses, 'the price of gas to domestic users is not market related' (Price Commission, 1979, p. 30). The situation in 1979 is reproduced summarily in Table 5.5. This shows that the cost of utilizing other fuels was 25–120 per cent more expensive than the cost of utilizing natural gas. Price hikes since 1979 have eroded this advantage, perhaps, but not by much – as the cost of alternative fuels has also risen. Price rises were politically unpopular in the past, and, as is so often the case with natural gas, political expediency took precedence over economic good sense.

Table 5.5 *Cost of domestic fuels compared to natural gas, classified by use, 1979 (cost of gas = 100)*

Use	*Electricity*	*Solid fuel*	*Oil*
Central heating	140–170	130–150	125–140
Cooking	160–200	—	—
Spacc heating	170–200	125–140	—

Source: BGC figures cited by Price Commission, 1979, p. 116.

Consolidation By 1979, there was every reason for British Gas to rest on its laurels. The conversion programme was finished and paid for. Natural gas, helped by advantageous North Sea contracts, had scored a spectacular success *vis-à-vis* competitive fuels. The BGC was meeting the rates of return specified for it, and was even at the point of being embarrassed by its richesse.

Yet there was no time for self-congratulation. The position that natural gas had won on the energy market would have to be consolidated. Two exigencies were of particular importance here: the widespread domestic use of natural gas could create future load factor problems; and there was a need to replace the gas consumed under the North Sea contracts with additional supplies.

With regard to increasing flexibility of gas supplies, the British Gas Corporation worked first off its knowledge of liquefied natural gas. Near the ends of the main transmission lines at Glenmavis in Scotland, at Partington in the Southwest, at Avonmouth, and the Isle of Grain, the BGC built extensive peak-shaving plants to liquefy surplus incoming gas and store it for periods of higher demand. At the same time it shifted the LNG terminal at Canvey Island to peak-shaving load factor use. The storage capacity of all these facilities is of the order of 52 million MMBtu at present. This policy was complemented by an interest in using depleted reservoirs for storage purposes, leading to the conversion of Rough field, a minor North Sea gas field, into a storage reservoir to take care of seasonal variations.

Not content with these accomplishments, British Gas has begun a series of *quid pro quo* negotiations with the operators of the Southern Basin fields, granting price concessions in exchange for more flexible load factors. This had led to considerable corporate investment in new compressor facilities and the like on these fields. It has also generated renewed interest in Southern Basin offshore acreage, as the companies see that the BGC is willing to pay more than previously for natural gas. Furthermore, the Corporation has structured the development of a major gas field discovered by its exploration affiliate, Hydrocarbons Great Britain – the Moracombe field off Liverpool – with an eye to increasing the flexibility of gas supplies in the UK.

Finally, considerable use has been made of interruptible supply contracts to industry. Although not favoured as a method of increasing load factors, about one-third of the gas supplied to industrial customers is on an interruptible basis (*Financial Times*, 24 January 1980, p. 32).

While the drive to increase supply flexibility has been a considerable success, progress in obtaining new gas supplies has been mixed. This has particularly been the case when the BGC has competed with

Continental buyers for Norwegian gas. As of this writing, the score stands at 2 to 1 in favour of the European Continental consortia. First, the Gas Council was interested in obtaining the associated gas from the Ekofisk complex and the non-associated gas from Albuskjell. This gas was the subject of intense bidding by both the Gas Council and a Continental consortium headed by Ruhrgas. According to reports, the Council's unwillingness to link the price of Ekofisk gas to the prices of oil products (most notably those of heavy fuel oil) led Norway to decide to market Ekofisk gas on the European Continent.

BGC was more successful in procuring the reserves from Frigg field, an immense structure straddling the Norwegian/British North Sea sector line that contains some 212 billion m^3 of gas. Here contract details have been sparse, but it is clear that Frigg gas was tied to oil product prices. The price for the gas from the British sector of this field is somewhat lower than that from the Norwegian sector. The gas is delivered to the BGC at Saint Fergus through twin pipelines run by Total, the Compagnie Français des Pétroles affiliate in the North Sea. Whether the escalation clause is tied to landed prices in Saint Fergus or to *ab platform* deliveries to the pipeline is uncertain. If the latter should prove to be the case, it is evident that, despite doubled or tripled oil prices, the BGC has struck a good deal.

It was largely this issue of *ab platform* versus beach prices that led to the failure of the third British attempt to obtain Norwegian gas, this time primarily from the Statfjord field. In early January 1981, the Norwegians sold Statfjord gas (together with supplies from such fields as Heimdal) to the Continent. British Gas Corporation bid hard for this gas, hoping to transport it via the Brent collection line (the FLAGS line). Although the price *ab platform* offered by the BGC was higher than the beach price minus transportation offered by Continental buyers, the fact that escalation terms would apply to the landed price of natural gas on the Continent together with considerations of landing the natural gas in Norway before shipping it south nudged the Norwegians into accepting the Continental offer.

British Gas has been a good deal more successful in obtaining additional supplies in the British sector. The Brent gas line will undoubtedly be used for gas gathering elsewhere in the North Sea: from the British portion of Statfjord, from the Cormorant fields, and from other discoveries in the Northern North Sea. These supplies have been procured at a price, however. Table 5.6 gives an impression of how these Northern North Sea gas prices compare to those from the Southern Basin. Quite a few oil fields have already been connected to the Frigg and Brent lines (Beryl, in the table, is perhaps representative of them), but prices to the BGC are high.

The increased prices paid for gas to replace the supplies currently being consumed has raised the question of what the proper price for natural gas in Great Britain should be. The Price Commission and other governmental bodies in the UK have argued that the price of natural gas sold in the UK should cover the long-range cost of the marginal supplies required to replace the gas consumed. In 1979, the Price Commission calculated that the marginal cost per therm for the domestic market was of the order of 27.0p and that the average price paid by the consumer was 19.5p/therm. Figures for the industrial and

Table 5.6 *Estimated price per therm: Southern gas fields vs. Northern gas fields (p.)*

	Price per therm	
Gas field	*Initial*	*1982 (est.)*
Southern fields:		
Average	1.2	3–6
Asking for new fields[a]	—	ca. 10
Northern fields:		
Frigg (Norwegian)[b]	8.8[c]	11.5
Brent	6.5	12.5
Beryl[d]	—	16.0
North Alwyn[e]	—	22–3
Asking for new fields	—	20.0

Notes: a *Source:* ML Petroleum Services as quoted in *Noroil*, December 1981, p. 27. According to source, up to 576 million m^3 obtainable at this price.
b Terms of delivery of British gas probably considerably different, with price closer to that of Brent.
c Price in 1974.
d Price on shore after delivery via Frigg line.
e Reported price for N. Alwyn associated gas (*The Oilman Newsletter*, 21 August 1982, p. 2).

commercial markets were 17.5p/therm and 14.5p/therm, respectively (Price Commission, 1979, p. 40). Although the assumptions of the Price Commission report are open to criticism, the general picture is clear: replacement costs for the gas consumed are far too high. Prices for natural gas should increase to enable the orderly development of this resource. This picture has not changed much since 1979; although domestic and industrial tariffs for natural gas have risen, so too have the costs of development. It would appear that the days of rapid British gas expansion are at an end.

The political 'rough and tumble': 1972–8

The 'ping-pong' nature of British policy on national industries is so widely known that it is rarely commented upon. Successive Labour governments have come to power vowing to nationalize industry. Successive Conservative governments have been returned vowing to denationalize. Beyond the rhetoric, until the election of the Thatcher government in 1979, little has happened. The steel industry – nationalized in 1949, denationalized in 1953 and renationalized in 1967 – was perhaps the most significant case of switches between Labour and Conservative government policies, but it was a relatively isolated phenomenon. It certainly did not carry over into the area of natural gas. The nationalized industries, moreover, have never really been employed as strategic instruments of industrial policy (Wilks, 1982, p. 11). There was, in fact, little strategy involved in governmental supervision of the BGC and its predecessor. As executives of the BGC have themselves put it, 'Governmental policy in general has been benign'. In the years to 1979, this statement was not inapt. So long as the Gas Council and the BGC could meet their specified rate of return and did not draw overmuch on the financing service available, they have been left in peace.

There have been exceptions to this, of course. Pricing of natural gas, particularly domestic pricing, has always been a sensitive political issue. The BGC fought and lost a battle against the application of the petroleum revenue tax to natural gas, whether from associated oil fields or not. Producer profits from all offshore contracts concluded after 31 May 1975 are subject to this tax – a factor not designed to make the producer prices to the BGC any cheaper.

A harder blow was perhaps the failure of the Gas Gathering Pipeline Scheme in 1978 and its replacement with private initiatives. This scheme envisaged public ownership of a gas-gathering line to the Northern North Sea of about 66 2/3 per cent, divided equally between the BGC and the British National Oil Corporation (BNOC). The balance of the investment would have come from the companies involved. For various reasons, primarily expense and the economics of the project, the Gas Gathering Scheme fell through. Its replacement with private schemes, favoured by the Thatcher government, has in fact meant that the gathering lines are built after the associated gas has been discovered. This has led in turn to increased natural gas prices because the gas contracts signed with the BGC are needed by the companies to finance the required gathering lines (*World Gas Report*, 17 August 1981, p. 8). Thus, rather than controlling a pipeline that can gather natural gas from the various North Sea fields at a cost-based price of some 10p/therm, the BGC is receiving delivery of transaction costs – the latter being hardly in keeping with Conservative ideology.

Similarly, Conservative policies seem to contradict each other. Thus, in opposition to the spirit of the Oil and Gas Enterprise Bill, the UK Chancellor of the Exchequer, responding to industrial complaints about high gas prices, froze the prices paid for natural gas to all customers buying over 25,000 therms per year at 30.3p/therm and rolled back prices to new contract customers from 48p to 30.3p per therm (*World Gas Report*, 22 March 1982, p. 11). This effectively cut non-BGC private interest in supplying natural gas to industry under the terms of the Bill.

The 'hiving off' of British Gas assets is not progressing well either, owing in part to lack of private interest in some of the oil fields being put up, and in part to skilful lobbying on the part of Sir Dennis Rooke, chairman of the Corporation. Clearly, passing legislation through the House of Commons is one thing; having the political will to push issues is quite another. More promising than breaking the Corporation's monopoly is the recently rumoured undertaking to sell shares of the BGC to the private sector. If true, such a privatization scheme is both a tribute to the integrated manner in which the BGC has built its monopoly position and an admission of failure of attempts to break up that monopoly.

Notes

1 For example, after consolidation, three of the newly constituted area boards found that there were over 400 tariffs in operation, excluding special contracts (Select Committee on Nationalized Industries, 1961, p. 43).

2 Other fields and dates: Dottie (1967), Deborah and Rough (1968) and S. E. Indefatigable (1969) – Peebles (1980) p. 31.

3 It is curious that this facet of the Southern Basin negotiations has not been commented upon elsewhere. Dam, for instance, notes the strange reluctance of the companies to come forward with detailed explanations of cost and risk involved in exploring for and developing these fields, but nowhere mentions that the Gas Council was in fact a member – in two cases, a very considerable member – of the producing groups involved. The two factors may have been interrelated. See Dam (1976) p. 81.

4 For example Dam (1976) p. 82:

> Exploration costs raise perhaps the sharpest issue. . . . Further exploration could be expected only if the costs were those that would be involved in exploring for and developing new fields. But attempts to extrapolate those costs from actual historical experience were frustrated by the companies' refusal to reveal their own cost data.

This failure could not have helped the companies' case *vis-à-vis* the Ministry of Power, particularly if the GC presented its historic evidence on risk.

5 The opportunity cost of this failure (and others) is elaborated upon at length in a somewhat hypothetical analysis in Davis (1981b) pp. 133–43.

6 The IRRs to industry from the gas fields are drawn from industry sources. The average rate of return to capital is from Blair (1976) p. 308.

7 For the opposite point of view, see Dam (1976) pp. 94–9. It is noticeable that few of the promising blocks issued in 1964 but not fully explored have been wholly relinquished. Exploration rates have fallen and the number of takers of blocks issued in the area have become fewer, but this could have more to do with the monopsony status of the BGC and the amounts furnished on current contract than it has with whether or not there was a significant amount of rent capture in the 1967–8 negotiations.

8 What is referred to in the first instance is a breaking of BGC's monopoly over pipeline transport. Should the British government wish to allow private firms to produce and market natural gas in the United Kingdom, it will be difficult for it to determine the rates/capacity that British Gas should charge for use of its transmission lines. This would have been much easier with the old area board system of LNG sales.

9 For a further explanation of these points see Davis (1981b) pp. 126–9.

10 Adelman (1972) p. 241. It should be noted that this criticism is only partially justified. Nothing is said about the length of contract, the terms of delivery or the use to which the natural gas is put. It should be noted, however, that this is not just a failure of Adelman, but is a general failure of such discussions. What is said about the relative prices of natural gas here is subject to the same caveat.

11 These are Beryl A, Montrose, Fulmar, Beryl B, Hutton, and N. W. Hutton.

6

Natural Gas According to the 'Plan': The Soviet Experience

Two issues

Any analysis of Soviet natural gas must necessarily begin with the discussion of two topics: the size and potential of Soviet reserves, and the criteria against which to judge the Soviet experience. The first of these issues is far less controversial than the second. But for the comparative purposes lying behind this volume, the second is by far the more critical.

Known Soviet gas resources are huge. No fewer than forty of the world's ninety-six giant gas fields listed in Table 2.1 in Chapter 2 lie within the Soviet Union. Estimates of the Russian share of the total world natural gas resources lie in the range of 36–40 per cent. The potential for finding still more natural gas is also vast. It has been estimated that some 37 per cent of the world's sedimentary basins – the geological structures necessary to contain oil and natural gas resources – lie within the Soviet Union. (Comparable figures for the United States and the Middle East are 2 per cent and 11 per cent, respectively; Goldman, 1980, p. 127.) Since both its known and its potential resource base dwarf those of other nations, the question of how the Soviet economy performs becomes of vital import, not only for developments within the Soviet Union itself, but also for those nations that are liable to become major importers of Soviet natural gas in the future.

To what degree is it right or fair to judge Soviet performance by Western economic criteria? The socialist command economy is by its nature considerably different from the market-oriented consumer-preference Western economies. As a command economy, the socialist economy rejects the very utilitarian basis of Western economic theory, the basis used by Western economists to evaluate the performance of the command economy. The Western economist rejects in turn the rational basis of the command economy except in so far as it fits into a utilitarian consumer-preference framework. To understand Soviet natural gas policies in their own terms, it is necessary to find a relatively 'neutral ground' from which to define 'rational' behaviour.

Continental Weberians may provide such a 'neutral' framework. These analysts restrict rationality to two ideal types: formal rationality (rationality based on unambiguous, internally consistent, quantifiable end–means calculation), and substantive rationality (rationality shaped by economically oriented social action within criteria expressing ultimate societal values). In terms of internal consistency, ambiguity and quantifiability, substantive rationality cannot live up to the criteria of formal rationality. In terms of societal criteria, formal rationality, while internally consistent, could be correct or erroneous according to ultimate social values.

Western economic rationality best fulfils Weberian rational requirements in a formal sense, while being less satisfying perhaps on the substantive plane. Its measures – price demand relationships, utilitarian consumer preferences, and the like – are unambiguous and relatively consistent. Soviet command economy rationality is perhaps more substantive in nature. The tools of measurement here – expressed in a series of non-market related measures: input–output tables, material balances, and engineering criteria – are ambiguous and, although quantifiable, are often inconsistent. They are, however, tailor-made to the criteria imposed by planner values. As such they are shaped by 'substantive rationality'.

What makes judgement of Soviet performance so difficult is the persistence of a 'market' form of rationality as the ultimate criterion. Thus, Robert W. Campbell states in his discussion of the issue:

> There is often a temptation both for Soviet writers and outsiders to think that the concepts and criteria of rationality of decision-making accepted in reference to market economies are not applicable to the Soviet-type economy. The mischief in this notion lies in its being about half right. The key to clarification of the issue is to see the Soviet economic system as one in which the leaders choose the economic goals while planners and lower level decision makers choose the means for achieving the goals. The leaders express ultimate aims in terms of preferences among alternative mixes and magnitudes for the bill of final goods. Once we accept the substitution of planners' preferences for household preferences, we can go ahead and reason about rationality in this setting in precisely the same way we do for any other economy. (Campbell, 1968, pp. 39–40)

In other words, Campbell argues for the substitution of planner preferences for consumer preferences and the retention of a market economic perspective as a basis of judgement. This argument contains two fallacies: first, that a set of planner preferences is interchangeable with a set of household preferences; and second, that the performance criteria of the Soviet bureaucrat are interchangeable with those of a Western economist.

In the Western world, the question of preferences is left to market forces; within the Soviet Union, preferences are the result of bureaucratic battles: struggles between ministries, Gosplan (the central planning body), party hierarchies and the like. Very much as with the thesis behind Arrow's Impossibility Theorem, the preference orderings of these bodies are probably so different that it would be impossible to generalize one set of preferences from them. This in itself is not a fatal flaw. The problem is that, for a Western economist, the determination of such preferences is perceived in political terms – terms in which he is largely uninterested:

> [I]t is a polycentric system in which many of the issues of energy policy get settled through bureaucratic struggles and the political processes of an oligarchic power system. It is only in these terms that we can interpret outcomes. . . . Unfortunately the working of these processes is often opaque, and an understanding of them is more likely to come via the skills of Kremlinology than those of economic analysis. (Campbell, 1980, p. 19)

This leads to judgements or arbitrariness or political irrationality when assessing the ends–means relationships through the eyes of Western economic theory.

Secondly, because of a Western preoccupation with efficiency in the market place, emphasis is removed from the major objective of the Soviet command economy: the fulfilment of the 'plan'. It is true that, in concern for plan fulfilment, many economic steps are taken that seem irrational or wasteful to Westerners. There is little doubt that capital investment in the Soviet Union seems unduly intensive and inefficient. Yet here it should be stressed that 'waste' is a two-way street. Thus, while Westerners describe horror stories of the exploration parties that drill shallow wells to achieve the plan-prescribed meters of total drilling year after year without finding any resources (which might lie in deeper sedimentary strata), Soviet economists point to the wasteful example of East Texas where, because of the American 'Law of Capture', some 25,000 wells were drilled on a field that could have been developed by 5–10 per cent that number with higher rates of recovery and greater maintenance of reservoir pressure (Campbell, 1968, p. 131).

Discussions of this sort soon prove to be fruitless, however. Clearly, the institutional constraints within which economic activity takes place can result in a waste of resources, whether one is discussing the experience of Soviet exploration parties or the 'Law of Capture'. More to the point would be a comparison of the Soviet economic performance in a field like natural gas to that of a large corporation. Both the large corporation and the Ministry of the Gas Industry (Mingaz) have specified goals, a set of plans to be fulfilled within a period of time. Both the firm and Mingaz are judged by the degree to which they can achieve these goals – a certain rate of return: rate of growth or market share on the part of the firm, a fulfilment of a plan on the part of Mingaz. The firm is perhaps the more autonomous, being able to set its own goals in conformity with the wishes of its shareholders. Yet the ministry also has a degree of freedom to adjust the plan goals set for it through ministerial and bureaucratic bargaining and 'log-rolling'. Just as corporations fall short of their prescribed goals, so the Ministry of the Gas Industry can fail to achieve the performance prescribed for it by the plan.

Here, however, the analogy stops short. The company is subject to the discipline of the market place. Irrespective of the degree of oligopolistic competition or the size of the firm, a failure to perform can have disastrous consequences for the firm itself. In this respect, bureaucratic disciplining (the Soviet counterpart of the market) comes but a very poor second.

Similarly, market prices are notably more effective in adjusting supply to demand than are administered prices – although administered prices can also have some effect on demand. Finally, the Soviet approach to economics makes application of an entire range of Western economic tools difficult if not impossible – tools such as interest rates, discount rates, and an emphasis on marginal instead of average costs. Where such tools are used, they have to be justified by roundabout methods. (The use of discount rates, for example, is justified through allowing for a reciprocal to the pay-back period – a reciprocal approximating that of an internal rate of return.) And they are used in addition to the Soviet matrix-based input–output planning system, to which is grafted a very 'engineering-oriented' notion of efficiency.

The Soviet performance

The dynamics of the Soviet natural gas industry are not essentially different from those in the West. In both East and West, the natural gas industry is plagued by the problems of field production, load factor, 'peak' and 'valley' consumption, and storage. Superficially, the manner in which the Soviet industry is organized is no different. The Soviet industry too has functional firms: exploration and production enterprises, transmission companies, distribution boards. Yet these enterprises are contained within one enormous ministry, the Ministry of the Gas Industry. Rather than distinct corporate entities hammering out contractual marketing relationships among themselves, there is a single unit, the component parts of which are subject to discipline and directives from above. There are undoubtedly gains in stability here, but these gains are made at the expense of efficiency and flexibility, which are replaced by the rigidity so common to bureaucracies everywhere. Instability becomes a question of bureaucratic ordering and infighting. Plan non-fulfilment – the pre-eminent criterion of performance in the command economy – results in loss of power in administrative reorganizations.

Despite these drawbacks, the Soviet record in finding natural gas fields, although the industry started almost inexplicably late, was impressive from the outset. Although they continued to focus on Baku/Azerbaijan, Soviet explorers expanded their activities in the early 1950s and made a number of large discoveries in already known petroliferous areas. In the North Caucasus, the Stavropol field, a giant encompassing 230 billion m^3 of reserves, was discovered in 1951. The discovery of this field, along with others in the same general region (Mirnoye, Maykop, Berezanskaya and Leningradskaya), led to the development of the Central Trunk System. This line was orginally completed in 1956 and comprised 790 miles of 28 in. pipeline. It has since not only been extended to Leningrad to the north and the Baku area in the south, but has been supplemented by four additional pipelines of up to 40 in. diameter. The Central Trunk System currently has a 40 billion m^3 annual capacity.

The Ukrainian systems were constructed at the same time as the Central Trunk System. The incentive to build these systems lay in the discovery of various small fields in the period 1949–56 (the first pipeline to Moscow was built in 1949). Things took a more serious turn with the discovery in 1956 of the Shebelinka field, containing 530 billion m^3 of gas, and with the subsequent discoveries of the Yefremovka and Krestishiche fields. Currently, three lines run from the Ukraine to Kiev. Varying from 40 in. to 56 in. diameter, they are capable of transporting 40 billion m^3 per year, not only to Kiev but to a range of other destinations.

Also of interest is the extension of the Bratstvo Export System through Eastern Europe to Czechoslovakia (and ultimately Austria). This 32 in. line has a current capacity of 28 billion m^3, slightly more than it had on its opening in 1967. It furnishes natural gas to East Germany and Poland as well as the original two parties.

Despite these finds during the 1950s, there could have been room to worry about the possibilities of finding yet more reserves. (Indeed, natural gas production from all the above areas is now in decline.) Instead, a stream of remarkable discoveries in Central Asia was to serve as a buffer and a base for a further 'take-off' that would utilize the still large reserves in the Urals, the Komi ASSR and Western Siberia. In Uzbekistan, the discovery of Shebelinka was paralleled by an equally large find, the Gazli field, with some 456 billion m^3. Uzbek reserves were supplemented by further discoveries in Turkmenia – the Achak field, the Gurgurtl field and finally the immense (876 billion m^3) Shatlyk field. This last named, found in 1968, was merely icing on the cake, as the great West Siberian fields had already been located. Nonetheless the Gazli, Achak and Shatlyk fields were the basis for the Central Asia/Central Russia pipeline system, which, out of a total production of 60 billion m^3 per annum, was delivering 58.5 billion m^3 to Moscow and central Russian industry as of 1978. The completion of this system of lines, totalling some 13,750 km, ushered in an age of expansion of natural gas to the Russian heartland.

In the Volga-Urals area, discoveries increased in 1966 with the find of the vast Orenburg field, the largest Soviet field with a favourable location – some 1,792 billion m^3 close to industrial centres, but outside the deserts to the south and the tundra swamps in the north. This field quickly became the hub of a series of domestic lines running to power plants, chemical processing centres and industry concentrations. By 1976 the Orenburg was connected by parallel lines to the Central Asia/Central Russian system. Two years later a line to Uzhgorod on the West Soviet frontier was completed. With an annual capacity of 28 billion m^3, it is designed to supply Eastern Europe with some 11.2 billion m^3 and Western Europe with approximately 12.8 billion m^3 each year.

Further north, in the inhospitable climes of the Komi ASSR and West Siberia, a string of gigantic finds provided further evidence of the resource depth of the Soviet gas reserves. The discovery of the Vuktyl field in the Komi ASSR increased this resource base by about 388 billion m^3. But this find was to be dwarfed by a series of discoveries beginning a year later in 1965, the well-known West Siberian finds. Four of the largest of these – the Urengoy field (3,900 billion m^3), the Yamburg field (2,500 billion m^3), the Zapolyarnoye field (2,000 billion m^3), and the Medvezh'ye field (1,500 billion m^3) – together account for about 40 per cent of Soviet natural gas reserves. The Komi field was put into production almost immediately. (Particularly desirable were the large amounts of condensate connected with the field.) In addition, it was made a base of the ambitious 'Northern Lights' pipeline system – a system that would carry some 20 billion m^3 of Komi gas plus some 30 billion m^3 of Siberian gas. This system goes to the Czech frontier but also intersects the Moscow–Leningrad pipelines and is expected to be a major source of energy for the second largest Soviet city. Current plans for Siberian fields encompass the largest (Urengoy) and the most accessible (Medvezh'ye) of the Siberian giants. One transmission system for the Medvezh'ye field is in operation. Yet another, connecting the Urengoy with European Russia and ultimately with Western European markets as well, is under construction. The total capacity of these systems is enormous.[1] In 1985, Urengoy alone is scheduled to produce 250–270 billion m^3 – an amount equivalent to the total production of the USSR in 1974![2] Figure 6.1 illustrates the interrelationship between the regions, the giant fields and the pipeline systems.

Nor are these oil and gas provinces the only ones of interest. In Yakukst, deep in the East Siberian lowlands of the Soviet Far East, lies another complex of structures. The major field developed to date is the Ust'-Vilyuv field. Distance and the difficulties of developing

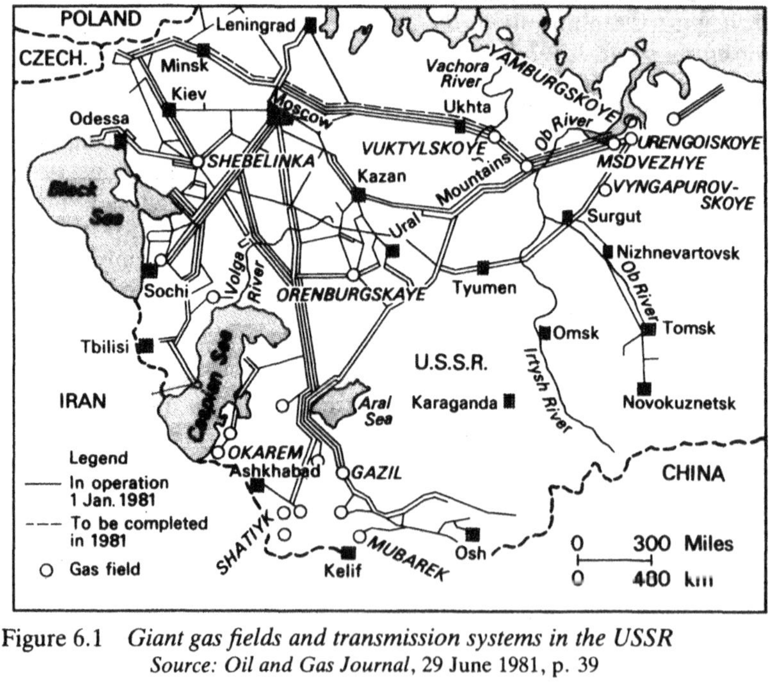

Figure 6.1 *Giant gas fields and transmission systems in the USSR*
Source: Oil and Gas Journal, 29 June 1981, p. 39

these structures admittedly create even more problems than the giants in Siberia. Considerable gas and oil also lie in Northern Sakhalin. Yet the remoteness of these resources has limited their use to cities in the Soviet Far East, most notably Khabarovsk and Nakhodka-Vladivostok. This has not precluded Soviet interest in different cooperative ventures with the Japanese, with a view to export to the Japanese market. Soviet-Japanese cooperation is already well underway in proving up some of the Sakhalin hydrocarbon resources.

Despite all the finds made in the past twenty years, the Soviet gas industry is not above criticism: 'The gas industry has consistently failed to meet its planned targets over the last decade' (Russell, 1976, p. 63). Much of this failure is the result of problems with pipelines and production facilities, which have proved more difficult than the planners expected. Critical to non-fulfilment of plans have been gas-processing facilities and compressors. Inadequate repair and maintenance of compressor stations and/or unsatisfactory preparation of natural gas have led either to underutilization of natural gas pipelines or to system breakdowns. They have caused delays or interruptions in delivery for several major pipeline systems, most notably the Central Asia/Central Russian system and the Orenburg systems. The exploration effort is also not above criticism. Virtually all of the major additions to Soviet gas resources have been located at a considerable distance from the centres of consumption. Planned increments to reserves in the older, more proximate gas provinces in the Ukraine and the North Caucasus have not been met. Furthermore, in many areas, extraction of natural gas has exceeded additions to reserves by factors of 2 to 5 (Russell, 1976, p. 64). Lack of adequate performance has made Mingaz the object of considerable criticism and the subject of bureaucratic reforms.

Despite these flaws, in terms of the sheer length of pipeline constructed and of the magnitude of the undertakings involved, the results are not wholly unimpressive. Table 6.1 demonstrates the extent of Soviet pipeline construction in the period 1950–80. If anything, these figures understate some of the problems and achievements. Much of this pipeline is based on new technology, often unproved when installed. Additionally, the areas of pipelaying are among the most inhospitable in the world: inaccessible, bitterly cold in the winter, and swampy in the summer. Temperatures drop so low in certain areas that high-pressure pipeline has been known to crack. In other senses, too, the record is somewhat more impressive than these figures indicate. Of the pipeline installed since 1975, 12,000 km consist of mammoth 56 in. diameter pipe. At current pressures, one of these lines can transport some 35 billion m^3 of natural gas. At plan-

Table 6.1 *Soviet transmission line construction: end 1950 to end 1980 ('000 km)*

Year	*Length of transmission lines (total)*
1950	2.3
1955	4.9
1960	21.0
1965	42.3
1970	67.5
1975	98.8
1980	132.4[a]

Sources: Russell, 1976, p. 67; Dienes and Shabad, 1979, p. 83; *Oil and Gas Journal*, 29 June 1981, p. 40.
Note: a of which 7,400 km not yet operating at design capacity owing to lack of compressor capacity.

ned higher pressures with laminar pipeline, they will be able to handle some 68 billion m^3 apiece if the gas is reduced in temperature to −30° C, as is now the plan (*Oil and Gas Journal*, 29 June 1981, p. 40).

The consumer markets serviced by these pipeline systems have, on the one hand, been relatively unaltered and, on the other, been revolutionized. As can be seen in Table 6.2, primary consumption of natural gas by sector has grown phenomenally. In the period 1960–

Table 6.2 *Natural gas share in primary energy markets of the Soviet Union, 1960–75 (percentage share of sector)*

Year	*Non-fuel use*	*Electrical generation*	*Industrial use*	*Housing and municipal*
1960	20.2	8.7	9.2	7.7
1965	40.6	19.2	17.9	19.7
1970	25.3	19.6	27.3	24.0
1975	44.4	18.0	31.9	28.4
Per sector growth in overall energy consumption, 1960–75	+940	+672	+384	+265
Total per sector growth in natural gas consumption, 1960–75	+2006	+1452	+1334	+978

Source: Campbell, 1978.

75, natural gas quadrupled its share in the Soviet housing and heating markets, tripled its share in the industrial sector and doubled its share in the electrical power industry. When the overall sectoral rise in total energy consumption is taken into account, 1975 consumption of natural gas alone accounts for anywhere between nine and twenty times the natural gas consumption levels of 1960.

Interestingly enough, however, despite this increase in sectoral consumption of natural gas, allocation of natural gas among the sectors remained relatively constant for this period: some 6–9 per cent of natural gas going to non-fuel purposes, 41–44 per cent to industry, some 15 per cent to municipal and domestic heating uses, with the remaining 29–30 per cent going to electrical generation. This distribution is particularly striking when compared to the allocation of natural gas by sector in the United States and the Federal Republic of Germany (Table 6.3). In the USSR, industrial, electrical generation

Table 6.3 *Allocation of natural gas consumption by sector: USSR, USA and the Federal Republic of Germany, 1975 (percentage of total)*

	Non-fuel use	*Electrical generation*	*Industry*	*Domestic/ commercial (USSR: domestic/ municipal)*
USSR	9.6	29.5	43.8	15.1
FRG	13.6	23.5	28.7	28.9
USA	0.2	18.8	36.8	44.4

Sources: Campbell, 1978; International Energy Agency, 1981a.
Note: National totals do not add to 100 per cent owing to losses and other uses.

and non-fuel use (83 per cent of the total) weigh very heavily when compared with these categories in the FRG (66 per cent of the total) and the United States (56 per cent of the total).

Although Soviet literature apparently has little information on the number of interruptible customers served, clearly the amounts consumed by electrical generation and heavy industry relieve the natural gas planners of load factor problems and the consequent problems of storage. This coincides with reports that Soviet gas planners are having difficulties finding exhausted reservoirs close to centres of consumption that can be utilized for off-peak storage purposes. As a result, use is being made of aquifers (into which natural gas is pumped, displacing water). About one-half of this gas is lost as 'cushion gas'; the other half can be recovered and delivered to distribution lines. An estimate of some 8 billion m^3 capacity for the period 1975–80 (Elliot, 1974, pp. 61–2)[3] is still only some 2 per cent of total annual consumption, a far cry from the 30 per cent storage capacity in the United States. Planners therefore use surplus gas extensively in power stations, and it is not unlikely that there are quite a few Soviet industrial plants with interruptible supplies. Whether or not this speculation is grounded in fact, it is clear that Soviet supplies could be employed more flexibly than they are at present.

Industrially, natural gas is proving to be quite a significant fuel. This trend was already notable by the 1960s, and is accentuated today. By 1965, Soviet planners estimated that replacing coal with natural gas led to the savings of 2–15 rubles per ton of conventional fuel in boiler usage. These savings were remarkable not only for their size but also for the fact that they were generally higher than savings incurred by substituting fuel oil for coal in underboiler use (Elliot, 1974, p. 237). Natural gas has been additionally helpful as a chemical feedstock. Its use as a fertilizer base is especially significant. By 1965, 100 per cent of all ammonia produced in the Soviet Union was based on natural gas as a feedstock. While perhaps not fully introduced into the steel industry by that year, natural gas accounted for close to 47 per cent of the fuel mix in steel smelters and 31 per cent of that for the production of rolled steel (Elliot, 1974, p. 237).

One should be careful not to read too much into the relatively low market share of natural gas in the residential heating sector. As of 1979 there were 54 million householders utilizing natural gas for heating purposes, but more than two-thirds of these were using liquid petroleum gases (LPG), certainly a more flexible fuel than piped natural gas. This relative lack of household use of natural gas is in part due to planning priorities that favour military and industrial applications of this fuel. In part it could also be due to the great demands that household consumption of natural gas places on storage capacity and load factors. As important as these concerns, however, is the political objective to heat homes through regional plans involving the heat cogenerated in electrical power plants. According to Goldman, as much as 45 per cent of Soviet urban homes should be serviced in this fashion today (Goldman, 1980, p. 143). The problem with such urban heating schemes is that they displace natural gas, which is best used for domestic heating in precisely those areas that are best for district heating schemes.

The problems of planning: energy balances

'One of the cornerstones of Soviet economic doctrine', it has been perceptively stated, 'is the requirement of proportional development of the economy, the need to ensure certain proportions and relations'

(Tretyakova and Birman, 1976, p. 158). As a major economic input to the 'plan', energy is a particularly vital component in the doctrine of 'proportional development'. Natural gas as a fuel energy is consigned to a set of material balances classified as 'extractive industries'. The role of the planner is to find the proper proportions. First he receives his directives; then he works out the resulting material balances as follows:

> The production programme of the fuel industry takes account of how economical individual types of fuel are, which factor necessitates the volume to be extracted. Since oil and gas are valuable raw materials and the demand for them virtually unlimited, the plan for oil and gas extraction is based on production capabilities. The need for oil and gas, including for export and domestic consumption, determines the lower limit of production.
>
> The plans for the development of oil and gas envisage:
> - the maintenance of rational balances between planned extraction and proven deposits of the minerals;
> - the widespread introduction of advanced highly efficient methods for working deposits;
> - maximum use of incidental minerals;
> - rational proportions in the dynamics of fuel extraction and its transportation;
> - the qualitative preparation of the gas and oil for subsequent refining. (Anchishkin, 1980, p. 199)

A sample of the energy balances matrix relating the availability of the fuels to their uses is given in Table 6.4. One does not have to be an expert in centralized planning to see how complicated this apparently simple idea of planning balances in fact is. To begin with, there is the question of determining 'rational balance', 'efficient methods for working deposits', 'rational proportions in the dynamics of extraction and transport', and other planning concepts. These concepts not only have to be operationalized without a market but they also have to be integrated with one another. A further problem, one left out of the citation by Anchishkin above, is the balance between these objectives and the demand for fuel from the consuming sectors. (This is alluded to in the fuel and energy balances shown in Table 6.4.) The problem is not made any the lesser by the magnitude of these balances and their internal consistency. What guarantee is there, for example, against a significant amount of double counting or special regional bureaucratic pleading in establishing the goals for the 'plan year'? The problem can be further elaborated through referring to the Soviet 'input–output' planning model in 1970. This model consisted

Table 6.4 *Fuel and energy balances (millions of tons of conventional fuel)*

	Report (base) year	Plan year
I *Resources – total, including:*		
1 Fuel extraction		
a hard coal		
b oil		
c gas		
d shale, peat and others		
2 Generation of hydroelectricity		
3 Imports		
4 Reserves at end of year		
II *Distribution – total, including:*		
1 Consumption – total, including:		
a for the generation of electricity and thermal energy		
b for production and technological purposes		
2 Exports		
3 Reserves at end of year		

Source: Anchishkin, 1980, p. 200.

of 6,727 direct input coefficients, of which 2,320 were calculated by technical–economic forecasting. For the fuel industries alone, there were seven sub-industries, which together accounted for 385 direct input coefficients, 60 of which were derived from technical–economic forecasting (Ellman, 1971, p. 80). (To make things even more complicated, input–output matrices, although they enable better planning vision, are often inoperable as planning instruments because of problems of non-comparability, differences in product prices and in classifications between the matrices and actual data used in drawing up plans.)

The problem of the Soviet planner approaches schizophrenia. The planner is aware of the directives concerned with his branch – here, the fuel and energy section of the state planning agency (Gosplan). But Gosplan has no executive power of decision, it merely serves as a source of information for the Soviet Council of Ministers, which retains decision-making powers over the energy and fuels branch. The planner is torn. He must elaborate targets set by the executive agencies through the use of his professional modelling expertise. His career depends on this. He must also contend with the problem of obtaining the proper balances necessary to achieve the target figures within the resources available. Finally, he must keep a 'weather ear' out for the political struggles among various ministries as to how his input–output/balances modelling is going to be formed (Katsenelinboigen, 1978, p. 13). The resulting balancing act can prove difficult even for the most hardened Soviet bureaucrat.

Gosplan is supplemented by the Ministry of the Gas Industry (Mingaz), an enormously large bureaucracy whose organizational structure is constantly evolving as various reforms are implemented. Russell (1976) described Mingaz in 1974:

> the main departments of the ministry are linked by operational divisions reflecting the territorial divisions of the country. . . . Each of the thirty-two regional administrations comprises its own enterprises dealing with: drilling, gas production, transport, gas processing plants, engineering and repair organizations, scientific research and planning institutes, supporting and servicing subunits. (pp. 224–5).

The number of enterprises engaged in a regional administration can be staggering. Russell (1976) names regions with as many as forty-five gas-producing enterprises apiece. Since too much regionalization deprives the gas industry of its necessary central coordination, some centralization is also present. The inter-regional gas transport system reports directly to the ministry. Other priority areas – storage, machine construction (reflecting a lack of necessary inputs from other economic sectors), servicing and repair work – have been made separate divisions of federal industrial boards. Additional divisions include: standardization of materials, standardization of production quality, and automation systems. Activities for foreign gas come under the purview of the Federal Board for Foreign Gas.

This organizational structure has probably been changed at least once since Russell's admirable description. This does not really matter much. The major point is that, while Gosplan represents the priorities of the Soviet economy as a whole, the complex bureaucratic system within Mingaz reflects the industrial dynamics described in Chapter 2. Particularly noticeable is how Soviet organization specialists, when deprived of capitalistic market indicators, resort to divisions and functional specialities that do not make much sense from a Western economic perspective but that reflect the engineering realities lying behind the Soviet natural gas industry.

Mingaz coexists uneasily with a series of other ministries with which it alternately cooperates or 'does battle': the Ministry of Electrical Power and Electrification (Minenergo), the Ministry of the Coal Industry (Minugol'), the Ministry of Oil Extraction (Minneft'), the Ministry of Oil Refining and Petrochemicals (Minneftekhim), and the Ministry of Geology (Mingeo). These ministries have an interest in natural gas variously as (1) a fuel and a product (Minneft' and Minneftekhim), (2) a fuel and potential competitor in consuming markets (Minenergo), (3) a competitor in consuming markets alone (Minugol') and (4) a resource to be explored for (Mingeo). Given these different interests (often divided within competitive ministries), it should not be too surprising that inter-ministerial battles are alluded to in Soviet sources. Such allusions are infrequent but they do occur. A conflict between the Minister of Geology and Minneft' was resolved with the removal of the minister in 1962 (Campbell, 1968, p. 35). Responsibility for offshore drilling is concentrated in Mingaz owing to lack of inter-ministerial coordination (Goldman, 1980, p. 126). Lack of coordination between the Ministry of the Petroleum Industry (Minneft', prior to 1970) and Mingaz has led to the flaring of up to 20 billion m^3 of associated natural gas per annum, equivalent to 7 per cent of total annual production (Goldman, 1980, pp. 47–8).

Neither are these conflicts limited to sister energy ministries. Access to the West Siberian fields, for example, has led to conflicts with the Railroad Ministry. The Medvezh'ye, the most accessible of these fields, could be reached by a river route – Ob River to Ob Gulf and then up to the Nadym River – a route open only three months per year. Alternatively, use could have been made of an abandoned railroad, the Salekhard–Igarka line, built by slave labour in Stalin's time but since abandoned as impractical. Here, ministerial logic triumphed over economic sense. The Railroad Ministry refused to rehabilitate the route, arguing somewhat tautologically that rehabilitation was made difficult by the route's state of disrepair. The Gas Ministry thereupon took over and repaired a vital segment of the railroad, an incursion into an unusual line of business for the gas industry, to put it mildly (Dienes and Shabad, 1979, p. 90).[4] Other problems also plague Mingaz. Compressors for the gas lines are delivered by the Ministry of Power Machinery. This ministry of course has other obligations, so there are seldom enough compressors to go around; moreover, those that exist are often inadequate for the jobs that they are to perform. Similar problems can be seen with regard to the provisions of pipeline. (These are problems to which we shall return later.)

These tensions between the central planning authorities responsible for the entire Soviet economy and Mingaz, between the various divisions in Mingaz, between the particular regional demands within the ministry, and the imperatives imposed by technology and central decision-making between Mingaz and its sister ministries and between Mingaz and other non-energy ministries is the stuff of Soviet natural gas production, transmission and distribution. A knowledge of energy balances and organizational structures is not enough for an understanding of Soviet natural gas policies. It must be supplemented by an appreciation of decision-making in the system.

The 'plan': bureaucratic bargaining and technical rationality

As emphasized in the introduction to this chapter, decision-making in the Soviet Union is an outcome of bureaucratic 'infighting' in an oligarchic system, an 'infighting' that is more or less opaque to the outside world. In lieu of fact, we must content ourselves with generalizations on decision-making within the gas industry.

As with most Western bureaucracies, the style of Soviet decision-making is one in which ministries, non-ministerial organizations and individuals play for stakes. In contrast to Western bureaucracies, however, in the Soviet Union it is the ministries that predominate in most decision-making processes. This is because the relationship between a ministry and an enterprise places the enterprise manager within a hierarchical structure, a structure that will determine that manager's chances for future promotion and advancement. This is reinforced by a general patron–client relationship between upper-level administrators and their middle-level subordinates, and between the middle-level administrators and their subordinates.

This form of organization, which still predominates despite the introduction of reforms, has two facets. The first of these is that conflicts are to a large extent confined to the ministries; and, given the hierarchical nature of the ministries, the conflicts are internalized. As a result, few administrators play the role of 'whistle blowers'. The exceptions to this generality are few and far between. The second facet of hierarchy is that there are apparently few inter-ministerial coordinating committees. What coordination does take place occurs at the upper levels of management. Unfortunately, there is little Western literature on inter-ministerial coordination in Russian energy policy. Such committees as do exist are mentioned infrequently and always in connection with a limited decision-making competence. Otherwise statements are general and non-specific:

> The Soviet economy relies much more on the administrative approach to coordination. Very little is left to the realm of lateral

> communications and interaction between units analogous to the firms of a market economy. In constantly reshuffling subordinations, changing systems of agglomeration, moving functions from one unit to another and so on, the Russians are trying to accomplish the same kind of results that growth in size of firm, changing specialization, and so on bring about in a market economy. (Campbell, 1968, p. 25)

Such administrative inflexibility (and apparent lack of coordination) can have disastrous consequences as regards natural gas. For example, the question of disposal or sale of associated gas should require extensive coordination between Mingaz, Minneft' and Minneftekhim. Issues of natural gas use in electrical power generation should involve coordination not only between Minenergo and Mingaz, but also between other organs such as Minngol'. Yet an examination of the manner in which inter-ministerial problems are solved indicates that there is little effective coordination (particularly in areas such as flaring of associated gas). Rather the evidence points to hierarchical structures within ministries that have their goals set for them by the planning process and bargain for bureaucratic stakes that plan-fulfilment or non-fulfilment can give them. This bureaucratic bargaining introduces further complications. Senior administrators are going to think twice about sharing information with counterparts in other ministries if the latter are going to use the information to 'capture' some of the authority of the sharers. Another element of confining coordination to the very top is that it places enormous burdens on competent top-level management in the first instance and on the Council of Ministers in the final instance. These administrators will tend to defend their narrow ministerial interests not only because of the 'stakes' involved and their client relationship, but also because information necessary to coordination will be hard to come by, as will be managers who are expert in the necessary areas of coordination.

There are advantages to this bureaucratic structure, primarily in that it shifts coordination and planning responsibility to those few top-level structures charged with such responsibility. This advantage is not to be underestimated in a 'command economy' such as that in the Soviet Union. But, seen with Western eyes, it is more than offset in terms of wasteful use of a depleting resource and duplication of functions in the ministries involved.

A second characteristic of Soviet decision-making in natural gas is the substitution of 'technical rationality' for Western 'economic rationality'. Complaints that the Soviet production, transport and sale of natural gas lack a fundamental rationality and are characterized by *ad hoc* material balances planning are only partially correct.

In the Soviet system, the planning role of 'capitalist' investment criteria is either replaced or supplemented by a form of 'technical rationality', an adaption of engineering criteria as a tool for ascertaining the appropriateness of ends–means relationships. (It must be emphasized in this context that this is not a reference to any technocratic elite; rather, 'technical rationality' indicates a decision-making style pervading natural gas industrial relationships in the Soviet Union.) What is meant by 'technical rationality', and how does it operate in this particular context? A few examples of how it functions should answer both of these questions.

First, in a non-market economy where prices are unreliable indicators of supply and demand, it is necessary to substitute some other quantifiable measure as a means of 'checking' on a possible maldistribution occurring through the Soviet price mechanism. In Weberian terms, 'technical rationality' is a means of imposing a formal rationality on a substantively rational economic system. Complementary to this function, engineering criteria are eclectic in nature, and their application can be altered or varied depending on how the criteria serve. This enables widespread borrowing between engineering disciplines and from abroad – borrowing seen not in terms of ideology but rather in practical terms of ends–means relationships. The very eclecticism and end–means relationship inherent in technical rationality enable Soviet planners to 'decouple' the use of formally rational Western criteria such as 'pay-back period' and interest rates for capital (used in weighing diameter versus compression characteristics in transmission pipe planning) from the substantively rational Soviet command economy. What appears to Western economists as a glaring inconsistency is seen by Soviet planners as just another project measure utilized to obtain the best engineering results and as a means to fulfil substantively rational goals.

Second, the very necessity of gathering data for production norms and plan specification will tend to emphasize 'technical rationality'. As stated previously, price data are unreliable. Data are necessary, but to be useful they must be logically interrelated, which Soviet pricing data often are not. The data used for planning should ideally be related to multiple variables, and engineering data fulfil this objective admirably. The result is that the data selected can be incredibly detailed in their specification: recovery coefficients, drilling footage ratios, physical output measurements, detailed cost improvement figures, and the like, and can be mixed in with more normal data such as fulfilment of plan, actual production figures, etc.

Third, engineering and technical data are understood more and more by Soviet decision-makers. Elite recruitment studies and analyses of the Soviet education system reveal the emphasis given to engineering (in the Soviet sense of the term). This emphasis is increasingly showing results within the bureaucracy at large, and even, to an extent, in the Communist party. (Recent studies have indicated, for example, that the route to the Politburo includes stints at both a technical school and a job as industrial manager.) As a result, formally rational arguments not deemed relevant to Marxism–Leninism are comprehensible to one's superiors and subordinates in the bureaucratic enterprise system.

In practice, technical rationality is manifest in a plethora of measurements that are meaningless and unconnected in terms of Western economic theory. Yet there is one major consistency with this approach, which accounts for much of the technical differences between the Soviet natural gas industry and that in Western Europe and the United States: an emphasis on efficiency in general and energy efficiency in particular as a means of measuring performance. Examples of this tendency are numerous. The Soviet 'love-affair' with turbo-drill technology, a technology that has proved very costly in other terms, could in part be based on an engineering fascination with drilling efficiency. The Soviet electrical power generating industry is also super-efficient by European/American standards. According to Campbell (1980, p. 27) the efficiency in Soviet electrical power conversion (power delivered), including the use of district heating, was .487 in 1975. Figures for Western Europe and the United States were .327 and .307, respectively. Similarly, energy balances are used for project planning. The question of whether to abandon a producing well, for example – elsewhere a matter of comparing cost and price – is decided in the Soviet Union through energy balances. The energy costs involved in continuing to produce from the well in question are weighed against the energy that the well contributes to the system (Campbell, 1968, p. 55). Such practices are also written about and resorted to elsewhere.

Curiously, energy efficiency is not a particularly marked concern outside the energy industry. This is particularly true of the industrial uses to which natural gas (and other fuels) are put:

> . . . Soviet efforts need to be directed primarily at efforts to save energy in industrial processes. I suspect that because of systemic differences, energy conservation in industry may be more difficult to achieve in the USSR than in market economies. In the US, the stimulus of higher energy costs leads businessmen themselves to seek out many ways to save energy. . . . In the Soviet Union the system is unlikely to show this kind of automatic response. . . . The Soviet rationing system probably leads to wastage of fuels in the nonhousehold markets. . . . Customers pad requests in the

> expectation that they will be cut which leads the rationers to arbitrary cuts, and so on in a vicious circle. (Campbell, 1980, p. 20)

Lack of efficiency is also characteristic of domestic heating, although for slightly different reasons. Orudzhev, a Minister of the Gas Industry, has been quoted as saying that, owing to declining housing standards, particularly in the quality of insulation, it takes 60 per cent more fuel to heat a house built in the 1970s than a similar dwelling built twenty years previously (Goldman, 1980, p. 143).

Another 'irrational' characteristic of the technical approach is the phenomenon of 'engineering around' a particular resource lack in the Soviet system. What makes this characteristic 'odd' in Western eyes is that it is not due to actual resource costs or to market price considerations. (After all, plastic products have replaced wood in the Western world in the name of cheaper products and decreased production costs.) Rather, this engineering effort is due to a bureaucratic inability to provide goods or services in the necessary amounts or quality. The best example of this tendency is the development of the turbo-drill. Without doubt, *the* major factor behind the development of the turbo-drill was the lack of drill pipe of sufficient quality to support conventional rotary drilling practices. Rather than reorganize the steel industry, it was easier to adopt another time-consuming drilling technology that, despite its theoretical merits, is increasingly acknowledged as a failure.

In all these cases, 'technical rationality' lacks the clarity and consistency of Western equilibrium economic theory. It is a form of rationality but, in the eyes of many, it seems riddled with 'rule of thumb' engineering considerations. More importantly, technical rationality is not the sole guiding principle here. As important as any sort of engineering consistency are the bargaining relationships and the procurement of enough energy inputs to fulfil goals set by the planners.

Natural gas as a panacea: internal energy balances and currency earnings

The general problems of Soviet energy balances and the drive to export natural gas are, not unusually, seen as interconnected phenomena. The Soviet Union, it is argued, needs its natural gas reserves developed, but this development is plagued by bottlenecks, most notably in the transmission networks required to bring natural gas to markets in Western Russia. A means of 'killing two birds with one stone', therefore, is to import Western pipeline technology and pay for this technology with natural gas. Perceptions such as these have in the past led American policy-makers to oppose the sale of petroleum technology to the Soviet Union. (There was a celebrated affair in the late 1970s over whether Dresser Industries would be allowed by the American government to set up a drill-bit factory on Soviet soil.) The change in American administrations and the conclusion of a series of major European–Soviet natural gas deals sharpened American determination to 'do something' about the issue of European dependence on Soviet gas and the Soviet use of foreign exchange. The resulting ban on the export of all pipeline equipment (but particularly compressor rotor technology), either directly by American firms or indirectly by non-American firms working under licence, was a direct attempt to sabotage a series of deals that at one and the same time were seen as posing a threat to Europe and as resolving one of the more difficult domestic Soviet quandaries – the procurement of sorely needed pipeline technology. After a brief trans-Atlantic fracas in the early 1980s, the Americans backed down from their direct ban.

The issue of European dependence on Soviet gas exports is dealt with further in Chapter 8. The question here is whether there is a direct connection between Soviet gas export policies and an energy 'crisis' within the Soviet Union.

To begin with, the problems plaguing the Soviet gas industry are multiple: a lack of skilled labour to accomplish plan objectives in Central Asia, Western Siberia, Yakutsk in Eastern Siberia; insufficient infrastructure to support natural gas production, treatment and transportation from these areas; an extremely harsh environment, which imposes seasonal swings on well-drilling and pipeline-laying progress; and capital equipment that is incapable of operating in the extreme cold of Western Siberia and Yakutsk. All of these factors have led to increasing costs in the additional marginal unit of natural gas. Indeed, costs have risen at such a rate that, despite the vast gas resources available, Western experts are beginning to question if coal is not becoming an increasingly cheaper alternative fuel, making oil and gas 'uneconomic in many of the uses which now absorb large amounts of these fuels' (Campbell, 1980, p. 29).

Apart from possibly very specialized areas (drill bits), Western technology imports have with one exception very little potential to resolve these crises. The exception concerns the areas of transmission pipeline and compressors. And here the picture is really considerably more complicated than one would otherwise believe. The overall dependence of the Soviet Union on imported oil and gas pipe of above 40 in. diameter has been variously estimated at around 56–68 per cent (Campbell, 1980, p. 209). The dependence on imports of wide-diameter gas pipe alone is probably considerably higher. Campbell (1980, pp. 209–12) points out that the Soviet Union has the ability to manufacture all diameters of pipe including the new laminar 56 in. line. The same can be said of compressors. In fact, the Soviet Union has more compressor capacity installed on its transmission lines than the American transmission company lines (if we are to take these as models of efficiency). If this is true, why is it that the Soviets are thought to be dependent on Western technology?

Official Soviet sources are in no doubt about the nature of the problem; it does not exist. *Oil and Gas Journal* (29 June 1981, p. 41) quotes F. Salmanov:

> [N]ow is a good time to state that the widespread view that Western technology is of decisive significance to the Soviet Union's oil and gas industry development is false. We can construct oil and gas pipelines very well on our own and faster and better than any other country in the world.

But are Western experts and Soviet officialdom talking about the same thing? Few will dispute the Soviet ability to construct their transmission lines. The question is whether deadlines and efficiency goals can be met at the same time. The evidence here, while not conclusive, is at least convincing: it is hardly likely that the ambitious goals can be met without some import of Western technology.

The problem is best revealed through studying the performance of existing Soviet gas lines. In the early years of the Soviet gas industry (the mid-1950s), several Soviet pipelines operated at no more than 45–50 per cent (Campbell, 1968, p. 153). Utilization of Soviet line as a percentage of its design performance factor is now considerably better, averaging around 80 per cent, but this level of performance is still relatively poor compared with performance factors in the West. The Soviet gas industry, with more installed compressors of horsepower and with pipeline of greater diameter, should be able to outperform the Americans with ease. Yet, in terms of cubic meters per kilometer distance, the work performed was in fact slightly less in the Soviet Union than in the United States in 1975 (Campbell, 1980, p. 297).

Although the lacklustre performance may be the result of a number of factors – lower operating pressures, corrosion in the lines, lack of storage capacity leading to higher degrees of 'line packing', clogging of lines with natural gas liquids – it is likely that difficulties with line pipe and compressors contribute substantially to planner headaches. The problems become even more severe when expectations of higher pipeline pressures and greater compressor capacities are taken into account.

Soviet expectations of future pipeline pressures of 1,469 and 1,763 lb/in^2 are considerably in excess of the upper limit of pressures currently being used (1,020 lb/in^2). Campbell estimates that most Soviet lines operate at pressures considerably under American pressures – some 700–1,000 lb/in^2 (1980, p. 208). The question remaining is whether Soviet line can stand up to these pressures, or whether imported line may be necessary for this purpose.

Problems with compressor capacity and reliability are if anything even more vexing. Currently, plants operated by the ministries of power machinery, ship building, chemical and petroleum machinery, and aviation all manufacture various types of compressor equipment. While this sort of multiple effort is not unusual given the great resources of a command economy to 'get things done', the resulting waste and lack of uniform specification must be high. Furthermore, these disparate efforts have not produced a single highly reliable standard compressor type. Availability of compressors for American transmission lines is around 95–97 per cent of planned availability; time budget figures for Soviet gas turbine, electric and piston-driven compressors show operation times of 68, 42 and 51 per cent, respectively (Campbell, 1980, p. 209). Operation time between breakdowns shows the Soviet-designed compressors at even greater disadvantage. The GPA-Ts-6.3 gas turbine is acclaimed a success with 1,970 hours of operating time before breakdown, whereas American units in the same class are reputed to run 25,000–40,000 hours before breakdown. Soviet compressors are heavier as well, and have the marked disadvantage that they are assembled at the construction site instead of at the factory where they are built.

These problems are compounded by the large compressor units needed for the new wide-diameter pipeline. Even given the presence of large gas turbines in the 16,000 and 25,000 Kw class for 48 in. and 56 in. lines, no fewer than three turbines have to be present at each station (one on line, one in repair, and one back-up). If the large units are not available, then the compressor station must use a multiple of smaller units, also in groups. According to Kurbatov, chief of the Main Siberian Pipeline Construction Administration, a standard number of smaller turbines will be around eight (*Oil and Gas Journal*, 27 June 1981, p. 43).

Thus there can be little doubt that Western technology can ease bottlenecks and aid in achieving planned goals. In this regard it is easy to be overcritical of the Soviet gas development effort. The West Siberian achievement alone has been compared to the simultaneous construction of no fewer than eight Alaskan Natural Gas lines. Given that, with a $47 billion price tag, American interests have yet to proceed with the actual building of this line, one must credit the Soviet planners with considerable energy and courage, all the more so given the hostile climate and the planned specifications of the transportation systems involved.

The connection between the import of Western technology and the export of natural gas must remain largely a matter for conjecture. There is little doubt that the Soviet planners needed Western technology in the 1960s and 1970s. Whether they depend on this technology today is another question. There is undoubtedly an element of convenience associated with the continued import of Western technology. It is better, easier to install, more efficient, and can be relied upon to be delivered more or less on time. Moreover, through association with a major Soviet export commodity, Mingaz no doubt is given preference in obtaining large amounts of foreign exchange. (This sort of preferential standing is not unusual among Soviet domestic and export industries.) If so, it would make sense to utilize this foreign exchange reserve to secure the equipment that will enable the generation of even more foreign reserves. For there is no doubt that money is there to be made. The title of this book, 'blue gold', is the Russian nickname for the fuel, a reference perhaps to the US $7.6 billion earned through Soviet gas exports to Western Europe in 1981. An increase of two to three times 1981 revenues should be 'gold' enough to justify future reliance on Western technology when needed.

Exporting natural gas

In the past few years the Russians have become not only the world's largest, but also one of the more capable exporters of natural gas. Exports started modestly enough to Austria in 1968, a development that led to some rearranging of internal West European trade. The Austrians were joined by other major importers: the Italians (1974), the West Germans (1975), and the French (1976), the last named through a complicated 'swap' arrangement with the Italians. Exports of natural gas to Europe can, with a reasonable degree of confidence, be expected to double in the next five years or so. As Table 6.5 demonstrates, this growth in gas exports to West Europe has been paralleled by a similar growth of exports to Eastern bloc. Soviet potential in export markets is tremendous. In addition to the pipeline exports listed in Table 6.5, various LNG projects have been under consideration involving Urengoy gas (with a US consortium headed by Tenneco), Yakutia gas (US and Japan) and possibly Sakhalin gas (to Japan).

The commercial and engineering skill with which these natural gas projects have been organized has been as impressive as the growth in

Soviet trade. This is particularly evident in the arrangement of the export trade to France. Here, a three-way swap agreement was made: Soviet gas sales to France were diverted to Italy; Dutch gas sales to Italy were diverted to France to substitute for the Soviet sales to Italy, and France and Italy settled accounts between themselves regarding the differences.

A similar deal was struck in Soviet trade with Iran. Iran, with the second largest exportable surplus in the world, is proximate to the Baku and Azerbaijan gas provinces, which are currently in the process of depletion. The Soviet Union concluded a pipeline trade (IGAT-1) with Iran (they also have one with Afghanistan), which involved some 8–9 billion m^3 per annum from 1970. This trade

Table 6.5 *Growth in Soviet natural gas trade, 1972–80 (billion m^3)*

	Western Europe		*Eastern Europe*		*Iran/Afghanistan*	
Year	*Exports to*	*Imports from*	*Exports to*	*Imports from*	*Exports to*	*Imports from*
1972	2	—	3	—	—	11
1974	5	—	5	—	—	11
1976	11	—	13	—	—	12
1978	24.6	—	9.2[b]	—	—	12
1980	28	—	28.5	—	—	n.a.[a]

Sources: 1972–6: Goldman, 1980, p. 94; other figures from the Gas Committee of the UN Economic Commission for Europe, *Annual Bulletin of Gas Statistics for Europe* (annual).

Notes: a Discontinued owing to dispute over prices in the case of Iran. Reputed that some of shortfall is being made up by additional exports from Afghanistan.

b Excluding GDR and Yugoslavia.

enabled Soviet use of infrastructure in the Azerbaijan area and compensated for falls in production in the area. The trade was not without problems; the Iranian Shah forced the Russians to increase their prices for natural gas in 1974 and again in 1977 (Goldman, 1980, p. 106).

Nevertheless, the success of this first trade led to IGAT-2, another 'wheeling-dealing' Soviet trade. Here the Soviets joined with the Czechs, Austrians, West Germans and the French. The idea was to 'swap' Iranian imports for supplies destined for Soviet consumption; these supplies would then be exported to Western Europe. The Soviet Union would be left with the earnings from the 'swap' plus transit fees. More particularly, IGAT-2 would deliver 17.2 billion m^3 per year from 1984 to 2001 to the Soviet Union, which would utilize this gas domestically. This would free some 13.5 billion m^3 per annum to send to Europe. The Czechs would take some 2 billion m^3, a portion of which would be a fee. The balance would be divided among Österreichische Mineralölverwaltung Aktiengesellschaft (Austria) – 1.86 billion m^3 per year; Ruhrgas (West Germany) – 5.7 billion m^3; and Gaz de France (France) – 3.8 billion m^3. The Soviets further contracted to build the 600-mile line in return for 1 million tons of crude. The terms of this swap have elicited admiration in the West: 'Note the beauty of the scheme. The Soviets would receive 1 million tons of oil for building the Iranian pipeline, 3.5 billion cubic meters in part as a transit fee, and the entire cost would be underwritten by the West Europeans' (Goldman, 1980, p. 81). With the overthrow of the Shah and renewed Iranian demands for higher natural gas prices for both IGAT-1 and IGAT-2, the two trades have been suspended. (The cif price of Iranian exports to the Soviet Union was around $0.72/MMBtu, and the cif price of Soviet exports to Western Europe was approaching $3.00/MMBtu; under these conditions, the Iranians began to get somewhat restless.)[5]

The question of the commercial terms behind Soviet exports is contentious. Throughout the 1950s, allegations were made that the Soviet Union 'dumped' oil on world markets at sacrifice prices. These assertions have been difficult to prove. It is similarly difficult to ascertain the degree to which the Soviet exporters today resort to major price reductions in their sales of natural gas. Table 6.6 gives the destination and prices of Soviet exports in 1979. For comparative purposes, the table also shows the average Western prices for the markets concerned. (Soviet prices here are computed at official dollar/ruble exchange rates.) It is understandable that Soviet prices to markets such as Austria and Finland are higher than prices to other markets. These markets are currently not tied into Western European supplies and there is little competition. The high prices charged to Germany seem a little more problematic, but could reflect the fact that the German contracts allow for escalation in prices with little or no lag. (Lagged price increases are a common feature of many international gas contracts; in this respect 1979 is not a typical year – prices for oil products rose significantly in the last two quarters.) When Soviet prices are compared to the unweighted average of other contract prices to the same markets it would appear, on the basis of Table 6.6, that Russian prices do not reflect a policy of dumping.

This conclusion must be amended in several respects. First, much of the payment for Russian gas is in the form of industrial goods, pipeline and other items. The commercial value of these items is difficult to reflect in cif import prices. Additionally, there is evidence that major contract prices in West European–Soviet trade are fre-

Table 6.6 *Soviet exports and price comparisons with average Western prices, 1979*

Exports to	*Start up*	*Exports in 1979 (billion m³)*	*1979 prices USSR (US$/MMbtu cif)*	*Av. Western price*
Eastern block:				
Poland	1965	4.0	2.25	—
Czechoslovakia	1967	7.2	2.20	—
East Germany	1973	4.3	2.30	—
Bulgaria	1975	3.2	1.95	—
Hungary	1976	2.8	2.50	2.25[a]
Yugoslavia[b]	n.a.	0.8	2.00[d]	—
Romania	1980	n.a.	n.a.	n.a.
Western Europe:				
Austria	1968–78	3.4	2.70[d]	—
Italy	1974	8.6[d]	1.15[d]	1.30
W. Germany	1975	7.4	2.50[d]	2.025
France	1976	2.7	1.25[c]	1.79
Finland	1974	1.1	2.23	—

Sources: Start-up dates: Tiratsoo, 1979, p. 350; 1979 prices: Segal and Niering, 1980, p. 374; trade data: Gas Committee of the UN Economic Commission for Europe, 1980, *Bulletin of Gas Statistics*.

Notes: a Price of Romanian gas exported to Hungary.
b Yugoslavia receives its natural gas transferred through Hungary.
c Price of Dutch gas to Italy (swapped for Russian gas to France).
d Estimates.

quently supplemented with shorter-term 'spot' deals that involve considerable volumes of natural gas at lower prices. Without further knowledge of the timing and other conditions of these sales, it is difficult to ascertain what effect they have on overall pricing behaviour. Finally, Russian gas figures given to the UN Economic Commission for Europe Gas Committee uniformly utilize the net heating values of natural gas, not the gross heating values (for definitions see Appendix II). To the degree that Russian sales to Europe are based on net heating values, there could be a considerable price reduction involved. An example might serve in this regard. The widely quoted 1982 price bases of $4.60/MMBtu for the mammoth deliveries of Urengoy gas through the Yamburg pipeline may not be strictly comparable to Western prices, which are at gross heating values. If gross heating values were substituted for net heating values as the pricing basis, the price of the natural gas involved would fall to about $4.08/MMBtu – a price level that would make it one of the cheaper sources of European natural gas in 1982. These are all rather hypothetical cases based on sketchy evidence. Yet if one or more of these conditions obtain, Russian natural gas, while not sold at 'dumping' prices, clearly underprices other natural gas in inter-European trade.

It is hard to ascertain definitively the degree to which *machtpolitik* enters into Soviet considerations when weighing commercial versus political factors. Clearly, politics are a factor – and one of which customers ought to be aware. The party cadre interviewed by *New York Times* correspondent Hedrick Smith had little doubt about the political merits of Western European dependency on Soviet supplies of natural gas. Referring to the occasion when Soviet supplies to East Germany were constrained because of differences between the Soviet Union and the GDR, the apparatchik stated:

> [T]hey [the East Germans] tried to threaten us but they could not threaten us because we have . . . the taps [pipeline valves] in our hands. That is very useful to have the taps in your own hands. We have the taps now for both Germanies in our hands. The more they take, the more we have in our hands. So they threaten us, the [East] Germans and to us such a threat was just a trifle. We stopped deliveries for awhile and they understood what it meant. (Smith, 1977, pp. 366–7)

Clearly, not too much should be read into such a statement. Russians see East Germany as a client state and amenable to their terms of trade. The situation with Western Europe is different; the Russians need the commodities involved. How much difference this will make is open to interpretation. Cost–benefit considerations will certainly enter into any Soviet political use of natural gas supplies to Europe. How much weight such considerations will be given is, of course, another matter entirely.

The Soviet Union: the 'steam-roller' of the natural gas world?

With all its imperfections and heavy-handedness, there is something awesome about the manner in which the command economy can muster resources to a definitive end. One is reminded of the way the Red Army 'steam-rollered' over the Nazi armies on the Eastern front, at enormous cost in Russian men and matériel but with clearly ascertainable results. It is understandable that a highly formal rational manner of thinking finds the efforts criticizable in terms of economic efficiency and coherency – let alone in terms of individual versus collective values. However, the Russian effort has little formal logic; it is based on substantive goals. In the words of Soviet commentator Gennadi Pisarevsky:

> President Leonid Brezhnev has laid special stress on the importance of developing Western Siberian gas deposits in the 1980s. He has emphasized that these deposits are unique and that the biggest of them – Urengoiskoye – contains such gigantic reserves that it can for many years meet the nation's domestic and export needs, including deliveries to capitalistic countries. (*Oil and Gas Journal*, 29 June 1981, p. 41)

The substantive goal here is reliance on Western Siberian gas reserves. One almost gets the impression that once the 'steam-roller' gains momentum it will be difficult if not impossible to stop. In this, the move towards natural gas in the Soviet energy economy is reminiscent of the move towards electrification, which started decades ago and is still underway, although with reduced priority. The resources that must be brought into play are enormous by any measure. The Soviet goal regarding gas deliveries from one province alone, the Tyumen region, is fully 1 trillion m^3 per year by 1990, equivalent to 170 per cent of total US annual consumption in 1980. In calculations of this order, the celebrated Yamburg pipeline, which has split Americans and Europeans, fades into insignificance.

It is understandable that this 'steam-roller' has sinister overtones in relation to other natural gas markets. The Yamburg line is only part of the issue. Security analysts in both Europe and the United States are right to worry about the beguiling effect of so much natural gas so close to Western European markets, but they perhaps logically confuse a Soviet domestic endeavour with Soviet trade policy. The Russian gas effort is primarily aimed at filling the objectives of the plan. Exports of natural gas are only one objective of the plan – and not the major one at that. The Soviet 'threat' should be seen in this light rather than as a determined effort to undermine the Atlantic Alliance.

In this regard it is vital that the differences between the natural gas behemoth and its capitalist counterparts be understood. To sum up, it might be said that the industrial dynamics are the same both East and West: the particular technological needs of natural gas production, transportation and marketing know no frontiers. Yet what we have characterized as market instability – the bargaining relationships between natural monopoly actors – is not really the major Soviet problem. In the Soviet case, the bargaining relationships are internalized within Mingaz, which orders them so that contractual rela-

Table 6.7 *Differing industrial characteristics: western vs. Soviet modes of organization*

Performance measures	*Western industry*	*Soviet industry*
Criteria for success	Profits	Achieving plan goals as specified for the enterprise
Investment measure	IRR, 'payout' net present values	Opportunity costs as represented by 'pay-back' period (although price is an unreliable measure)
Planning measures	Private industry: 'target rate of return' Public: various public measures regulate allowed profits	Use of material balances methods combined with computers and tools of input–output analysis
Choice of technology (general)	Technology fit economic criteria	Use of engineering criteria to make choice/efficiency, etc. Occasional use of investment measures/interest on capital (transmission lines)
Well spacing/ production rates	Use of inter-temporal discount/interest rates	Project-making institution works out engineering variations, which are weighed in oportunity cost terms (pay-back) before selection of definite plan
Well abandonment	Comparison of prices and avoidable costs	Energy costs in continuing production versus energy won through production
Marketing	Determined by price Marginal costs/end-use value determine marketing strategy	Use of average costs versus other fuels Non-reliability of prices Use of plan goals (material balances) to determine strategy

tionships, vertical integration and the other mechanisms so critical to other markets become less important. Finally, because of the different underlying rationality behind the Soviet command economy, the ends and means of the Soviet planning process are different from those in the West. Some of these differences are outlined in Table 6.7. Deprived of the rational cohesion of a set of prices and of the interest rates provided to Western industry, the Soviet system appears *ad hoc* and bureaucratic.

Soviet pricing is unreliable as a planning indicator. Although it is arguable that price increases, such as that of 1967, have had an impact on consumption patterns (Kelly, 1978), the Soviet system does not rely on the pricing mechanism to allocate fuels to consumer markets either directly or indirectly. In the West, price ties seemingly unrelated phenomena – field abandonment and market demand – together in a rationally united whole. Such is clearly not the case in the Soviet Union, nor is it intended to be. The result is that, despite an impressive ability to mobilize resources, misallocation and bureaucratic infighting will continue to plague the Soviet system because of the substantively rational manner in which the system is designed.

Notes

1 The discussion of field discoveries and establishment of transmission systems is drawn from various sources: Dienes and Shabad, 1979, pp. 79–95; Elliot, 1974, pp. 54–6; Economist Intelligence Unit, 1975, pp. 14–17.
2 'Pipelines hold key to Soviet gas production', *Oil and Gas Journal*, 29 June 1981, pp. 39–40.
3 This figure is obtained through the addition of Elliot's active storage figures for 1974 – and including the planned expansion figures given in Elliot.
4 In fairness to Soviet planning, it should be noted that only a 65-mile segment of the railroad was ever restored. Since then, a 400-mile Surgut Urengoy railroad line has been built, a line that serves a series of oil fields (in addition to gas fields) en route.
5 According to *World Gas Report* (25 April 1983, p. 4), the price for Iranian gas eventually rose to some \$2.66/thousand ft^3 in 1980. At the time that the Shah was deposed, the Iranians were demanding some \$3.80/thousand ft^3. The new regime, failing to get this price, broke off negotiations and deliveries under IGAT-1 ceased. IGAT-2 is also in abeyance, although 40 per cent of the required pipeline has been laid and 30 per cent of the construction work completed.

7

Continental Europe: Nationalism and Constraints

Two paradoxes

In terms of population, of closeness to natural gas reserves and of a modern capital-intensive industry, the European Continental market is an ideal target for the expanded use of natural gas. Compared with the Soviet Union or the United States, the geographical juxtaposition of European resources to markets is a gas man's dream. With a population of 250 million, located largely, though not exclusively, in densely populated urban corridors not far from Holland's gigantic Schloctern natural gas field, Europe has a vast potential, unequalled perhaps anywhere else.

Paradoxically, despite all these advantages, it is arguable that this potential has been realized in the past, and it may not be realized in the future. The market dynamics of natural gas, as noted elsewhere, reward economies of scale. Yet the 'European gas market' as such does not exist. What exists is the *sum of national markets*, all organized according to individual national historical experience and political–economic preferences. To take a parallel, imagine the United States not as one market but as the sum of the markets of each state. Here Texas, Pennsylvania and California would each possess their own national laws and their own national supply policies rational to the end of each state. There would be no coordinating juridical instance, no common sovereign power to adjudicate in the event of disagreement among the states.

What is plainly irrational on the large US market is rational in Europe. Here it has been natural for the Netherlands (Texas?), West Germany (California?) and Italy (Pennsylvania?) each to possess its own market and organize it with relatively little thought to how this organization might affect the natural gas markets of its neighbours. It is the purpose of this chapter to show how, on a single continent, such a variety of individual national solutions to the problems of natural gas market organization has come to exist.

There is another paradox to the European market, interlinked with the first. Despite their diversity, the national Continental markets do have a larger international logic. To return to our analogy, both California and Pennsylvania in the US depend on natural gas 'imports' from Texas in much the same manner as West Germany and France depend on imports from the Netherlands. Inter-state commerce in the United States, however, occurs within the jurisdiction of the national government, whereas trans-border commerce in Europe involves the cooperation of up to a dozen sovereign states. This trading dimension presents very real problems of international market organization and of conflicting national jurisdictions and legal practices. Yet the Europeans have worked out a solution to this, too, preserving the interests of both the gas exporters and the gas importers, while retaining the essential market stability for all the parties involved.

This is the kernel of the second European paradox: how can a series of diverse, 'segmented', national markets achieve the high degree of international natural gas trade (both from European sources as well as from the Soviet Union, North Africa and Iran) that currently exists while remaining non-integrated in any but the most superficial sense? This paradox and its ramifications for the future of the European markets constitute the focus of Chapter 8.

A wealth of national experiences

Progress towards general European economic integration differs widely from sector to sector. With varying success, the European Communities have pursued common policies in trade, agriculture, money and fisheries. Perhaps the most intractable commodity area within the Communities is that of natural gas. Here, rather than the European Communities, the main international organization involved in the coordination of national natural gas policies is the United Nations Economic Commission for Europe (UNECE). The Gas Committee of this organization nevertheless has a largely toothless informational role and confines its activities to the issuance of periodic statistical and technical reports.

Why have the Europeans been so uninterested in – one might even say resistant to – the idea of greater policy coordination? One could point out that diversity in natural gas policies is partly due to the fact that some European countries are net exporters (Norway, the Netherlands), others are net importers with some national natural gas production of their own (West Germany, Italy, France), and still others are totally dependent on outside sources of natural gas (Belgium, Switzerland). While appealing, this argument is tenuous. That some countries are net exporters, others net importers, and others completely dependent on imports could equally well be an argument for rational integration. Yet it does not happen; the nations involved, as said, plan their markets with only minimal attention to the effect of their actions on others. In 1979, for example, the Netherlands suddenly announced that it was not renegotiating the contracts it had concluded with its importers. This unilateral decision left the other parties in a real quandary about how they were to find substitutes for Dutch sources of supply.

National differences and the lack of impetus towards integration, we shall show, have their roots elsewhere: in the varied national political–economic rationales of industrial market organization, in the differing historical experiences that most European nations have had in the coal and town gas industries, in the individual national responses to oil multinational corporations, and in their dissimilar technical and marketing basis.

National economic ideologies of industrial market organization in Europe by themselves constitute a major reason for national irreconcilability. Just as some countries generally prefer extensive government control of fiscal and monetary affairs and others emphasize the role of the private sector, so do these preferences spill over into individual policy areas like natural gas. In Denmark, for example, the formation of a state-owned gas company, the Danish Oil and Natural Gas Company (DONG) has been an arduous process, with much pushing and pulling between the political left and right over whether the creation of a state company represented excessive government interference in the market place. This view is simply not comprehensible to the Norwegians, whose gigantic Statoil is playing an increased role in furnishing the European market with natural gas. Neither would it be understood in France or Italy. Other governments are less convinced of the merits of state enterprise, prominent among them being West Germany.

Another source of national diversity is the historical experience of the individual countries with coal markets and associated manufactured gas. Blast furnace gas, coking gas, water-based gas and town gas, as we have noted, are largely local monopolies. In both the United States and Great Britain, the transition from locally manufactured gas to natural gas had organizational consequences long after the change came into effect. In the United States, suspected price-fixing among the transmission companies and local manufactured gas utilities was one of the reasons for the passage of the Natural Gas Act of 1938. In Britain, the Gas Council and twelve area boards were established not to coordinate the development of North Sea gas by virtue of their monopoly of the British market, but to rationalize and ensure the survival of an industry in its death throes. Similarly, the dying town gas industry on the Continent led to differing national solutions that, slightly modified, still exist today. (Natural gas was introduced at a much more recent date in European markets and the after-effects of town gas are thus more obvious.) One need only look at the origin of the current state companies in the Netherlands and France to see the consequences of this transition. In the Netherlands, Gasunie (the sole gas transmission company) is owned jointly by Dutch State Mines (the old national coal firm) and the private Royal Dutch/Shell–Esso consortium known as Nederlandse Aardolie Maatschappij (NAM). The French Gaz de France was a consequence of the nationalization of gas works and an attempt at rationalization; today Gaz de France, together with privately owned Société Nationale Elf-Aquitaine (SNEA), dominates the French natural gas market. In West Germany, state involvement in coal took a slightly different turn. The major transmission company, Ruhrgas, was essentially a state-industry coalition of interests between the German coal industry, Shell and Esso (the Dutch exporters), and various utilities.

Of slightly different significance, though interrelated with policies towards coal and town gas, are national attitudes towards the multinational oil corporations. These vary from active cooperation with a private 'national champion' (as with NAM) in the Netherlands, through a 'hands-on, hands-off' strategy (West Germany), to active policies of responding through the erection of state companies (Italy and France). Italy in fact has utilized its domestic reserves of natural gas to build the Ente Nazionale Idrocarburi (ENI) into a national state-owned alternative to the oil companies.

Finally, the technical and marketing differences between national markets are very real. This can be most easily seen in the relative market shares of domestic/residential, commercial, industrial and electrical generating sectors for the respective countries (Table 7.1). The Netherlands are the undoubted champions in utilizing natural gas in all non-transportation energy sectors, which is hardly surprising. However, there is enormous variation in market penetration among the other three nations. The German market share for natural gas in electrical generation is more than twice that of Italy and four times that of France – countries that are more reliant on coal, oil and nuclear power. Italy, on the other hand, follows the Netherlands in the supply of the domestic residential markets, which are notoriously difficult to service due to load factor problems. Yet the figures are to some extent deceptive. In quantity terms, the West Germans utilized only slightly less natural gas in electrical generation in 1977 than did the Dutch. Furthermore, while the market share enjoyed by Dutch gas in this sector has since fallen to under 50 per cent, the German use of natural gas (and thus the market share of natural gas) has actually increased. The essential point here is that each of these national markets has achieved its own equilibrium and that these equilibria are different.

The availability of storage capacity is further evidence of differing national equilibria. Italy has storage roughly equivalent to 32 per cent of its national annual consumption, and is followed closely by France. West Germany has only a fifth of the storage capacity of Italy in terms of annual per cent of consumption. The Dutch have little storage; rather, Gasunie relies on the unique productive qualities of the Groningen field – increasing field production in the winter, decreasing it in the summer. This practice has led to some second thoughts, however (N. V. Nederlandse Gasunie, 1979). *Ceteris paribus*, one would therefore expect that residential markets would be emphasized less in West Germany than in the other three countries – an expectation borne out by the figures in Table 7.1.

Table 7.1 *Market shares of natural gas: West Germany, the Netherlands, Italy, France, 1977 (percentage share)*

	Electrical generation	*Domestic/ residential*	*Commercial/ public service*	*Industrial*
West Germany	17.1	17.8	22.9	22.5
Netherlands	77.7	69.9	96.1	52.8
Italy	8.2	26.7	3.8	25.1
France	3.6	11.4	50.8	16.0

Source: International Energy Agency, 1981c.

The markets still incorporate remnants of manufactured gases, but once again in varying quantities and differing manners. Italy, West Germany and France produce a considerable amount of manufactured gas through blast furnaces (furnace gas), coking ovens, refinery operations, and even town gas works, but only a fraction of this gas is ever used outside its industry of origin. In West Germany, coal and steel manufactured gas use amounts to some 4.7 per cent of total manufactured gas/natural gas consumption. The planning and structure of the French manufactured gas segments is somewhat different. In France, town gas is no longer supplied to residences, a use that is still retained in Germany.[1] The planning and structure of the Italian industry in this regard differs from that of both France and Germany, while in the Netherlands the outside use of manufactured gases is negligible.

This basic pattern of diversity applies to the entire European Continent. In order to keep our analysis within manageable limits, it is necessary to focus on the most significant cases. Accordingly, this chapter will discuss only the four major natural gas markets on the European Continent: those of the Netherlands, West Germany, Italy and France.

National markets: the Netherlands

Any discussion of the Continental natural gas markets must begin with the Netherlands. Not only is natural gas extremely important to the Dutch economy (providing roughly 7 per cent of national income and some 20 per cent of government tax revenue), but Dutch gas exports account for roughly 50 per cent of all natural gas consumed in Western Europe. In a very real sense, the growth of the European natural gas market can be traced to the 1959 discovery of the huge Schloctern field in Groningen.

The Dutch, too, began their sally into gas with the production of town gas from coke and other manufacturing processes. During the 1940s came the discovery of two smallish natural gas fields: one at Coervorden (1943) and the other, the more important of the two, at Wijk (1949). These finds were succeeded by the discovery of a series of other onshore gas fields: Staphorst (1950), Wanneperveen and Tubbergen (1951), Denekamp (1952), Rossum (1955), De Lutte (1956) and Schoonebeek (1958) – all close to the West German frontier (Peebles, 1980, p. 121).[2] Conversion to natural gas or 'reformed' natural gas was therefore well underway when the Royal Dutch/Shell–Esso exploration and operating consortium, Nederlandse Aardolie Maatschappij (NAM), struck natural gas at the Schloctern gas field in Groningen in 1959. The significance of this field was, however, slow to dawn on Dutch authorities. Initially the reserves were assumed to be of the order of some 60 billion m^3. In 1962, the Dutch Ministry of Economic Affairs declared the reserves to be 500 billion m^3; in October 1963 the figure was revised upward to 1,100 billion m^3. It is currently estimated in oil company circles that the Schloctern field contains in all some 2,000 billion m^3 of natural gas (Peebles, 1980).

While not as significant as the 'oil shocks' of 1971 and 1973–4, the discovery of Schloctern revolutionized the European energy equation of the 1960s. The field was so big that the Dutch decided to use it not only to spur a widespread national conversion to natural gas, but also to export it. Thus the orderly introduction of Schloctern gas to both the Dutch and Continental markets took on a critical significance. How was it to be accomplished?

Dutch society has not inaccurately been characterized as 'a type of institutionalized Gemeinschaft society reinforced with utilitarian services; institutional patronage, and the inducement of emotional solidarity' (Scholtern, 1978, p. 4). In such a society, the role of corporatist bodies becomes very important, their function being to arrange for conflict resolution/consensus maintenance in industrial-governmental relationships.[3] It follows that large, well-defined organizations – such as NAM – would exercise considerable power. To many observers, indeed, Dutch gas policy is NAM policy ('NAM runs everything around here', is not too untypical of the more casual comments one encounters). Clearly, this statement is an oversimplification. The Dutch administration of natural gas has been characterized by gradual change, which, although preserving NAM's status as the most important corporate entity in Holland, has also eroded many of its powers. While the most important of these incremental steps were hotly debated in the Dutch context, few of them were noticed in the world at large. The result has been that NAM and Gasunie (which took over the functions of the old state Gas Board) perform an unusual private enterprise/state enterprise function that cannot be found elsewhere in the world of natural gas.

The corporate arrangement of this complicated compromise is illustrated in Figure 7.1. The 'Maatschap', the critical government holding *vis-à-vis* production of Schloctern (Groningen) gas, is not a corporate entity but a curious anomaly, a 'financing partnership' in which the old national coal firm, Dutch State Mines (DSM), would share the costs and profits incurred through Groningen production. DSM is today, with the winding up of coal production in the Netherlands, also a holding company that, in addition to its 40 per cent share of Gasunie, is the full owner of chemical firms, building materials industries and a ceramic business. Given these diversified interests, plus the very real differences in the energy content of Schloctern gas *vis-à-vis* that of other fields (its calorific value is only about 80 per cent of that of Ekofisk natural gas), it was probably well that Dutch State Mines could rely on NAM when it came to the wholesale conversion of the Dutch energy economy to natural gas.

After Schloctern was discovered, the Dutch authorities resolved to introduce this gas gradually, so as not to dislocate existing energy markets. This decision had two unavoidable consequences. First, it meant that the sales of Dutch gas were dependent on the sales of other forms of energy. Second, it made it necessary to expand Dutch natural gas exports rapidly. Both of these consequences have since

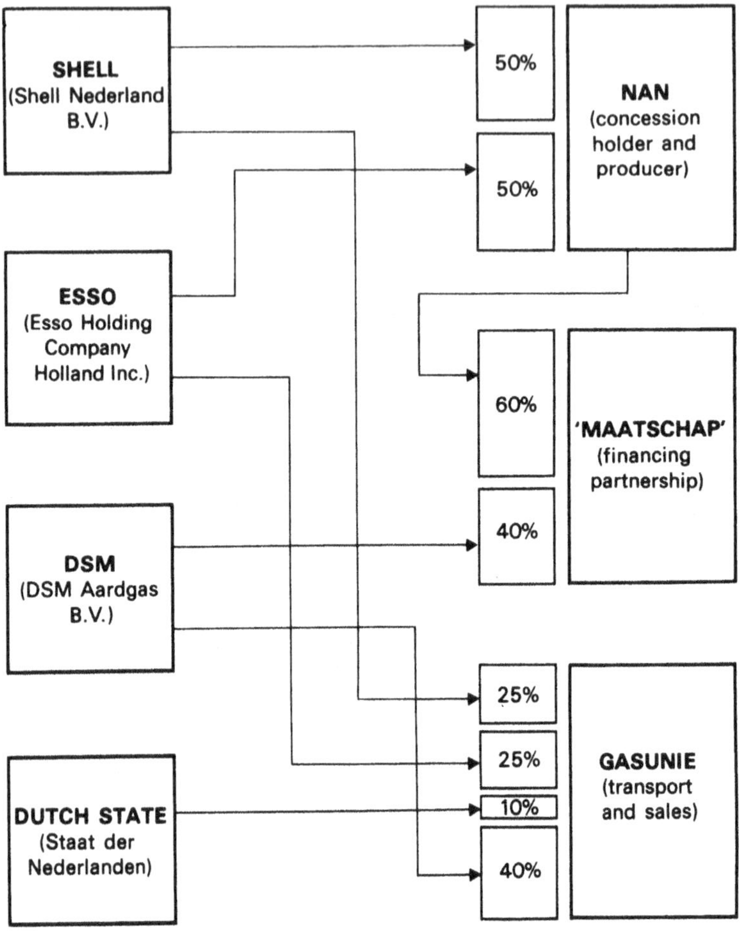

Figure 7.1 *Groningen gas: corporate arrangement*
Source: Peebles, 1980, p. 126

been subjects for debate in domestic quarrels with gas utilities and in difficult international negotiations with other European countries.

Domestically, Gasunie negotiates with the distribution companies. In February 1963, these companies formed a grouping known as SROG, whose purpose was to set up the organizational arrangements for gas supply. In 1972, SROG was dissolved and replaced by VEGIN, the national association of gas boards. The marketing price of natural gas established by these distribution companies is exemplified here by the prices for home heating use. Table 7.2 shows how the price of natural gas, first set competitively with coal, shifted in 1965 after the introduction of Groningen gas to a price competitive with home heating oil, a position it has maintained to the present. (The price of gas rose on 1 January 1979 to some $3.961/MMBtu, and has increased since then.) This interesting shift is remarkable not only for itself, but also in its contrast to Table 5.5 in the British chapter.

Prior to the discovery of Groningen gas, the question of marketing and profit-sharing had been thoroughly discussed by an industry–government Natural Gas Advisory Committee. This committee recommended in 1953 that Dutch gas should be marketed so as to render the maximum 'net profit' to the Dutch economy. Subsequently, the state Gas Board was awarded monopsony powers for the purchase, transmission and sale of natural gas to industry and local distribution companies (Peebles, 1980, p. 122). This arrangement, covering the

Table 7.2 *The home heating market: competitive fuels (US $/MMBtu)*

		Fuel	
Year	*Gas*	*Coal*	*Oil*
1963	1.002–1.170	0.919–1.337	0.919
1964	1.005–1.173	0.921–1.340	0.670
1965	0.538	0.916–1.333	0.666
...	—	—	—
1976	2.816	n.a.	3.550

Source: Peebles, 1980, p. 137.
Note: Conversions to energy equivalents by me; exchange rates are those for the years concerned.

small gas fields discovered prior to 1959, was certainly not sufficient to handle the quantities involved in the exploitation of the Groningen find.

In arranging for the production, transmission and marketing of Schloctern gas, another equilibrium was achieved. This was guided by two principles upon which all parties could be agreed: first, the natural gas should be sold on a 'commercial basis consistent with the greatest possible benefit to the national economy', and, second, the gas 'should be introduced as smoothly as possible without undue dislocation of the existing energy market' (Peebles, 1980, p. 125). This translated in practice into the dissolution of public ownership of the transmission lines as established in 1954, the active participation of both NAM and the state in the transmission and distribution of natural gas, and the passive participation of the state in natural gas production at Schloctern and in the determination of gas export policies.

The partner chosen for state participation was, as already mentioned, the state firm Dutch State Mines, which, as the major existing producer and distributor of coke oven gas, 'would be affected by the introduction of Groningen natural gas into the market' (Peebles, 1980, p. 125) and would hence have somewhat similar interests to those of Shell and Esso, whose oil products would also compete with natural gas. Natural gas supplies to industry are undertaken directly by the joint state–industry transmission company Gasunie. Since 1967, gas prices to major industrial suppliers have been directly linked to heavy fuel oil prices (not unlike what happened in the residential heating market, as described earlier). Gasunie maintains a 3,450 km main net, a 6,250 km regional net, and supplies natural gas to 193 distribution companies, 22 electrical companies and some 403 industrial clients (Communauté Européenne et les Entreprises Publiques, 1979, pp. A-106 – A-107). The gas is marketed according to strategies worked out by Gasunie, the Department of Economic Affairs (the ministry mainly responsible for energy policy) and VEGIN. Goals are specified in an annual plan, the Plan van Gassfzet, and selling prices must be approved by the ministry. This procedure does not wholly alleviate dispute, for there are very often conflicts between VEGIN and Gasunie, 'especially regarding price setting' (de Man, 1982, pp. 14–15).

A direct consequence of this gas marketing policy was the reasonably rapid build-up of natural gas exports to other nations in Europe. The Dutch strategy presents an interesting contrast to the British, whose solution was both not to export it and to sell natural gas cheaply – or, more pejoratively, to 'dump' it – on the domestic energy market. A consequence of the Dutch 'orderly marketing' was that not only had quantities of natural gas to be disposed of elsewhere – and here there were ready purchasers in West Germany, Belgium, France and Italy – but the gas was also marketed at a higher price. The export of Dutch natural gas was to be carried out by a firm called NAM/Gas Export, which acted both for Royal Dutch/Shell–Esso and for the state participants in Gasunie. This company, with assistance from the Internationale Gas Transport Maatschappij (another Shell/Esso company established to develop gas and distribution ventures outside the Netherlands), continued to function in this role until replaced by Gasunie in April 1975 (Peebles, 1980, pp. 127, 140). The pricing/escalation terms of these early exports are illustrated in Table 7.3.

Another contrast between the Dutch solution and the monopsonistic/nationalistic British solution was that the considerable rent element in Groningen gas went not to the consumer, as in the UK, but to private industry and the state. NAM, in fact, made considerable

Table 7.3 *Initial export prices and subsequent price increases: Netherlands to Continental markets (US $/MMBtu)*

Date	*Country importing*	*Reported initial price*	*Price in 1976*[a]	*Price in November 1981*[e]
1966	Germany[d]	n.a.	1.062	4.06[f]
1966	Belgium	0.332	1.062	4.06
1967	France	0.367	1.062	4.06
1970[b]	Germany	0.385[b]	1.062	4.06
1969–70	Italy	0.437[c]	1.062	4.06

Source: Valais and Durand, 1975.

Notes: a Prices are average price for all Dutch gas in 1976. In October 1976 negotiations were undertaken to increase prices and tie them to competing oil product (low-sulphur fuel oil) prices.

b Price is on contracts begun in 1966 but renegotiated.

c No escalation clause.

d This is by no means the first import of natural gas into Germany. The Dutch were exporting considerable quantities to Energieversorgung Weser Ems in Germany before the development of Schloctern.

e Tied to fuel oil prices and more or less uniform. The prices are *ab* border and consequently are adjusted for the cost of transportation to national markets.

f Excluding Rheinisch-Westfälisches Elektrizitätswerk and Energieversorgung Weser Ems, which had not signed contracts.

profits in the early years. The Dutch manner of resolving the issue of rent capture has been incrementalist, relying on the establishment of a consensus and then only in one area at a time. This is illustrated in Table 7.4. Thus the smaller onshore fields still are obliged to share only some 10 per cent of their profits before taxes with the state; and it was not until 1976 that the tougher continental shelf conditions were extended to new licences onshore with the granting of the Twenthe concession to NAM. The NAM Groningen concession itself was unaffected by the tough terms meted out to holders of newer North Sea licences on fields such as Placid and K14 as these came onstream from the Dutch shelf. Rather, the Dutch state decided to alter 'financing arrangements' within the Maatschap, changing these from the earlier 50:40:10 (NAM:DSM:state) to 15:85 (NAM:DSM/state) in 1971, and further hardening the terms in 1974 – some eleven years after the first gas had been sold from Groningen. The exact nature of this financing relationship is unclear, but it is thought to be a profit-sharing arrangement, the exact particulars of which have not been publicly explained.

More recently, with renegotiated Continental (see later in this chapter) and domestic prices, the state is planning to levy a further 'windfall tax' on the difference of the price at which the natural gas is

Table 7.4 *Rent capture and Dutch natural gas: a chronology*

Year	*Area*		
	Onshore		*Offshore*
	Groningen/Schloctern	*Others*	
1948	—	10% profit share to the government	—
1956	—	Calculations of profit shares specified (levied before taxes and therefore deductible)	—
1963	10% profit share after taxes. DSM a 40% share of profits. Gas to be sold by Gasunie owned 25:25:40:10 by Shell, Esso, DSM and state – each gets respective share from the Maatschap.	—	—
1967	—	—	Continental shelf opened. State receives surface lease payments, royalties of 10–16%, a profit share of 50% (from which corporate taxes can be subtracted), and 40% participation rights for DSM.

1971	85–15% financial sharing between state and NAM in Maatschap	—	—
1974	95–5% financial/profit-sharing in Maatschap	—	—
1976	—	—	Revised surface lease payments. Royalties on subsurface value. State profit share raised to 70%. Participation rate raised to 50% and revised to include petroleum
1977	—	NAM given Twenthe concession on same terms as continental shelf in 1976	—
1981–2	Attempt to levy windfall profits taxes on differential between reference price (ca. 19 Dutch cts/m^3) and the price realized by seller.		

sold and a reference price of some 19 Dutch cents m^3. The negotiations over this tax have been prolonged, and its exact status as of this writing is unknown.

Although Groningen still is the major supplier of natural gas to the Dutch and European markets, gas from Schloctern is rapidly being supplemented by natural gas from the Dutch North Sea. The latter will, however, never be a sufficient replacement for Groningen gas despite the remarkable rate of growth in North Sea production (see Table 7.5). An additional problem with North Sea gas, as indicated earlier, is that its Btu content is considerably higher than the Btu content of Schloctern (or Groningen) gas. Roughly speaking, Groningen gas has a gross thermal content of some 840 MMBtu/thousand ft^3, offshore gas a content of some 1,077 MMBtu/thousand ft^3; and the other onshore fields are somewhere between these two extremes (Peebles, 1980, p. 124). These differences have caused technical problems, and not only in the Netherlands. The international transmission of Groningen gas through the 'L'-gas system (for low thermal content) has had to be supplemented by the building of a new 'H'-gas

Table 7.5 *Dutch gas production, 1960–81 (billion m^3 at 15°C)*

	Onshore	*Offshore*	*Total*
1960	383.9	—	383.9
61	476.4	—	476.4
62	538.1	—	538.1
63	602.7	—	602.7
64	875.8	—	875.8
1965	1,817.6	—	1,817.6
66	3,585.0	—	3,585.0
67	7,888.2	—	7,888.2
68	16,076.1	—	16,076.1
69	25,263.4	—	25.263.4
1970	37,196.0	—	37,196.0
71	51,102.1	—	51,102.1
72	65,877.2	—	65,877.2
73	78,085.1	7.4	78,092.5
74	92,730.6	10.4	92,741.0
1975	97,057.6	960.9	98.018.5
76	101,981.6	3,092.3	105,073.9
77	97,682.1	4,825.8	102,507.9
78	86,058.2	6,293.5	92,351.7
79	85,367.2	10,925.4	96,292.6
1980	78,208.9	12,102.0	90,310.9
81	70,928.3	11,798.3	82,726.6

Source: Netherlands Ministry of Economic Affairs (1981) Annex 8a.

system (high thermal content) to handle North Sea shelf gas from Norway and the Netherlands (Tiratsoo, 1979, p. 255).[4]

With the expansion of international gas trade and the depletion of Groningen reserves, the Dutch government is disinclined to renew the existing contracts and has begun the process of renegotiating them. This turnabout has caused considerable worry in Western Europe. Figure 7.2 shows the rapid decline in natural gas exports that will occur on simple termination of the Dutch international contracts. Efforts are currently underway to ease this downturn by rewriting these agreements to allow for more flexible terms stretching over a longer period of years. The Netherlands is also engaged with its Continental trading partners in contracting for additional gas supplies

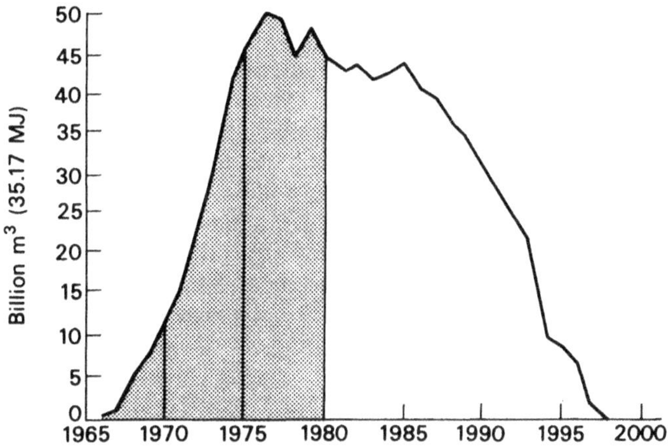

Figure 7.2 *Projected decline in Dutch exports: 1980–2000*
Source: N.V. Nederlandse Gasunie, 1979, p. 6.

from elsewhere in Europe (Norway and the Soviet Union), from Africa (LNG from Algeria and Nigeria), from the Middle East (Qatar) and from North America (the Canadian Arctic).

National markets: West Germany

Some 43 per cent of Dutch exports go to the Federal Republic of Germany, the country with the largest domestic West European gas market. The West German market is very complex and a thorough description of its particulars is beyond the scope of this book. There are, however, several features of the market that are of special interest in this context, and it is to these that we will turn.

Georg Küster, in a widely cited essay, has succinctly expressed German attitudes to general economic policy. In characterizing Germany's economic policy through the years 1950 to 1970, he writes:

> Common to the structure throughout was a pluralist view of society and a neglect of social antagonisms. Society was seen as a collection of interests and groupings of equal rank, no group being able to dominate another, each group being limited by countervailing power (Küster, 1974, p. 67).

In practice, this has led both to a strong belief in the efficacy of the private *geordnete markt* (orderly market), and to a surprising degree of government intervention in the energy field. How can these seemingly disparate policies of orderly market competition and state interference be reconciled?

One area where it had been difficult to obtain an orderly market was coal production and marketing. As early as the late 1950s, the German hard coal industry was in considerable trouble. Two different government approaches to saving the situation were tried. The first was to discriminate against other fuels, particularly oil. In the period 1965–71, companies were made voluntarily to restrict their imports of fuel oil. The supply of fuel oil to the electric power companies was contingent upon the consent of the Economics Ministry. With a view to security of supplies, the government required importers to stockpile oil, a requirement that effectively excluded smaller importers (Küster, 1974, pp. 74–5). Excise taxes were levied on heavy and light heating oils. A protective tariff (ostensibly to encourage domestic production of crude) also raised the price of imported oil to the consumer. A special tax on heating oil was introduced in 1960 and is still in force today (Grayson, 1980, pp. 147–9). This discriminatory approach was complemented by a conciliatory attitude. Thus the major companies were encouraged to cooperate with one another, although small competitors were excluded from such cooperation. To accomplish this, the oil industry was exempted from the German Cartel Act. It was in this period of concerted policy-making that the fundamental structure of German gas imports was built.

Dutch gas exports to West Germany did not occur in a vacuum. There was quite considerable domestic natural gas production in Germany in 1964 of around 1.9 billion m^3, which rose to just under ten times that amount fifteen years later, due largely to discoveries in the Weser-Ems and the Ems estuary (Tiratsoo, 1979, pp. 89–91). There also was considerable legal precedent on the books about how that natural gas should be marketed, mostly dating from a law promulgated by the Reichstag in 1935. According to that law, the various

Länder in the Federal Republic had wide discretionary powers in planning for the transmission and distribution of natural gas and in implementing prices. When the federal government took control of most of the policy-making in the energy area (covering coal, oil and refining policies), much of the power of the *Länder* became a 'dead letter'. It is interesting in this context that one of the few remaining areas of *Länder* discretion was licensing policy and rent capture. In the areas of the Federal Republic formerly governed by the Prussian state, a royalty of only about 17 per cent is demanded (in addition, of course, to corporate taxes). But in Lower Saxony and in Bavaria, two of the prime gas-producing areas, there are other sets of rent capture mechanisms. This regional independence has prevented the federal government from imposing 'windfall profit' taxes on domestic oil and gas production. It is estimated that this cost the state some DM 916–1,577 million in tax revenues in 1975 alone (assuming that all windfall profits could be captured through taxes on the economic rent) (Grayson, 1980, pp. 167).

Against this background, the introduction and build-up of natural gas to the West German market was a carefully balanced affair, taking into account both the interests of the companies and the interests of the German state. In practice, this led to a high degree of vertical integration between the producers of natural gas and the transmission companies in West Germany, and between the oil companies and the major coal and electrical utilities in running the German transmission lines.

As Table 7.6 demonstates, the oil majors hold a tight control over natural gas production in the Federal Republic; Shell and Esso alone account for more than 50 per cent of the total. If one excludes Wintershall AG, production by firms owned by West German interests amounts to less than 5 per cent of the total.

A multinational dominance is even more pronounced in the ownership of the German transmission companies. These consist mainly of the major holders of the individual pipeline companies, which control the individual pipelines. The ownership of these major transmission companies is illustrated in Table 7.7. Although ownership patterns have shifted considerably since 1975, the oil majors have a very dominant position, particularly again Shell and Esso. These transmission companies are not all equally important, however; it is estimated that, directly and indirectly, Ruhrgas may account for 67 per cent of the German market, and Shell/Esso own about 40 per cent of Ruhrgas through their individual holdings and their Brigitta partnership.

In general, the companies listed in Table 7.7 also have major purchasing interests in Holland. Thus Deutsche BP (Gelsenberg) is an importer of Dutch gas in its own right, and Brigitta (Shell/Esso in

Table 7.6 *German domestic gas production: corporate shares*

German company	*Ownership*	*Production and market share*			
		1975		*1977*	
		billion m³	*%*	*billion m³*	*%*
Oil majors total		15.2	78.9	15.3	79.4
Gewerkschaft Brigitta Gewerkschaft Elwerath	Shell/Esso Shell/Esso	10.47	54.5	10.5	54.4
Mobiloil AG	Mobil 100%	4.3	22.2	4.3	22.3
Deutsche BP (Gewerkschaft Norddeutschland)	BP 100%		—	0.02	0.1
Deutsche Texaco	Texaco 100%	0.43	2.2	0.5	2.6
German firms total of which:		4.02		4.0	
Wintershall	BASF[a] 100%	1.85	9.6	2.06	10.6
Preussag		0.42	2.2	0.37	1.9
Deilmann Ag		0.08	0.4		—
Gelsenberg (Veba)		0.223	1.2	0.25[b]	1.3
Other		1.45		1.32	
Total majors + German firms		19.28	100	19.3	100

Sources: 1975 figures: *Petroleum Times*, 12 November 1976, pp. 31–3; 1977 figures: from industrial sources.

Notes: a Badische Anilin- und Soda-Fabrik AG.
b Merged with Veba.

Germany) purchases natural gas from its partners (Shell/Esso) in the Netherlands plus the other members of Gasunie. This is not always the case, however. Rheinisch-Westfälisches Elektrizitätswerk (although it owned a major share in Gelsenberg in 1972) is a minor direct purchaser of Dutch gas, as is Vereinigtes Elektrizitätswerk Westfalen (a sister electrical company).

In sum, while precise data are hard to come by, there can be little doubt that a consequence of the German style of 'orderly markets' has been to let the oil majors play a vital role in the production and transmission of natural gas in West Germany. Nor has time significantly altered this picture. When the German state changed course and attempted, through the structuring of the national state-owned giant Veba, to go directly into the West Germany energy market, it appeared that existing natural gas marketing arrangements would be

Table 7.7 *German transmission companies: corporate ownership (estimated share)*

Transmission company	*1980 ownership*			*Estimated market share[a] %*
	German co.	*%*	*Oil MNC*	
Ruhrgas	Shell AG	15.0	Shell/Royal Dutch	50.0
	Brigitta	25.0	Shell/Esso	
	Deutsche Texaco	0.4	Texaco	
	Deutsche BP (Gelsenberg)	25.31	BP	
	Veba	5.0	—	
	Mobil	7.4	Mobil Oil	
Thyssengas	Stichtung Administratie Kantor Thyssen	50.0	—	8.4
	Deutsche Shell	25.0	Shell	
	Esso AG	25.0	Esso	
Deutsche Erdgastransport GmbH	Deutsche Shell	25.0	Shell	
	Esso AG	25.0	Esso	3.3
Gasversorgung Süddeutschland	Brititta	25.0	Shell/Esso	6.9
	Cities and *Land* Baden-Württemberg	75.0		
Erdgasverkaufs-GmbH	C. Deilmann AG	—	—	2.9[b]
	Deutsche BP	n.a.	BP	
	Deutsche Schachtbau	n.a.	—	
	Gewerkschaft Elwerath	n.a.	Shell/Esso	
	Mobil	25.0	Mobil Oil	
	Wintershall	n.a.	—	
Bayerische Ferngas	Four cities	n.a.	—	n.a.
Salzgitter Ferngas	Ruhrgas	24.0	See under Ruhrgas	n.a.
	Holding Co. (Ruhrgas owns 20%/ nominates management)	24.0		
	Brigitta	12.5	Shell/Esso	
	Lower Saxon utilities	26.0		
	Münster Erdgasverkaufs GmbH	12.5		
German electrical utilities	Vereinigtes Elektrizitätswerk Westfalen (VEW)	13.0		
	Rheinisch-Westfälisches Elektrizitätswerk (RWE)			
	Energieversorgung Weser Ems (EWE)			

Source: industry sources.

Notes: a The position of these firms as owners of transmission lines is unclear. The consumption attributed to them is based on their imports into Germany. They are probably owners of shares in the various pipeline companies as they negotiate independently with Gasunie. The figures given for their consumption are probably low.

b Based on estimated production of these companies.

more or less thrust to one side. In 1969, with government encouragement, Veba, together with eight German companies (including Gelsenberg, Preussag and Deutsche Schachtbau) founded an international oil exploration firm called Deminex.[5] Gelsenberg, a large German energy group, was the next Veba acquisition. Veba purchased 48.3 per cent of the share capital of Gelsenberg from the electrical holding company RWE in late 1973, and soon held a majority in that company. In September 1974, Veba acquired Deutsche Schachtbau and Preussag AG. The outcome of these moves was that the state acquired substantial energy interests and, for the first time, became (through Veba) a major shareholder in Ruhrgas, some 30.31 per cent. These shares were, however, later transferred to the oil major British Petroleum in a massive deal in which BP refineries and guaranteed supplies of crude were exchanged for shares in Ruhrgas and other Veba interests. (The resulting widely changed interests are those represented in Table 7.7.)

It seems clear that the West German market has needed the expertise that large firms so often possess. One of the main reasons is the mix of gases used to supply the German market. A large portion of West German gas is in fact coking gas from the Ruhr. Added to this is the natural gas from Groningen, with an entirely different range of thermal values, natural gas from the North Sea, and natural gas from the Soviet Union. All this gas must be purified, reformed or upgraded at various plants throughout West Germany, most notably in the Ruhr, which is the coking gas centre. Furthermore, most of the international gas contracts into which the transmission companies enter have a very high load factor (some 90 per cent). The general load factor for distribution firms, on the other hand, lies at about 45 per cent; consequently, the demands on both transmission lines and distributors is large.[6]

National markets: Italy

Italy has the third largest natural gas market on the Continent, after West Germany and the Netherlands. Like West Germany, Italy is both a considerable producer of natural gas in its own right (generating some 13 billion m^3 in 1978), and reliant on natural gas imports (about 14.3 billion m^3 in 1978) to satisfy domestic demand (Tiratsoo, 1979, p. 96). The character of the Italian natural gas market organization, however, provides an interesting contrast to the Dutch and German experiences. It is dominated almost totally by the two concerns AGIP and SNAM, affiliates of the state-owned and -run Ente Nazionale Idrocarburi (ENI).

The reasons for this contrast are to be found in the economic background of the Italian state. Private initiative in industrial development in Italy has historically always been weak, requiring a large state-directed sector as an initiator and supporter of structural change. As a nation with its own set of infant industries exposed to other more mature competitors, Italy has been well aware of the imperfections of the free market as an engine for industrial growth (Prodi, 1974, pp. 45–57). The oil industry, with its well-defined oligopoly markets, was therefore a 'natural' target for Italian state initiative.

The Azienda Nazionale Generale Italiani (AGIP), the first company of what was later to become the ENI group, was founded in 1926; its *raison d'être* was to conduct and promote exploration, production, refining and marketing of hydrocarbons both in Italy and abroad (Grayson, 1980, pp. 107–8). Despite its mixed record abroad, with the onset of the Second World War further complicating matters, AGIP was a success at home, particularly in the Po valley, which gave this firm its first secure source of energy. Here, at Cortemaggiore, accumulations of what was practically pure methane were discovered in 1949. These finds, and some smaller ones before them, led to renewed oil industry interests in this area. In the ensuing political fracas, the head of AGIP, Enrico Mattei, capitalized on the Cortemaggiore find not only to secure exclusive exploration rights to 55,000 km^2 of the Po valley for AGIP, but also to obtain the sole right to run pipelines for the natural gas concerned. In a major move in 1953, the Italian government created ENI as a holding company under Mattei, assigning it all governmental interests in the hydrocarbon industry (Penrose, 1968, p. 142).

These moves had two enormous consequences for natural gas in Italy. First, in granting Mattei and the ENI exclusive rights to the exploitation of Po valley natural gas, the Italian government also assigned him the right to all the rent involved in natural gas production. (The Italian government still has one of the more advantageous tax and royalty systems in Europe – applying largely to offshore areas.) Second, in assigning the newly created ENI subsidiary Società Nazionale Metanodotti (SNAM) exclusive rights to the transport of this gas, the Italian government placed the Italian natural gas distribution markets firmly in the hands of ENI. In 1977, fully 94.2 per cent of all natural gas produced in Italy (including that of the offshore shelf) and some 97.2 per cent of all natural gas sold in Italy passed through ENI affiliates.

Impressive as these figures are, however, it should be noted that they exclude considerable ENI/private capital cooperation. ITALGAS, a large Turin–Rome gas manufacture and distribution firm, which sells both manufactured gas and some 857 million m^3 of natural gas in 136 communities, is a case in point; although owned some 34.38 per cent by SNAM, it is also widely held by a large number of other enterprises.[7] Similarly, AGIP, although capitalizing on its concession in the Po valley and enjoying preferential treatment in other onshore and offshore licensing rounds, dominates but does not wholly exclude the oil companies. In 1981, AGIP, possessed twenty-six of the thirty production concessions related to offshore Italy; the French Elf possessed the rest. In the exploration phase, AGIP is either on its own or is operator of some ninety individual/joint ventures out of 156 currently covering the Italian offshore (*Noroil*, May 1982).[8] Exploratory efforts show every promise of being able to maintain the current domestic production rate of 13 billion m^3 per annum for some time to come.

Yet because the Italian market, much like the West German, has expanded far beyond the total annual production capacity of the Italian fields, ENI has also been a leader in procuring natural gas imports for its Italian customers, currently a little more than half of the 28–30 billion m^3 consumed annually. Here ENI has explored several options. First, it negotiated with the Soviet Union (in the 1960s) without, however, securing any immediate results. ENI also capitalized on its relationship with Esso and the Libyan government to establish an LNG trade, which at its peak amounted to imports of some 2.4 billion m^3 per annum. Through its affiliates SNAMPROGETTI and SAIPEM, ENI has additionally pioneered deepwater pipe-laying technology in building the first Trans-Mediterranean Pipeline. The company has not, however, been fully successful in these efforts owing to the general uncertainty prevailing in the international trading of natural gas (see Chapter 10). More fruitful have been imports of Dutch and Soviet natural gas – some 7.04 billion m^3 and 5 billion m^3, respectively, in 1980 (note these are affected by a swap arrangement with France).

As a national holding company, ENI has been plagued by political impediments. Despite effectively capturing all the rents from its Po fields,[9] it is currently running at a loss. In part this is due to price regulation in the Italian oil products markets (as in Germany and Holland, the price of natural gas is generally linked to the price of oil products, with perhaps a slight lag of a year or two); in part it is due to difficulties of management and political interference. The problems of ENI's chairman have been likened to 'a man who is trying to run the National Aeronautics and Space Administration at a profit with some responsibility for Indian Affairs and economic development and he is working with a regulatory structure as backward as those we find in Latin America' (*Business Week*, 3 March 1973, p. 68; quoted in Prodi, 1974, p. 280). Because of this, ENI and the Italian gas industry generally are more dependent than most national firms/markets on who is operating what company and what their political connections and interests are. Italy is fortunate, nevertheless, in that its exhausted fields have been increasingly used for storage, giving the gas industry there considerably more flexibility than is the case elsewhere in Europe.

National markets: France

Perhaps nowhere else in Europe has a working synthesis between state and industry been more successful than in France, a synthesis that also extends to the organization of the French gas industry. Terms such as '*étatism*', '*planification*' and the like have developed a symbolism of their own, reflecting a state-management style with healthy doses of Colbert and Descartes. It has been estimated that roughly one-half of the investments in the private sector in France were financed directly or indirectly by public funds. Such integration makes the separation of private and public sectors highly difficult, with constant movement out of one sector and into the other and vice versa.

The oil and gas industry provides an excellent illustration. The largest French oil firm, the Compagnie Française des Pétroles (CFP) was at first privately held, but after 1931 some 35 per cent of its shares and 40 per cent of its voting rights were acquired by the state. Another company, the Centre de Recherche de Pétrole du Midi, was a privately owned exploration company supported by government funds when it struck the first French gas find at Marcet in 1939; in order to exploit this find, it was replaced by the Régie Autonome des Pétroles (RAP), the first state-owned company to have a direct industrial and commercial role in the development and sale of hydrocarbons (Grayson, 1980, p. 76). The Vichy government gave large concessions to a competitor to RAP, the Société Nationale des Pétroles d'Aquitaine (SNPA). This firm, which was to draw on private capital in pursuit of a common national effort to find hydrocarbons (and was accordingly held 51 per cent by the state and 49 per cent by individual investors), was the one to find the giant gas field at Lacq in 1951.

In the post-war period, a new body, the Bureau des Recherches des Pétroles (BRP), 100 per cent state-owned, was founded, initially to finance further exploration in French territories, but later to function as a holding company. It took over responsibility for RAP and controlled the government's 51 per cent shareholding in SNPA. In

1975, the state companies, excluding SNPA but including the state-owned refining and marketing companies, were united to form the Entreprise Recherches des Pétroles (ERAP). The following year ERAP and the SNPA were merged into the Société Nationale Elf-Aquitaine (SNEA). The state froze some 18 per cent of its 70 per cent share in the new grouping, with the thought of eventually selling it to the public.

Table 7.8 gives the ownership of the major gas fields, and their division into public and private spheres. While not all fields have been listed in this table (such as the ones at Valenpoulieres, Proupiary, Auzas, Bazordan and Germ in the Southwest; Schaeffersheim in Alsace, and some very small finds in the Paris area), 60 per cent of all gas produced in France is from the Lacq structures, with most of the balance from the Meillon field (Société Nationale Elf-Aquitaine, *Annual Reports*, 1977–9; Tiratsoo, 1979, pp. 87–8).

The state was also active elsewhere on the French energy scene. The coal industry was nationalized in 1945. The following year the gas industry (largely based on town gas) was nationalized, and control was vested in the state firm, Gaz de France (GdF).

Table 7.8 *Natural gas production companies: Southwestern France*

Company	*Ownershp*	*Division of capital* *Public* %	*Private* %
The Lacq fields:			
Société Nationale Elf Aquitaine (Production)	SNEA(P)[b]	67	33
The Meillon fields:			
Société Nationale Elf Aquitaine (Production)	SNEA(P)	67	33
The Saint Marcet field:			
Société Nationale Elf Aquitaine (Production)	SNEA(P)	67	33
The Parentis oil field[a]*:*			
Esso REP[c]	Esso SAF[d] SNEA(P)	10	90

Source: Hervieu, 1969, p. 58, updated through private correspondence with SNEA.
Notes: a Associated gas.
b Société Nationale Elf-Aquitaine (Production).
c Esso Recherches et Exploitation Petroliers.
d Esso Société Anonyme Française.

The French gas industry presented two major problems. One was how to supply the markets of France with natural gas from the Southwest and town gas from the East. The other was reconciling the gas producers with the terms of sales to a national monopoly. As a further complicating factor, LNG imports from Algeria began in 1964 (although in a much limited form), Groningen gas imports commenced in 1966, and Soviet gas imports in 1976.

The solution to these problems lay in joint ownership of the major gas trunklines on the one hand, and a high degree of market-sharing among ERAP and SNPA (SNEA after 1976) and Gaz de France on

Table 7.9 *Ownership of French transmission lines*

Area	*Company*	*Ownership*	*Ownership* %	*Public* %	*Private* %
Southwest zone	Sociéte Nationale des Gaz due Sud-Ouest (SNGSO)	SNEA(P)	70	67	33
		GdF	30	100	
Centre of France	Compagnie Française du Méthane (CeFeM) – the network of CeFeM is rented by GdF to CeFeM	GdF	50	100	
		SNEA(P)	40	67	33
		CFP	10	35	65
Rest of France	GdF	GdF	100	100	

Source: As for Table 7.8.

the other. To begin with the first: ownership of the major transmission lines can be divided into three categories – the French Southwest, the Centre and the rest of France. The lines of compromise here were as follows: ERAP and SNPA were given majority ownership of the transmission lines in the region of their discoveries, and SNPA and Gaz de France shared ownership of lines in the rest of France. With the formation of SNEA, interests stand as given in Table 7.9.

On top of the regional differences characteristic of pipeline ownership and control, the marketing of natural gas was additionally divided according to the size of customers (Table 7.10). Transmission companies were generally allowed to keep larger industrial customers, while Gaz de France concentrated on smaller customers. Over time, thanks to its imports of natural gas, Gaz de France has become more and more important as the major distributor in France. The company currently serves some 8 million customers of various sizes.

Table 7.10 *Marketing natural gas*

Zone	*Client category*	*Company*	*Ownership* Public %	Private %
Southwest (Société Nationale des Gaz du Sud-Ouest – SNGSO)	• Public utilities • All industries outside towns • Industries of more than 15 miothermies[a] per year inside towns	SNGSO and CeFeM	77 81	23 19
Centre of France (Compagnie Française du Méthane – CeFeM)	• Direct clients of SNEA(P)	SNEA(P)	67	33
Rest of France	• Any use	Gaz de France	100	

Source: As for Table 7.8.
Note: a 8 miothermies = 898.3 MMBtu.

Gaz de France has also virtually completed a long conversion process from town gas to natural gas in all the areas it serves.

Gas production from the Lacq field is in decline. As with Germany and Italy, the French have begun to rely more and more on imported supplies from the Netherlands, Algeria and, more recently, the Soviet Union for future consumption. Table 7.11 shows this trend.

Table 7.11 *French production and imports, 1970–80 (billion m^3)*

	1970	*1977*	*1978*	*1979*	*1980*
Production	7.1	7.86	8.06	7.95	7.58
Imports	3.46	15.4	16.91	18.15	19.9
Imports as percentage of consumption[a]	34.5%	71.5%	72.1%	70.2%	76.0%

Source: Gas Committee, UN Economic Commission for Europe, 1981.
Note: a Includes balance after exports.

Conclusion

This brief highlighting of the basic features of the four most important Continental European markets leads to the question whether these markets have any political similarities. In fact they do – although the pattern is difficult to discern immediately and then only through looking at these markets in terms of the framework discussed in Chapter 3. The three elements to consider initially are limitation of actors, flexibility of terms of exchange, and time guarantees on the length of the business relationship.

State restriction of the number of actors is usually least where the exploration and production of natural gas is concerned, as outlined in Table 7.12. Given the risk involved in these activities, and the necessity for multiple geophysical interpretations for success in

Table 7.12 *Exploration/production regimes: the Netherlands, West Germany, France and Italy*

Country	*Description of actors and regime*	*Rent capture mechanisms*
Netherlands	Private companies encouraged to explore on own. At production stage, state participation necessary. (Informal NAM dominance?)	Flexible royalties. Differing tax/state participation regimes for onshore/offshore activities. Participation through Dutch State Mines.
West Germany	Private sector activity. With few exceptions, activities of oil majors predominate.	Royalties 17% onshore (offshore indeterminate). No oil-gas profits taxes.
France	Exploration open to both public and private firms but high degree of state intervention (government prohibits issuance of production permits to firms without substantial public holdings).	Production over 300 million m^3 subject to 5% royalty. (Local authorities may also levy royalties.) 23.5% depletion allowance before payment of corporate taxes.
Italy	Onshore – outside AGIP/ENI areas – access to Italian nationals or companies registered in Italy. Offshore, ENI allowed 25% of acreage on a preferred basis, balance available on application.	Royalty of 2.5–22% onshore, 5% offshore (excluding first 200 million m^3). Four echelon taxation scheme.

Source: United Nations Centre on Transnational Corporations, 1980.

exploration, one can argue that such a policy is rational. The character of national regimes at this stage is in notable contrast to that of the regimes for transmission and distribution, as can be seen in Table 7.13; restriction of numbers is most significant at the transmission stage. (It should be noted that the legal restrictions covered in these tables can be a poor reflection of political realities, which often favour a 'national firm' in ways both discrete and indiscrete.) With the notable exception of West Germany, transmission lines are closely controlled by government-owned companies. France shows the possible middle ground: what is not controlled by Gaz de France is controlled by SNEA. In Italy, the major competitor to SNAM, the

Table 7.13 *Concentration in transmission and distribution stages: the Netherlands, West Germany, France and Italy*

Country	*Transmission*	*Distribution*
Netherlands	Gasunie dominates	Municipalities
West Germany	Multiple transmission companies	Municipalities
France	Gaz de France dominates, though in conjunction with SNEA in pipeline companies SNGSO and CeFeM	Gaz de France monopoly in residential supplies. Shares in industrial sales with owners of transmission companies
Italy	SNAM predominates almost entirely	With the exception of some municipalities, SNAM predominates in all markets

firm ITALGAS, in fact is controlled about 34 per cent by SNAM shares.

One can reasonably conclude that it is through the control of these 'downstream markets' that the second condition of stability elaborated in Chapter 3 is achieved, that of flexible terms of exchange. On the European Continent this is interpreted as pricing natural gas at parity with other fuels – most notably oil products. That this can in fact be done is due in part to the tight state control over transmission lines. It is due perhaps even more to producer ownership (or representation) of the transmission lines. Thus, Holland's Gasunie contains NAM (the producer of Groningen gas), the German transmission companies encompass some of the major German natural gas producers, SNGSO and CeFeM in France contain SNEA as well as

Gaz de France, and Italy's ENI owns both AGIP, currently the major Italian natural gas producer, and SNAM, the national transmission and distribution monopoly.

Likewise through their respective laws of contract and their ownership/participation in the industry, European governments ensure stable business relationships over time. Here the pattern is less one of 'regulation' in the US sense and more one of direct intervention. (Germany and Belgium are exceptions.) This desire to ensure stable relationships and a continuing supply of natural gas has led the national governments to champion their leading natural gas corporations (public or private) in their search outside Europe for additional supplies of natural gas. Such international endeavours involve perils not encountered in national (or for that matter European) markets.

What none of these similarities reveals, however, is the wide variation of fundamental philosophies. This chapter has been sprinkled with terms such as 'Gemeinschaft society', '*geordnete Markt*', '*planification*', the need for a 'state-directed sector'. Ultimately it is these philosophies and the very real differences behind them that separate European markets so effectively and limit European natural gas cooperation to within national frontiers. The German distribution firm Gas-Union, for example, is a cooperation among five municipalities and Ruhrgas in the distribution and storage of natural gas among them. In Austria, the oldest gas market in Europe, the Austria Ferngas company was formed as recently as 1962 by three gas companies as a cooperative effort to supply natural gas to upper Austria. Since its inception, it has been joined by five distribution companies and the Republic of Austria; it now balances local production against imported supplies for all the parties involved. Both Gas-Union and Austria Ferngas are successful examples of cooperation. But both exemplify cooperation within frontiers – not without. Austria Ferngas does not coordinate deliveries for neighbouring German utilities. These must rely on German firms for their gas – irrespective of the merits or perhaps convenience of supplies close at hand across the frontier.

Notes

1 Town gas in domestic use amounts to 0.3 per cent of all gas use in the French domestic sector. The figure in Germany is 6 per cent.

2 Peebles' description of the Dutch natural gas industry is excellent and well worth reading.

3 According to Scholtern (1978), in 1975 there were some 402 advisory committees for these purposes: 90 per cent of these committees contained at least one departmental official in his formal role; 59 per cent contained over 50 per cent departmen-

tal officials. Some 58 per cent of all parliamentary candidates ascribed their election either to their expertise in advisory matters or to their interest organization affiliations.

4 The Dutch have solved the problem internally by distributing North Sea gas solely to industrial customers through a separate transmission/distribution system.

5 This encouragement included a government subsidy for overseas exploration of some DM 575 million for the period 1969–74, and a further DM 800 million for the period 1975–8, a subsidy that consisted of a loan to cover 75 per cent of exploration expenses (and which did not have to be repaid in case of lack of success) (Grayson, 1980, p. 150).

6 It is probable that the average load factor of imported natural gas has fallen somewhat and will continue to do so as a result of greater flexibility in the delivery terms of the now renegotiated Dutch contracts.

7 State holdings in ITALGAS are 34.38 per cent (SNAM), 7.11 per cent (IRI) and 7.18 per cent (Bank of Italy).

8 Note that AGIP has preference in new areas opened up for bidding and can acquire 25 per cent of the areas for its own use before the balance is offered up for bidding for foreign companies. Currently Total, Elf and Conoco are also exploring for oil. Gulf, Hudbay, Texaco and Burmah Oil are also in the rush for applications.

9 Grayson (1980, p. 118) quotes estimates, for example, of profits in natural gas of some $480 million in 1974, while there were overall group losses of some $89 million. This allows for considerable cross-subsidization.

8

Unity in Diversity: Inter-Nation European Gas Trade

The instruments of control

In 1969, Professor Peter Odell, in his inaugural lecture at the Department of Economic Geography at Erasmus University, pointed to the critical role that Royal Dutch/Shell and Esso had played, through their Dutch joint venture NAM, in the organization of the gas trade among the nations of Western Europe. They alone, he argued (1969, p. 11) possessed the requisite knowledge of how to dispose of Groningen gas: '[T]hese were business organizations immediately able, through their international structures, to collect and process the immense amount of data required for this exercise.' The provocative nature of this statement leads to an interesting question. In the previous chapter it was argued that national European markets are so different that their full integration is close to an impossibility. Here there is a contention that two multinational corporations from one country were responsible for structuring the European market. Can both assertions be right?

Seemingly contradictory arguments can often be resolved upon further examination. In Europe, as we shall show, the international supplies of natural gas are in fact superimposed on national marketing arrangements. This is not a new phenomenon. The import of natural gas by most of the nations of Europe in large quantities, first from Schloctern in Holland, then from Algeria, and finally from the North Sea and the Soviet Union, has presented existing Continental gas interests with some very complex problems. The first to confront them was Shell–Esso's NAM. NAM had to determine how the immense resources in Groningen could be sold throughout Europe on the most favourable terms and at the best possible price. How was this to be done? The solution was bounded as follows:

> Maximum profits after taking into account the following constraints: the need to avoid frontier price discrimination in export sales in light of the Rome Treaty and other international obligations of the Dutch Government; and the desirability of earning sufficient foreign exchange by sales abroad to satisfy the Dutch Government that the country's new resource was to play an effective role in strengthening the economy. (Odell, 1969, p. 11)

Intriguingly, NAM's own interests, as characterized by Odell, coincided in many important respects with the national interests of the individual European energy markets. The buyers of Groningen gas clearly had interests tied to their own domestic markets: West Germany was concerned in particular with the future of the Saar coal industry (and the not unimportant manufactured gas industry in the Ruhr); France was concerned with the viability of its own high-cost natural gas reservoirs at Lacq (and with the coal industry as well); Italy's ENI was concerned with the rent captured from its own natural gas fields in the Po region. All of these concerns were met by the NAM marketing policies (Odell, 1969, p. 15). Only Belgium would have been pleased with lower prices for Dutch natural gas.

The solution set adopted by all these parties, NAM included, was to price natural gas close to the prices of alternative fuels, primarily heating oil. This was easier said than done. Such a solution had to be stabilized in some manner; there had to be a means of enforcement against cheating. NAM found such a means through the control of the main transmission lines into Europe.

Later, as Groningen gas (and NAM control) declined in importance, this pattern of control changed slightly. Other European transmission companies united through ownership of these 'common carrier' transmission lines to manage the distribution of new natural gas in their respective national markets. Here too, however, the control of pipelines remained the 'key' to control of intra-European trade in natural gas, although the single actor of NAM had been supplanted by a group of firms. In both cases, pipeline contracts provided the legal instruments for the exercise of control.

European transmission lines are operated under two forms of contract. First, there are contracts of supply, in which the owner of the pipeline is also the owner of the natural gas transmitted through the line (such is the case in the United States and for landed gas in Great Britain). Second, there are contracts of transmission, in which the pipeline carries gas owned by others as well as itself (at times): 'shareholders of the pipeline company as well as third parties' (Organization for Economic Cooperation and Development, 1969). Both forms of contract are common in intra-European natural gas commerce and were critical to both NAM's strategy and that of its successors.

For the purposes of this chapter, the common carrier transmission line is much the more interesting of the two types. While the details of the contractual arrangements for such lines are highly secret, the general organization of the pipeline consortia can be deduced from publicly available information. Table 8.1 illustrates the organizational set-up of a typical common carrier arrangement involving shipments from exporting country X through its transmission company X_a to and through country Y, and its transmission companies Y_b and Y_c, to country Z, with its transmission company Z_d. Trade therefore runs from country X to country Z through country Y. These national transmission companies are likely to have national subsidiaries in the

Table 8.1 *National transmission companies and international pipeline consortium: countries X, Y and Z*

Transmission companies	*Pipeline consortium composed of national pipeline companies*				
	X_1	Y_1	Y_2	Z_1	Z_2
X_a	+	–	–	–	–
Y_b	–	+	–	–	–
Y_c	–	–	+	–	–
Z_d	–	–	–	+	+

Notes: + indicates active ownership/participation/operator status (responsibility for managing line).
– indicates passive interest/passive shareholding–voting rights/contract of access to pipeline transportation afforded by company concerned.

other countries to share in the ownership and control of the national segments of the international line running through this country. Thus there will be national pipeline companies X_1, Y_1, Y_2, Z_1, Z_2, where X_1 delivers to Y_1 and Y_2, and Y_1 and Y_2 deliver to Z_1 and Z_2 respectively, there being two transmission lines in countries Y and Z. The plus signs in the national pipeline company segments of the international consortium indicate that the transmission firm has an active interest in the pipeline company concerned; perhaps it even runs the pipeline on behalf of the other shareholder. The minus signs indicate non-active interests, ranging from a completely passive financial interest (and voting share exercised through the national subsidiary involved) to a mere contract guaranteeing access to the common carrier. To sum up, the national segments are run by interlocked national companies with the transmission companies of the country of passage normally operating the line on behalf of their partners, who keep an eye on them and provide similar services when the line passes through their own countries.

There are variations to the common carrier arrangement illustrated in Table 8.1. Most common are instances in which trunklines are operated by the natural gas buyer rather than a consortium. (There are probably aspects of this in the Gasunie–Distrigaz–Gaz de France arrangement for Groningen gas; here apparently Shell–Esso owned significant shares in both the Dutch and Belgian lines and Gaz de France purchased the gas at the frontier, although prices were based on delivered prices in France.) Irrespective of the particular arrangement, the resulting consortia are in a position to exchange information and coordinate policies, both directly for the active members of the specific consortium involved, and indirectly for those members who might be part of one consortium but not others.

Control of natural gas supplies in Europe through international transmission lines has historically involved both conflict and cooperation. This can be seen, for example, in the early pipeline system dominated by NAM. While other national transmission companies at first cooperated with NAM in selling natural gas in their domestic markets, this cooperation faded with the entry of newer, less expensive sources of supply. Thus, in much the same manner as in the United States, where transmission line monopsony power over field gas prices diminished upon the entry of new pipelines, so did NAM's attempt at exercising a monopoly over European natural gas pricing diminish upon the entry of cheap Soviet natural gas in the period 1968–73. The later 'H-gas' pipeline system, which supplemented the NAM Groningen system, was less an individual effort at enforcing a single international pricing policy on Europe than it was a cooperative effort by the national transmission companies to wield bargaining power *en bloc vis-à-vis* both Soviet and Norwegian newcomers to the Continental market, and to maintain respective national patterns of selling the natural gas involved.

In both systems, pipeline cooperation was a superimposed order. NAM's early pipeline policy was not resisted by its customers because the high prices charged suited their existing national natural gas interests. As these interests changed, NAM's policies had to change as well. This is also true today. With the import on a large scale of Soviet and Norwegian natural gas, international cooperation has tended to stop with the border delivery to the national transmission company involved.

During recent years, a new element has appeared that threatens to challenge this carefully balanced system of control: the signing of big new Soviet contracts for significantly increased supplies of natural gas. These contracts have a new political dimension as well, and have caused awkward questions to be raised about European cooperation in the event of a Soviet termination of supplies. Here the question of pipeline control has confronted yet a third problem – that of security of supply.

Control of Groningen natural gas

The export of Dutch gas to the various markets in Europe took differing organizational forms, forms that corresponded to differing exporter/importer preferences. NAM exercised perhaps the most control over gas exports to Germany, the largest export market. This was done through multiple ownerships not unlike the pattern in Table 8.1. The NAM companies held major shares in the German transmission companies: Ruhrgas, Thyssengas and the Deutsche Erdgastransport GmbH (DETG). This is illustrated in Table 8.2. These German

Table 8.2 *Shell–Esso interests in major West German transmission companies (percentage share)*

	German transmission companies		
NAM partners	*Ruhrgas*	*Thyssengas*	*DETG*
Shell	27.5	25	25
Esso	12.5	25	25

transmission companies owned and operated the major common carrier trade lines into Germany: the Nordrheinische Erdgastransport GmbH, the Mittelrheinische Erdgastransport GmbH and the Süddeutsche Erdgastransport GmbH. As can be seen in Table 8.3, the German affiliates of the oil companies also owned direct shares in these pipeline companies. (This table is perhaps more notable for what cannot be ascertained than otherwise.) While it would be instructive to know more about the power these arrangements conferred upon Esso and Shell (and NAM), it is probably safe to assume that at best they gave the two firms a controlling interest in the pipelines, at worst they gave them veto powers. Thus, there has been a high degree of vertical integration with regard to Dutch exports to West Germany. Shell and Esso have been involved in all stages of the trade, reaching even in some instances to ownership in distribution companies.

The German firms purchasing the gas have been organized into consortia (Gas-Union) that have had a major role in negotiating the contracts concerned. Table 8.4 lists the major trades in NAM gas

Table 8.3 *Ownership of pipeline companies importing Dutch gas to West Germany*

Pipeline company	*Ownership*	*Percentage share*
Nordrheinische Erdgastransport GmbH	Ruhrgas	n.a.
	Thyssengas	n.a.
	DETG	n.a.
Mittelrheinische Erdgastransport GmbH	Ruhrgas	n.a.
	DETG	n.a.
	Exxon[a]	16.7
	Shell(?)	n.a.[b]
Süddeutsche Erdgastransport GmbH	Ruhrgas	n.a.
	DETG	n.a.
	Exxon[a]	25.0
	Shell(?)	n.a.[b]

Notes: a Exxon holds this share in its role as major shareholder in Esso AG.
b It is not certain that Shell owns a share; it is included here on the assumption that the Shell–Esso partnership elsewhere in natural gas production and transmission in Germany also obtains for the transmission lines concerned.

with Germany, together with the average annual amounts specified in the contracts and the dates of initial signing and revisions of contractual terms. This degree of vertical integration has no doubt aided the orderly marketing of natural gas within Germany. In practice, it has meant that NAM has been able to adjust its prices, which initially were too high, to allow long-range market penetration elsewhere in Germany.

Schloctern gas has also been sold to other countries, primarily Belgium, Italy, France and Switzerland (see Table 8.5). Here, with the notable exception of Belgium, the problems have been somewhat different. NAM had to sell to a monopsony transmission company (Gaz de France, SNAM, Swissgas) in which there was little if any Esso–Shell holding. This led to considerable contractual negotiating difficulties. At one point, for example, the French claimed that NAM demanded control over the prices at which Dutch gas would be transferred to French transmission companies or consumers in the cases where Gaz de France had direct responsibility. The details of the contracts are of course a secret, but adjustments in price did occur at least once upon demand of the purchasing transmission companies.

For both the German and French–Italian trades, however, NAM experienced the common difficulty of maintaining its control over

Table 8.4 *NAM gas exports to West Germany*

Purchasers	*Average quantities p.a. (billion m³)*	*Exports beginning*	*Length of contract (years)*	*Contract signatures/ changes/ amendments*
Ruhrgas	6.68	1968	25	1966/1970/1980
Thyssengas	4.12	1967	25	1966/1980
Brigitta[a]	2.50	1972	20	——/1980
DETG[a]	1.55	1969	20	——/1980
Energieversorgung Weser Ems (EWE)	0.90	1966	20	——/1980?[b]
Rheinisch-Westfälisches Elektrizitätswerk AG (RWE)	0.75	1974	20	
Vereinigtes Elektrizitätswerk Westfalen (VEW)	1.30	1971	20	1971/1980?[b]

Source: 'Etat des contrats internationaux de gaz naturel à long terme au ler Mars 1977', *Le Pétrole et le Gaz Arabes*, vol. 9,199, pp. 21–3. Updated by the author. Hereafter this source is referred to as *Le Pétrole et le gaz*.

Notes: a Majority ownership Shell–Esso firms.
b Not included in compromise of 1980.

prices. Much in the same manner as the pipeline monopsony positions *vis-à-vis* Permian Basin producers in the USA crumbled upon entry of other competing pipelines, so too did NAM's pricing position begin to develop holes *vis-à-vis* its foreign customers just two to three

Table 8.5 *NAM gas exports to non-German markets*

Destination		*Average quantities p.a. (billion m³)*	*Exports beginning*	*Contract length (years)*	*Contract signatures/ changes/ amendments*
Company	*Country*				
Distrigaz[a]	Belgium	7.0	1966	25	1967/1968/1980
Gaz de France	France	9.25	1967	20	1966/1968/ 1972/1980
SNAM	Italy	6.25 (peak)	1974	20	1969/1980
Swissgas	Switzerland	0.50 (peak)	1974	20	1969/1980

Source. Le Pétrole et le gaz Arabes, pp. 21–2.

Note: a At time of initial contract signature, about 33% of Distrigaz was held by Shell–Esso.

years after the major commencement of its exports in 1966. In April 1969, Soviet Foreign Trade Minister, N. S. Patolichev, in discussions with Karl Schiller, Economics Minister for Germany, advanced the proposal that Soviet gas be exported to West Germany at a considerably lower price than competing Groningen gas. This discussion led to a great deal of debate in German gas circles. The domestic producers feared the impact of any future deal on their status. But this was nothing beside the fear felt by the NAM partners. These partners, wrote an expert on Soviet gas trade,

> judge that any reduction in the sales price of Groningen gas which the prospects of Soviet gas sales might entail, would have to be matched by a comparable reduction in sales price quoted France and Belgium. Moreover, sharp increases in natural gas supplies are likely to bring about reductions in the demand for fuel oil. (Ebel, 1970, p. 154)

Although the proposed German deal did not immediately go through, according to at least one authority (Odell, 1979, p. 50) Dutch gas pricing policies were changed in 1971, partially in response to impending Soviet imports.

Nor was the Soviet threat to NAM marketing policies unforeseen. In the same year that the meeting between Patolichev and Schiller took place, NAM had already consented to reduce the base price of its gas exports to France, Belgium and Germany by about 10 per cent to enable the customers concerned to market their natural gas in less accessible areas (Hervieu, 1969, p. 86). The main occasion for this price reduction was an Italian flirtation with the Soviets, and the signing of a contract between Italy and Libya for the import of LNG to La Spezia. There can be little doubt that these alternative sources of supply were desirable to Italy. NAM, confronted with the choice of either marginally dropping contract prices or losing future markets, opted for the first alternative. In 1969, the Italians signed two contracts, one for Dutch gas and one for Soviet gas. Some six years later the price of Soviet natural gas imported into Italy was reckoned to be of the order of \$0.64/thousand ft^3 (1975 prices). Dutch gas, on the other hand, which was sold to the French under a complicated 'swap agreement', was priced at about \$1.13/thousand ft^3. Even given the difference between other possible terms of sale that might be involved, this seems a significant variation in price.

Clearly, to retain continued stability of international gas markets in Europe, pipeline control would have to be extended to other sources of supply. It is to this further extension of control that we shall now turn.

H-gas: the new pattern of trade

'Ruhrgas thinks it dominates the European market', a Dutch authority stated recently to me; It doesn't. Gasunie does. Ruhrgas only thinks that it dominates the market.' Perhaps the most revealing thing about this remarkable statement is Dutch insecurity about their pre-eminence on the European market, despite the fact that they still supply the vast bulk of natural gas traded. None the less, with the development of the Norwegian North Sea fields and the commencement of Soviet imports, the 'centre of gravity' in European gas markets has moved eastwards to West Germany and towards Ruhrgas control/operation of the immense North–South and East–West transmission systems.

As indicated in the introduction to this chapter, these new transmission lines were based on a cooperative effort not only of Shell and Esso but of the combined national transmission companies of the major gas markets in Europe. In part, this effort was based on a perceived need to maintain a common front *vis-à-vis* the new entrants to the European natural gas market: the Russians and Norwegians. In part, the effort was aimed at re-establishing the orderly marketing achieved by NAM. Finally, the cooperative effort was prompted by the very real technical problems that use of these various new gases involved for the national markets concerned.

This last point requires a brief explanation. Groningen gas (L-gas) is not really interchangeable with other European natural gas (H-gas). Its calorific value is of the order of 8,400 kcal/m^3, as opposed to the some 10,000 kcal/m^3 values of H-gas. Soviet interest in natural gas exports to Europe and the discovery of not inconsiderable amounts of high-calorific gas in the Dutch and the Norwegian North Sea fields confronted European experts with the problems of conversion between gas qualities. It was decided to transport H-gas separately and mix it with L-gas and other lower calorific gases for local distribution. The North Sea gas promised to come on stream in the mid-1970s. Combined with initial Soviet imports in the period 1973–5, this furnished an incentive to build gas transmission networks in which Gasunie had only marginal interests. Even before Groningen gas exports peaked, plans were underway to transmit supplies of H-gas to European markets.

The first significant discoveries of H-gas were from the Dutch North Sea. Although traded domestically, this gas is also the basis of a not inconsiderable inter-nation trade, as illustrated in Table 8.6. Given NAM's predominance in the Dutch market, it is tempting to ask to what degree the contractual terms of the natural gas in Table 8.4 differ from those for NAM. The following conclusions seem fairly certain: NAM is not exporting any natural gas from the North Sea, preferring to dispose of it through Gasunie to the Dutch electrical generation and industrial markets. Natural gas exported to Europe is likely to be used for the same purposes (particularly given the number of electrical utilities purchasing the gas). The price of the natural gas is on a Btu basis, probably at the same level as that of NAM gas exports, although it is quite likely that other terms of sale (load factor, for instance) may differ significantly. Finally, it should be noted that the gas to be exported is transported through offshore lines at least partially owned and controlled by Dutch State Mines/Gasunie. This could well hypothetically give Gasunie (and, through Gasunie, NAM) some indirect voice in marketing terms both in Holland and elsewhere on the Continent.[1]

Table 8.6 *Dutch North Sea gas: sales to Continental markets*

Producing company	*Destination*		*Average quantities p.a. (billion m^3)*	*Length of contract (years)*	*Contract signatures/ changes/ amendments*
	Country	*Company*			
Petroland (Total)	France	Gaz de France	0.60	20	1970/1980
Amoco	Germany	VEW	0.60	20	1970/1980
Placid International	Germany	Ruhrgas VEW EWE	3.25 (peak)	30	1973/1980

Source: Le Pétriole et le gaz Arabes, p. 22.

As indicated above, pipeline construction had to proceed quickly if H-gas was to be marketed in an orderly manner. Gasunie completed an H-gas transmission line to the Dutch city of Maastricht on the southern frontier in 1974–5. A German–Italian consortium (Ruhrgas–ENI) built a continuation of this line across Germany and through Switzerland (where the Swiss have a 51 per cent share in Transitgaz SA, the pipeline company in Switzerland) to Italy. This line, dubbed the Trans-European Natural Gas Pipeline (TENP), started operations in 1977. The TENP is additionally connected to Emden through a Northern German spur and through Emden to the Norwegian natural gas from Ekofisk, Albuskjell, Cod and (in the near future) Statfjord and Heimdal. It therefore carries Dutch gas to Switzerland and Italy, carries North Sea gas from Germany to Switzerland and Italy, and distributes H-gas throughout the length of Germany (Tiratsoo, 1979). The other leg of the North–South system is constituted by the Société Européenne de Gazoduc Est–Ouest

(SEGEO), a joint Gaz de France–Distrigaz venture that transmits H-gas (including Norwegian gas) from the Maastricht area through Belgium to France (see Figure 8.1).

The East–West system is similarly designed to handle imports of Soviet gas. In this system, too, Germany and Ruhrgas play a critical role. Originally it consisted of two spurs, one 36 in. line running through Germany from the Czech frontier, the other crossing Austria (the Trans-Austrian Gasline (TAG) – an ENI, Gaz de France and ÖMV joint venture) into Italy. The German spur is built and run by

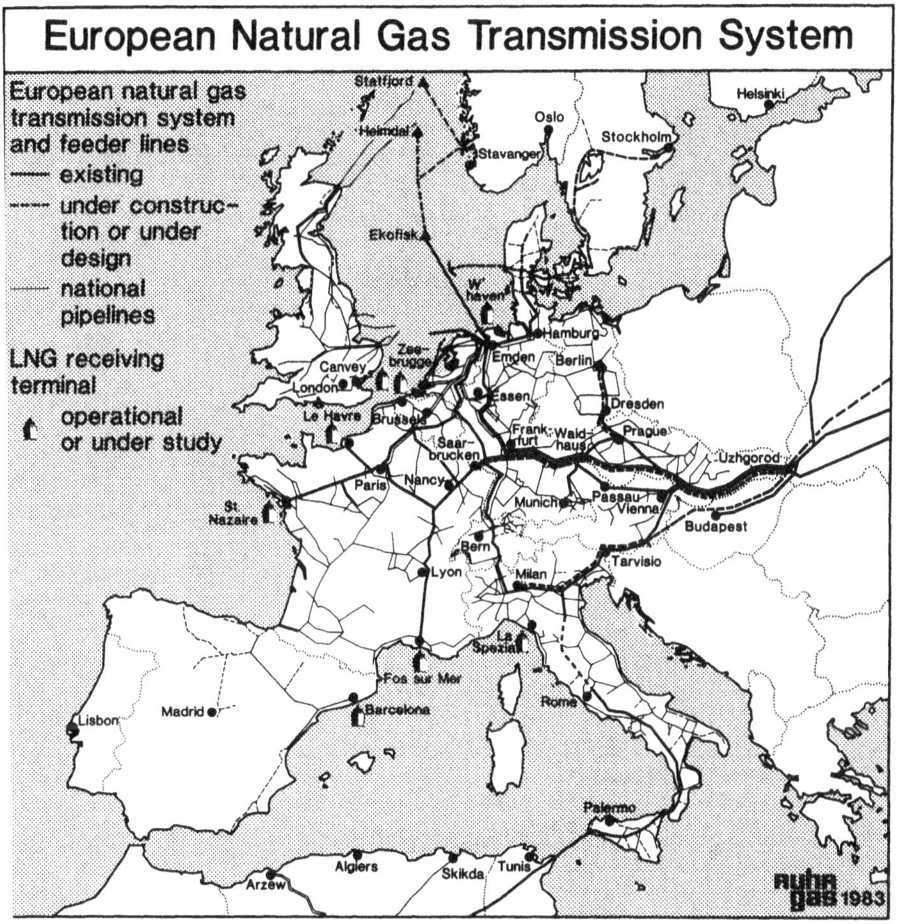

Figure 8.1 *West European H-gas grid*
Source; Ruhrgas 1983.

Table 3.7 *Purchasers of North Sea/Soviet gas correlated with transmission system*

Purchaser	*Amount (billion m^3)*	*Source*	*Transmission system utilized*	*Deliveries commenced*
Gaz de France	0.6	Petroland (Dutch)	North–South/SEGEO (after 1977)	1976
VEW	0.6	Amoco (Dutch)	North–South/TENP	1976
VEW	3.25	Placid (Dutch)	North–South/TENP	1976
Ruhrgas				
EWE				
Ruhrgas	6.8[a]	Ekofisk (Norway)	North–South/TENP	1977
Gaz de France	1.26[b]		North–South/SEGEO	
Distrigaz	1.26[b]			
Gasunie	1.26[b]			
Brigitta	0.72	Albuskjell (Norway)	North–South/TENP	1977
Ruhrgas–Thyssengas	0.63			
Distrigaz	0.45	Albuskjell (Norway)	North–South/SEGEO	1978
Gaz de France				
Gasunie				

Ruhrgas Gaz de France Distrigaz Gasunie	0.30	Tor (Norway)	North–South/TENP North–South/SEGEO	1977
SNAM/ÖMV	8.9	USSR	East–West/Trans-Austrian Line	ÖMV (1968/1975) SNAM (1973/1979)
Ruhrgas	6.0 (peak)	USSR	East–West	1973
Gaz de France[c]	4.0 (peak)	USSR	East–West/substitute for Dutch gas with Italy/TENP	1976
Gaz de France Ruhrgas SNAM ÖMV	11.5 21.1 6.7 ?	USSR[d]	MEGAL/TENP/North–South – (Trans-Austrian) – (West Austrian)	— —

Notes: a Of which some 4.1 billion m^3/year supplied under contract with base price of $0.52/thousand ft^3, the balance under contract with base price of $1.84/thousand ft^3.

b Base prices as under (a) although divided 1.36 billion/yr/lower base 0.90 billion m^3/yr/higher base price.

c Supplied under swap arrangement with Italy (Italy receives Soviet gas, GdF receives Dutch gas in equivalent portions).

d Recently contracted.

Ruhrgas, with the possible participation of other European transmission companies. Paralleling these lines are the proposed future lines of the Mittel-Europäische Gasleitungsgesellschaft (MEGAL), the dashed lines on the map in Figure 8.1. When completed, this transmission system will be the largest ever constructed in Europe, with two lines of 48 and 36 in. in diameter, respectively, the first running through Germany, the second dipping through Austria before re-entering Germany. Both will be connected in mid-Germany and run to join the TENP and the French system to the west (Tiratsoo, 1979).

Exact answers about system usage and interested corporate parties are unattainable. Table 8.7, however, which correlates the production and sales contracts from the Netherlands, Norway and the Soviet Union elsewhere on the Continent, probably gives a good approximation. Around the fringes of these pipeline systems are the various international LNG trades: Libya/Algeria–La Spezia (Italy), Algeria–Montoir, Alegeria–Fos-sur-Mer, Algeria–Le Havre (France) and Algeria–Zeebrugge (Belgium). These LNG trades can probably also be relied upon to contribute to the European network. Further, within a year or two the Algeria–Italy Trans-Mediterranean Pipeline will be completed to its northern terminus Minerbo, making this gas, too, available for the net. This deal alone can be expected to contribute up to 20 billion m^3 of Algerian gas per annum.

While corporate overlapping in the ownership, participating and operating interests in the H-gas transmission networks may not be as striking as that for NAM and L-gas, it is safe to assume that it is still significant. Table 8.8. gives a tentative impression of how national companies operating interlocks on common carrier lines can give rise to allegations that the companies concerned form a 'cartel'. Some of the probable participating interests are more likely than others. Thus Gaz de France and Distrigaz almost certainly participate in the TENP lines through Germany, as both companies receive Norwegian gas, which is carried part of the way by the German H-gas lines. Their participation in MEGAL is more problematic, in that it is assumed that such participation is necessary for the conveyance of the Soviet gas that both companies have expressed an interest in purchasing.

If we restrict our definition to the known national participations in lines and to the known operators of the national segments of these lines, what form of corporate interlocks occur and how frequently? Table 8.9 gives the answer based on the data in Table 8.8. Even excluding all probable participations, the results are impressive: Gaz de France interlocks no fewer than seven times with the other five major transmission companies, Ruhrgas has six interlocks (probably more), and so forth. It would seem that these interlocks necessarily aid in the coordination of European natural gas planning regarding

Table 8.8 *National transmission company operating interlocks and possible participatory interests*

National segment	*National transmission companies in each system*				
	TENP	*SEGEO*	*MEGAL*	*TAG*	*WAG*[a]
Netherlands	Gasunie	Gasunie	pp	pp	pp
West Germany	Ruhrgas	pp	Ruhrgas	—	Ruhrgas
Switzerland	Swissgas	—	pp	—	pp
Italy	ENI	—	pp	ENI	pp
Austria	—	—	pp	ÖMV	ÖMV
France	pp	GdF	GdF(P)	GdF(P)	pp
Belgium	pp	Distrigaz	Distrigaz(P)	pp	pp

Key:
Where operator status is definite – name of the transmission company.
Where participating interest is definite – name of transmission company followed by (P).
Where participating interest is probable but uncertain, no company given. Indicated by pp.
Note: a Western Austrian Gasline (portion of MEGAL system).

both present and future supplies. This close cooperation is in contrast to the differing national marketing arrangements noted in the previous chapter.

Despite this integration, the tightness of control exercised by international transmission lines for H-gas is no more than that exercised by NAM over Dutch gas in the late 1960s and early 1970s. In part this is due to the location of the significant Norwegian natural gas fields, which are equidistant from the Continent on the one hand and the United Kingdom on the other. This has deprived the Continental transmission consortium of any monopsony advantage in negotiating

Table 8.9 *Joint operating/known participation interlocks: H-gas network*

	ENI	*GdF*	*Ruhrgas*	*Distrigaz*	*Gasunie*	*ÖMV*	*Total*
	nos. of interlocks						
ENI	—	1	1	—	1	1	4
Gaz de France (GdF)		—	1	2	1	2	7
Ruhrgas			—	1	1	2	6
Distrigaz				—	1	1	5
Gasunie					—	—	4
ÖMV						—	6
Total							32

for gas supplies from this source. In part it is due to other factors. Here, some examples could be instructive.

In 1979–80, BP Norway signed a natural gas contract with BP Germany covering sales of natural gas from the Norwegian Ula field on which BP was operator. The terms of this contract were reputedly very favourable to Norway and created a minor fracas among the better-established transmission companies. Apparently there was nothing the consortium could do to block the sale of BP gas, as BP Germany was a considerable shareholder in Ruhrgas. That the contract came to naught was not due to a refusal of access of BP gas to the Continental transmission network; rather it was sabotaged by the German government's antitrust division, which declared the contract against the public interest. As a pacifier, BP was offered a share in a (now-dead) proposed LNG import project. This case is of considerable interest. Not only does it constitute proof that access to transmission lines in Europe is relatively free, but it also demonstrates that the political power of the transmission companies can be as big a hindrance to 'competition' as access to transmission lines. At any rate, the BP officials involved in this case are not in doubt about what went wrong: blame, in their eyes, 'does not lie in Bonn but in Essen, the headquarters of Ruhrgas the leader of the European "cartel"' (J. D. Davis, 1982).

Pipeline conflict with the other major non-communist producer of H-gas, the Netherlands, has to date resulted in one victory for free access and one defeat. The victory concerned the sale of Dutch North Sea H-gas from the L-10/L-11 Placid fields. Here Placid (and the Continental transmission companies) opposed a Dutch law that attempted to confine the sale of North Sea gas entirely to Holland. Placid, as the operator of the fields concerned, refused to sell the gas to Gasunie; instead it signed a contract with several West German utilities (see Table 8.6). The Dutch contested this contract. The European Commission, acting on Placid's behalf, threatened to take the case to the European Court, charging that the Dutch law was discriminatory as it interfered with the free movement of goods and services. The case was never brought to trial. Instead, all three parties agreed that one-half of the amounts in question should go to Gasunie and the other half to the transmission lines of the utilities concerned (Woodcliffe, 1975).

This victory (or half-victory) was soon counterbalanced by protracted new negotiations over the price of Dutch natural gas on Continental markets (J. D. Davis, 1983).[2] These negotiations (or, more precisely, renegotiations) commenced in 1976 and lasted until 1980. In 1976, Dutch gas accounted for roughly 74 per cent of all internationally traded gas on the Continent. It was felt that prices for Dutch natural gas were far lower than the oil product prices to which they were indexed. Renegotiations occurred in two phases. First, there was a preliminary round in which agreement was reached on the following bases:

- the price of Dutch natural gas fob should be 1 Dutch cent below the price of low-sulphur fuel oil (unspecified type and quantity);
- prices of the natural gas should escalate at a rate 80 per cent of that of the fuel oil concerned;
- the prices of natural gas would lag behind the prices of the relevant fuel oil by ten months.[3]

The Dutch were not wholly content with this progress, a discontent that strengthened with the significant price increases of the period 1978–9. A second attempt to renegotiate the contracts further with the ten major international purchasers failed when the Dutch negotiators found that they had no legal means to bring the companies to the negotiating table. A move to gain sympathy for Dutch objectives from the governments of Germany, Belgium, Italy and France also failed. These governments had no interest in revising domestic gas prices upward, quite the contrary. The Dutch government thereupon introduced legislation that in lieu of other progress, threatened to increase gas export prices unilaterally (Davis, 1983).

It is uncertain to what degree the Dutch actually meant business. The purchasers of Dutch gas finally entered into negotiations with the Dutch. After prolonged discussions, it was announced that the negotiations were successfully concluded on 28 August 1980. The Dutch procured terms such as the following:

- the price of Groningen quality gas was made equivalent to the then existing price for low-sulphur fuel oil on an energy content basis;
- indexation to the low-sulphur fuel oil concerned would be on a 95 per cent basis as regards future price increases of the fuel oil;
- the time lag between price increases on the fuel oil market and the consequent price increases for natural gas was reduced to five months.

It is typical of such negotiations that not all the gains were in favour of the Dutch. First, not all the purchasers of Dutch gas agreed to the new terms. (Two major German utilities refused their consent.) Second, in return for the higher prices, the purchasers were to get natural gas on better terms. The load factors concerned dropped significantly. Furthermore, the conditions of future deliveries were made much more flexible in terms of time. (Contracts due to expire in the 1990s were therefore extended, although the quantities delivered were to remain the same.) The upshot of these new terms was that the Groningen field became a 'swing field' – a field that provided increased flexibility of supply to purchasers. Given this new role, it is clear that the new prices are still low. With Groningen gas, the utilities and transmission companies could aim at more premium markets, supplement existing high load factor gas contracts and in the last instance utilize Groningen gas as a strategic reserve should future supplies (for example Soviet supplies) suddenly be terminated. (One can even advance the claim that Dutch gas, given these factors and the prices charged for natural gas elsewhere, is far underpriced.)

Other examples of pipeline control 'failure' could no doubt be adduced – the recent Distrigaz and Gaz de France LNG deals with the Algerians are especially relevant in this context[4] – but there seems little point to any further enumeration. What is significant in terms of control is that the transmission nets have been able to achieve the sort of stability described in Chapter 3. On the one hand, they have been able to maintain a rough parity with oil products on the selling end of the natural gas business; on the other, they have maintained average field/landed prices that allowed for continued market expansion throughout the 1960s and 1970s (an expansion that would have continued but for the price explosion of 1979–80). This has been no mean achievement. Tables 8.10 and 8.11 illustrate how it has been accomplished. Table 8.10 gives the producer price for all non-Soviet supplies for 1978 and estimations of prices for 1982 and 1990. For 1982 these amount to an average of $4.102/MMBtu. On the sales side, Table 8.11 gives the estimated costs of disposing of the natural gas on various Continental markets. Working backward from these, there appears to be about a $0.104/MMBtu difference between the $4.102/MMBtu actually paid to producers and the $4.206/MMBtu that the transmission companies could have afforded. This probably accrues to the transmission companies as a form of 'rent' and of course is the target of much producer–transmission company bargaining as regards future marginal supplies to the markets concerned.

The third phase of control – the Russians fill the 'gap'

Implicit in our discussion of pipeline control has been an important difference between the early NAM transmission system and the more recent H-gas system. The old L-gas NAM system was unique in that it involved both vertical integration and a significant control over

Table 8.10 *Price evolution. Groningen vs. other non-Soviet traded natural gases: Continental Europe*

	1978		*1982 (est.)*		*1990 (est.)*	
Exporters	*price ($/MMBtu cif)*	*% non-Soviet import*	*price ($/MMBtu cif)*	*% non-Soviet import*	*price ($/MMBtu cif)*	*% non-Soviet import*
Algeria:						
LNG (France & Spain)	1.76	5.7	5.50	4.5[a]	5.50	6.2[f]
LNG (Belgium)	—	—	5.50	—	5.50	9.4[f]
Trans-Med. Pipeline	—	—	5.22[b]	—	5.22	21.5
Norway:						
Ekofisk	1.91	20.7	3.87[c]	23.7	3.87	16.2[g]
Other	—	—	5.50[b]	—	5.50	9.4[g]
Netherlands:						
Pipeline	1.65	64.8	4.06[d]	67.6[e]	4.06	33.6
Libya:						
LNG	1.06	8.8	4.62	4.2	4.62	3.7[f]
Average price per MMBtu	1.66	100.0	4.102	100.0	4.671	100.0

Notes: a Estimated 50% contract deliveries to Spain and France combined.

b Estimated price from industry reports on current negotiations.

c $3.37/MMBtu for Ekofisk gas based on 2:1 high-sulphur/low-sulphur fuel oil ratio and 75% of Rotterdam spot market prices; added to this are $0.50/MMBtu handling and transport to Emden.

d Based on 95% Rotterdam spot market price of fuel oils averaging 2% sulphur content. Price will vary with transport differential.

e Percentage increase due to make up reported Italian and French shortfalls.

f Deliveries estimated at 75% of those specified for Algeria, 80% of those specified for Libya (this is to allow for the production factor vs. contracted-for volumes).

g Based on 7 billion m^3 from Heimdal, Statfjord, etc.; 12 billion m^3 from Ekofisk.

Table 3.11 *Continental natural gas markets (excluding the Netherlands), October 1981 (US$/MMBtu)*

	Housing premium market	*Industrial markets premium*	*Industrial markets Low premium*	*Interruptible supplies*	*Average price*
Percentage share of market	22.8%	22.1%	30.8%	24.3%	
Oil product substituted	Gas oil	Fuel oil 1% S_2	Fuel oil 3.5% S_2	Fuel oil 3.5% S_2	
Prices at which oil products are substitutable for natural gas (maximum utility price for natural gas)	7.95	4.95	4.65	4.15	5.347
Storage/transportation	(1.20)	(0.50)	(0.50)	(0.05)	
Distribution	(2.15)	—	—	—	
Transmission company:					
Price to utility (costs)	4.60	4.45	4.15	4.10	4.306
Transport/storage	(0.10)	(0.10)	(0.10)	(0.10)	
Maximum price to producer	4.50	4.35	4.05	4.00	4.206

Source: Bonfiglioli, 1980, pp. 55–8.

Note: Assumptions have been changed: first, the prices for oil products (3rd line) are those of Rotterdam, September–October 1981; second, the housing, low-premium fuel and interruptible shares are changed to allow for the exclusion of the Netherlands. Housing market enjoys a $0.50 premium above gas oil; other markets are Btu values with plusses to industrial markets.

transmission lines. Shell and Esso were involved both in Groningen gas production and in the major German and Belgian transmission companies; the sale of natural gas was almost an intra-firm sale. Control here was much sounder in its fundamentals than the present form of consortium control, in which the parties have little or no ownership over the sources of supply. The bilateral consortium purchaser–seller relationship has been effective to date in negotiating for, purchasing and disposing of significant new supplies of natural gas, but it does not involve actual management of the gas production in either Norway or the Soviet Union. Of necessity this 'bilateral monopoly' relationship is less steady than one in which vertical integration is present. (This is underlined in our discussion of instability in Chapter 3.) While not particularly worrying in the case of Norway, it can be argued that such a relationship with the Soviet Union is more questionable. This view brings into discussion yet a third aspect of pipeline control, that of possible consequences of dependency on one's supplier.

As with H-gas, the issues behind Soviet gas purchases did not originate in the Soviet Union. Rather they are rooted in political events in the Netherlands. The Dutch announcement that the contracts signed with Continental buyers in the 1960s and 1970s would not be renewed on expiration was a shock to Continental purchasers. Based on the contract information in Tables 8.4 and 8.5, there would have been virtually no natural gas delivered to these purchasers after 1994 with the exception of small amounts to be made up from the period the contracts had run. Fortunately the Dutch have been flexible in changing the termination dates and the terms of these contracts. This will ease future transition to other forms of natural gas. None the less, if nothing else changes, exports are scheduled to fall to some 33 billion m^3 (29 million tons of oil equivalent – mtoe) in 1990 and fall even further in subsequent years, so that by the year 2000 the quantities traded would be less than 10 billion m^3 (less than 8.4 million mtoe).

The search for supplies to fill the hole left by the phasing out of Dutch gas has as a result focused on several suppliers, most notably the Soviet Union. Here the critical issue is whether the European transmission firms can be as resourceful in assuring security of supply as they have been in organizing the inter-European market. It is difficult to answer this question definitively at present because of the political squabble between Europe and the United States over the desirability of the natural gas contracts. The US government, never very enthusiastic about these trades, exploited the Polish crisis of 1981–2 to argue against any Soviet imports to Europe. European governments under attack reacted predictably. The result has been the politicization of such a natural gas trade. There is nothing inherent to the European organization of natural gas imports that supports an alarmist view of future Soviet imports. This is, however, perhaps more the result of slower growth in European natural gas consumption than of any virtues of the European transmission companies. As we shall note, there are very real problems of inter-market coordination yet to be solved – a coordination that may be necessary to achieve future security of supply or minimize the impact of any Soviet cut-off of natural gas supplies.

The quantities of the natural gas at issue are substantial, and the gas is offered on attractive terms *vis-à-vis* alternative suppliers. The exact terms of the recent contracts are, of course, secret, and even if public would be hard to interpret given the diverse means of payment by European purchasers. Table 8.12 gives the quantities of Soviet natural gas imported by country in 1980 and further specifies the amounts involved in the contract negotiations of 1981–2. Should all the options in Table 8.12 be exercised, future contractual amounts of Soviet natural gas will approach the 84 billion m^3 per annum level, twice the anticipated Dutch levels of exports to European markets in

Table 8.12 *Soviet gas exports to Europe: purchases by country in 1980, quantities negotiated in 1981–2, and total available in 1990 (billion m^3 (st.))*

	Amounts purchased in 1980 (1)	*Quantities contracted for 1981–2* (2)	*Future options* (3)	*Maximum available by 1990* (cols 1 + 2 + 3)[b] (4)
Belgium	0	0	2	2
France	5	12	0	17
West Germany	10.7	22	0	32.7
Italy	7	7	8	22
Netherlands	0	0	2	2
Others[a]	2.5	4.8	1	8.3
Totals	25.2	45.8	13	84.0

Sources: Stern, 1982; cited in World Gas Report, 10 May 1982, and corrected after conversation with Mr Stern (30 January 1984). Contract termination dates are from Tiratsoo, 1979, p. 250.

Notes: a Spain, Switzerland, Austria, Denmark, Sweden, Luxembourg. This list excludes Finland, which is a major customer. This is done for comparability with Tables 8.13 and 8.14.

b Contracts under which gas was purchased in 1980 run out as follows: Austria, 2001, France, 1996; West Germany, 1995; Italy, 1994.

1990. Yet it is unlikely that the European buyers really need all that Soviet gas. In order to achieve the consumption of the amounts shown in the 'contracted-for' and 'optional' columns, it would be necessary for Continental natural gas demand to increase at a compounded rate of 3.2 and 4.2 per cent, respectively, for the eleven-year period 1979–91. Even as the European–Soviet negotiations were progressing, Continental demand sagged. The optimistic scenarios predicting demand increases of 2–4 per cent compounded per annum until the year 2000 have been replaced by more sober estimates. The International Energy Agency, in its 1981 report on European energy policies (IEA, 1981b), uses an average rate of demand increase of 2.1 per cent. This estimate could also be too high – although it is utilized for the material in Table 8.13 – since natural gas sales in the period 1980–2 have already slumped. Dutch sales in 1982 to the rest of the Continent were about 70 per cent of the 1979 levels. Concern at loss of governmental revenue and at a similar slump in domestic sales of natural gas have caused Dutch officialdom actively to consider additional export contracts in the 1990s. Sales figures elsewhere have similarly slumped.

Table 8.13, based on IEA European demand assumptions and currently contracted-for amounts from North Africa and the North Sea, demonstrates a second and complementary factor: Soviet gas supplies will not only be 'squeezed' by lack of Continental demand, they will also be pressured by gas from other potential suppliers. Far from needing some 84.0 billion m^3 of Soviet natural gas in 1990 (as predicted in Table 8.12), European demand, expressed as anticipated consumption minus anticipated production and imports from non-Soviet sources, could amount to some 33.9 billion m^3 – which is only slightly more than the 25.2 billion m^3 imported from the Soviet Union in 1980.

It is however, possible that, with falling oil prices, economic recovery may occur and demand for gas recover significantly. It would therefore be wise to examine the worst possible case in terms of dependency. For the sake of argument, it will be assumed that all Soviet imports in Table 8.12 actually do materialize for reasons of price and delivery terms. Should this happen prior to 1990, it would signify a 59.6 per cent French dependency on Soviet natural gas, a 60.9 per cent dependency for West Germany, and a 53.6 per cent dependency for Italy (see Table 8.14). European tendencies to minimize such dependencies are questionable. For example, the German claim that a Soviet cut-off would affect only 5 per cent of German energy consumption is misleading, given that in 1980 natural gas provided some 20 per cent of all fuel for electrical generation, 30.5 per cent of the final end-use energy consumption of industry and

Table 8.13 *The Soviet Union as a 'supplier of last resort': demand for Soviet supplies as a consequence of European supply/demand balances in 1990 (billion m³ (st.))*

	IEA estimated 1990 consumption (1)	*IEA estimated 1990 production* (2)	*Resulting balances (minus in brackets)* (2) – (1)
Major Continental markets:			
Belgium	18.02	—	(18.02)
France[a]	28.52	4.64	(23.88)
West Germany	53.68	17.64	(36.04)
Italy	41.04	7.88	(33.16)
Others[b]	27.28	7.77	(19.51)
Major European producers:			
Norway	—	30.0	23.9[c]
Netherlands	35.4	61.97	26.6[d]
North African exporters:			
Algeria	—	—	43.7[e]
Libya	—	—	2.5
Balance to be covered by USSR			33.91

Source: International Energy Agency, 1981b.

Notes: a Estimates based on anticipated production and consumption figures available to me.

b Includes Spain, Switzerland, Austria, Denmark, Sweden and Luxembourg, but excludes Finland, Greece and Turkey.

c Including Sleipner deliveries minus Frigg deliveries to Great Britain.

d Exports 30.58 billion m³; imports to cover consumption.

e Includes LNG contracts, plus Trans-Med and another 10 billion m³ through a potential Trans-Med 2 or a SEGAMO line.

16.8 per cent of commercial and domestic end-use. Even a small shortfall in energy furnished to these sectors would have a significant effect on economic activity. An ideal sector to wean away from dependency on natural gas is electrical generation, as many electrical plants can be converted to dual burners relatively cheaply. Table 8.14 therefore includes the percentage of natural gas consumption utilized in electrical industry. It also shows current interruptible supplies and storage expressed as a percentage of annual consumption in order to give an idea of how flexible the various natural gas markets in fact are. It is tempting to add up the annual consumption percentages for interruptible supplies, storage and potential electrical savings and define the problem of 'worst case' dependency away. Thus Belgium's market shares in these areas add up to a total of 71.8 per cent, quite adequate to take care of an 11.1 per cent shortfall in Soviet natural gas supplies. Unfortunately, this is grossly misleading. Some portion of the interruptible supplies (it is not known how much, but possibly half) is also dedicated to the electrical generation market. Thus any exercise along the lines described involves a good deal of 'double counting'. Additionally, the interruptible supplies and storage facilities listed in the table are already dedicated to national load

Table 8.14 *Market flexibility and Soviet dependency: natural gas (percentage of annual natural gas consumption)*

Country	*Anticipated 1990 'worst case' Soviet dependency*[a]	*1980*		
		Interruptible supplies	*Storage*	*Electrical generation*
Belgium	11.1	27	29.0	15.8
France	59.6	15	29.0	7.0
West Germany	60.9	11	6.3	24.2
Italy	53.6	16	32.0	10.6
Netherlands	5.6	7	—[b]	21.6

Source: International Energy Agency, 1982.
Notes: a Defined as column (4) in Table 8.12 divided by column (1) in Table 8.13.
b The Groningen field is used for storage needs.

factor/marketing purposes. Any use of these facilities for security of supply would wreak havoc with company marketing plans and customer supplies. For example, to conscript 92 per cent of the Italian storage facilities to ensure security of supply would definitely damage sales in the prime domestic and commercial markets. Finally, the figures in the table are not comparable. The 1990 figures cover a much greater consumption of natural gas than do the 1980 figures. Table 8.14 is primarily of interest as an illustration of how hard it would be for national markets operating on their own to 'ride through' a sudden Soviet termination of gas supplies.

Can the international transmission nets make up for lacks in domestic market flexibility? Two factors are of importance in answering this question: there must be alternative sources of supply that are capable of delivering widely varying amounts of natural gas; and (as a necessary consequence of the first) the transmission net must be capable of extraordinary 'swapping' and 'moving' of supplies in large volumes to the markets that most need these supplies.

All alternative suppliers, with the exception of one (the Groningen field), operate at high load factors, at times as high as 90–95 per cent. These suppliers are thus incapable of delivering the large quantities of natural gas that would be needed in the case of a major Soviet termination of natural gas trade. This is particularly true of North Sea and North African supplies at present. With regard to the one exception, the Groningen field, we have noted that the Dutch exchanged higher prices for more flexible contracts (for the purchasers) and a lower load factor. Groningen has been developed in such a manner that the field makes an ideal 'swing' supplier, starting up and closing down production and transmission within minutes or hours of notification. The Dutch thus possibly have the capacity to fill a Soviet shortfall. The problem here is that Groningen gas is L-gas, while Soviet gas is H-gas. The use of Groningen as an emergency back-up is therefore curtailed to the degree that distribution lines and customers can adapt to differing gas mixtures from the H-gas/L-gas, H-gas/other gas and pure H-gas mixtures that typify many markets. The answer to this puzzle is not obvious, but would prove interesting.

Conclusion: will the new dependency diminish control?

The ability of the European transmission net to swap quantities of natural gas is unchallenged. There are constant small net movements of natural gas between countries that are not explicitly covered in a major contract. There are also some larger examples of swapping. The Italians receive some 5 billion m^3 of Soviet natural gas destined for France, while the French are paid with Dutch gas originally contracted for Italy. The Dutch have additionally been capable of securing emergency supplies to the French when the latter faced curtailment of Algerian supplies. None the less, the quantities of natural gas involved in both these cases (some 5 billion m^3 annually) are relatively small when compared to a cut-off of Soviet supplies of some 80 billion m^3. Furthermore, the Dutch do not have inexhaustible supplies of H-gas or L-gas. Clearly, even given the best of efforts, the ability to compensate more than marginally for a shortfall in Soviet gas supplies is non-existent in today's European transmission network.

While there need be nothing inherent in the European organization of natural gas markets that necessarily precludes sizeable Soviet imports and a degree of security of supply, there is some room for considered concern. Although there is reason to believe that Soviet exports to Europe will be nowhere near the order contemplated in 1982, yet it would be a mistake simply to assume that a slump in demand and an increase in alternative suppliers will take care of the problem. Should Soviet exports assume a larger role than seems likely, the reorganization and tighter integration of European gas networks could well be a necessary object of inter-governmental policy-making consideration. In Europe, as elsewhere in the world, natural gas is too important to be left to the companies alone.

Notes

1 In this respect, the Dutch (through state participation on offshore lines) have done themselves one better than the British Gas Corporation, in that they are capable of appropriating some of the economic rent from the gas production in the North Sea through basing their price inflation clauses on *ab platform* rather than landed prices.

2 The following draws heavily on Davis, 1983, pp. 9–16.

3 The details here are deliberately vague. It is rumoured that the fuel oil concerned was industrial fuel oil with 2 per cent sulphur content.

4 Both these LNG deals were concluded with the assistance of the governments concerned and involved prices for LNG significantly above the then going prices for pipeline natural gas in Europe. As these deals involved natural gas destined primarily for sale in Belgium and France respectively – and thus did not involve extensive use of international transmission through H-gas lines – there was relatively little that the consortium could do.

9

Liquefied Natural Gas – Unfulfilled Promise

Auspicious beginnings

Although methane was successfully liquefied in the mid-nineteenth century, it was not until the *Methane Pioneer* eased its way out of Lake Charles, Louisiana, with a cargo of liquefied natural gas bound for Canvey Island in 1959 that liquefaction appeared suddenly feasible as a means of transporting natural gas over long distances that could not be covered by pipelines. In its quiet way this pioneering voyage was as revolutionary as was the laying of the first all-welded transmission line between northern Louisiana and Beaumont, Texas, in 1925. Scientists and engineers had long been interested in the potential commercial applications of liquefaction technology, and LNG facilities had been previously used for gas storage in Cleveland, Ohio. But it was not until the Lake Charles voyage and the British Gas Council decision to import liquefied methane from Algeria that LNG 'took off'.

The advantages of liquefied natural gas are many. As mentioned earlier in this book, the liquefaction of methane reduces its volume by a factor of 600–620. This greater density enables large quantities of gas to be both transported and stored in relatively confined areas. Furthermore, it liberates the natural gas trade from pipelines. Liquefied natural gas can be conveniently transported by sea in large supertankers, and also over land by lorry and rail; there have even been plans to carry natural gas in liquefied form in jumbo jets.

Two or three advantages stand out most particularly. First, reduction in the volume of natural gas has obvious storing and peak-shaving advantages. LNG facilities free distribution firms from the necessity of finding depleted reservoirs for storage purposes or from storing natural gas in expensive underground caverns or huge above-ground tanks. These advantages have also led to the widespread use of LNG as a means of supplying peak loads as desired, since the quantities of gas sent out to customers can be easily and flexibly controlled.

The use of LNG also frees producers from the need to separate out various natural gas liquids. These liquids can create difficulties for the pipeline transportation of methane and are often removed before the gas enters a pipeline system. This removal process can entail considerable inconvenience in out-of-the way fields. With LNG, these liquids can be included with the methane, transported to the proper destination, and then separated upon reception.

Finally, LNG holds an important key to utilization of natural gas resources previously inaccessible to the main consuming centres of the fuel. In the 1960s and 1970s, it was frequently pointed out how LNG transportation between the Middle East and Europe and between Venezuela and the United States could prevent the enormous waste incurred through flaring of associated gas. LNG transportation could also, as we have indicated, benefit the receiving country by not only contributing to base load needs, but also as an aid in peak-shaving, since it is delivered in a form suitable for immediate storage.

Such were the advantages of LNG that a bright future was promised for it almost from the outset. As recently as 1977, two experts predicted that by 1985 some 22 per cent of US gas supplies, 23 per cent of European gas supplies, and 86 per cent of Japan's gas supplies could come from LNG imports to those markets (Daniels and Anderson, 1977).

While LNG trade has expanded, it has not done so at these predicted rates. Table 9.1 enumerates the current and potential LNG trades into OECD markets, which account for virtually all LNG traded today. (There are two contemplated contracts between South Korea and Taiwan and the producing states of SE Asia, which are not included in Table 9.1.) While this list looks impressive at first glance, only twelve of the trades mentioned are currently operational. The balance are either 'under construction', 'tentative', 'suspended', or 'possible'.

What accounts for this disappointing performance of a technology that was seen as having so much potential? Two major reasons may be advanced: the first concerns the technology of LNG; the second concerns the markets to which LNG is supplied and the risk distribution between natural gas producers (who must invest in costly liquefaction facilities) and LNG consumers. With regard to the first, the problems were signalled early on. The earliest commercial use of LNG tanks for peak-shaving purposes – in Cleveland, Ohio, some few decades ago – came to an abrupt end when one of the tanks

Table 9.1 *Present and potential OECD LNG gas trade (billion m^3 p.a.)*

Export scheme	*Status*[a] *(1982)*	*1980*	*1985*	*1990*	*1995*
Algeria–USA (Distrigas)	Operational	1.1	1.4	1.4	1.4
Algeria–USA (El Paso)	Suspended	—	10.0	10.0	10.0
Algeria–USA (Trunkline)	Operational[b]	—	4.5	4.5	4.5
Sub-total		1.1	15.9	15.9	15.9
Indonesia–USA (Pac Indonesia)	Possible	—	—	5.5	5.5
Nigeria–USA	Tentative	—	—	—	8.0
Chile–USA	Tentative	—	—	—	2.5
Trinidad–USA	Tentative	—	—	—	7.0
Argentina–USA	Tentative	—	—	1.5	1.5
Canada(Arctic)–USA	Tentative	—	—	—	3.0
Sub-total		0	0	7.0	27.5
TOTAL into USA		1.1	15.9	22.9	43.4
Brunei–Japan	Operational	7.8	7.2	7.2	—
Alaska–Japan	Operational	1.2	1.3	—	—
Abu Dhabi–Japan (ADGLC)	Operational	2.7	2.9	2.9	2.9
Indonesia–Japan (Badak & Arun)	Operational	11.9	10.5	10.5	10.5
Indonesia–Japan (Arun)	Under construction	—	4.5	4.5	4.5
Indonesia–Japan (Badak)	Under construction	—	4.6	4.6	4.6
Malaysia–Japan	Under construction	—	8.4	8.4	8.4
Sub-total		23.6	39.4	38.1	30.9
Australia–Japan	Possible	—	—	8.4	8.4
Canada–Japan	Possible	—	—	4.1	4.1
USSR–Japan	Possible	—	—	4.2	4.2
Abu Dhabi–Japan (ADNOC)	Tentative	—	—	3.0	7.0
Qatar–Japan	Tentative[c]	—	—	8.0	8.0
Thailand–Japan	Tentative	—	—	4.0	4.0
Sub-total		0	0	31.7	35.7
TOTAL into Japan		23.6	39.4	69.8	66.6

Table 9.1 *Present and potential OECD LNG gas trade (billion m^3 p.a.)—continued*

Export scheme	*Status*[a] *(1982)*	*1980*	*1985*	*1990*	*1995*
Algeria–UK	Operational[d]	—	1.1	1.1	1.1
Algeria–France (Le Havre)	Operational	—	0.6	0.6	—
Algeria–France (Fos-sur-Mer)	Operational	5.28	3.6	3.6	3.6
Algeria–France (Montoir)	Operational	—	5.4	5.4	5.4
Algeria–Spain	Operational	—	4.5	4.5	4.5
Algeria–Belgium	Under construction	—	5.0	5.0	5.0
Libya–Italy	Suspended	2.10	2.5	2.5	—
Libya–Spain	Operational	—	1.1	—	—
Nigeria–Italy	Possible[c]	—	—	1.4	1.4
Nigeria–Belgium	Possible[c]	—	—	0.9	0.9
Nigeria–France	Possible[c]	—	—	1.8	1.8
Nigeria–Germany	Possible[c]	—	—	2.3	2.3
Nigeria–Netherlands	Possible[c]	—	—	1.0	1.0
Nigeria–Spain	Possible[c]	—	—	0.6	0.6
Cameroon–Europe	Tentative	—	—	5.0	5.0
Total into Europe		7.38	23.8	35.7	32.6
TOTAL trade with OECD		32.08	79.1	128.4	142.6
Total net into OECD		30.98	77.8	115.9	127.1

Sources: 1980 figures: Cedigaz as cited in *Noroil*, October 1981, pp. 34–5; rest of table: International Energy Agency, 1982, pp. 121–2.

Notes: a 'Possible' indicates signed contracts or letters of intent; 'tentative' indicates that one or both sides have bruited such a project, but that no firm decision exists to commence the project.
b Deliveries have not commenced.
c This volume may be divided between Japan and Europe.
d Contract has expired, but could be renewed.
e Volumes may be doubled if the Nigeria–USA contract is not approved.

collapsed in 1944. The LNG contents spilled out, re-gasified and ignited; 130 people perished in the resulting inferno, the heat of which proved sufficient to ignite buildings 800 feet distant. The above-mentioned Lake Charles voyage preceded an earlier attempt to ship natural gas from Louisiana to Chicago on a cattle barge. The barge had proved unable to take the extreme cold of the liquefied methane – the tank walls cracked and warped – and the Comstock

Corporation, the corporate sponsor of the idea, then ran into economic difficulties and abandoned the effort in 1952. The *Methane Pioneer*, the first successful LNG carrier, re-entered the LNG trade renamed the *Aristotle* and experienced no end of difficulties until finally retired as a storage vessel in 1975 (L. N. Davis, 1979, p. 51).

These technological problems were complicated by the vast disparity in the markets that the LNG trades were designed to service. These deals attempted to unite export and import markets that never before had been interconnected. As we have noted, the history of international natural gas trade outside of the USA–Canada relationship has been a very short one, so there has been no international precedent on which to base trading relationships. This has proved an immense handicap, and most of the proposed LNG trades have fallen through. As of this writing there have really only been two markets that substantially rely on LNG and that have adapted successfully to the LNG trade: France and Japan (and even here there have been difficulties). This general question of why LNG has fitted into certain national markets and not into others is in itself quite interesting; we shall attempt to answer it in the second half of this chapter.

LNG technology: the 'teething problems' of an industry

The advances in the commercial technology of LNG have been significant. Currently there are three major types of liquefaction plant heat-exchanger technology. The first, the cascade cycle, is a series of compression refrigeration cycles, each cycle rejecting heat onto the warmer cycle above it (cascading).[1] The second, the auto-refrigerated cycle, is a variant of the cascade cycle, but uses the gas liquids themselves as a part of the refrigerant process (this method uses 20 per cent more fuel but costs 20 per cent less to build). The third variety, the mixed refrigerant cycle, is a variant of the auto-refrigerated method and is the most widely employed technique at present. Today's liquefaction plants are built on a scale (and cost) previously unimaginable, being capable of liquefying 1.1 billion ft^3 per day, or 400 billion ft^3 per annum.

Technology for LNG carriers has similarly come a long way. Currently there are no fewer than twenty-one different designs for LNG ships. On 1 June 1981, of a total of fifty-seven carriers, some twenty-nine with a capacity of 120,000 to 135,000 m^3 apiece, were in service and of another fourteen on order, thirteen belonged in the 120,000 m^3 class. Each of the larger types carries enough natural gas to heat approximately 2.5 million homes on a −5°C (22°F) day (Davis,

1981a). Completing the LNG 'chain' are the reception terminals with their docking areas, re-gasification, storage and distribution facilities.

In short, the industry has made considerable progress since the days of the *Methane Pioneer*. This progress has not been without costs. The heat-exchanger technology utilized in liquefaction facilities is to the domestic refrigerator what the atomic bomb is to the Stone Age axe. A modern liquefaction plant often embodies 'state of the art' technology, with the latest engineering and scientific advances. The same is true of the insulation and design used in LNG carriers and in their reception and storage terminals. As with developing technologies elsewhere, LNG facilities have been both expensive and plagued with the problems of technological infancy.

This is especially true of liquefaction plants.

> typical problems include clogging of cooling water intakes by sand, seaweed, jellyfish or debris; failures in the blading in turbines, compressors or pumps; fouling of heat exchangers . . . due to build up of algae, clogging of spray rings for cooling the storage tanks, or bearing failures in pumps. (Office of Technology Assessment, 1980, p. 100)

These and other difficulties have often meant that the liquefaction plants do not live up to their design performances. The ratio between design throughput and actual throughput (the 'production factor') for these plants on average, to quote Bodle and Smith (1980), is as follows:

> Commencing at about 41 per cent in the first year, the average production factor gradually increases to about 74 per cent in the fifth year. Limited data beyond the fifth year show average production factors of 75 to 80 per cent . . . The most recent plants have shown a decided improvement in early performance. Hopefully the trend will hold.

This is not a small problem. Much more energy (normally feed gas) is lost in the liquefaction process if the plant is not operating to design expectations. Besides, as has been noted, a liquefaction plant with accompanying infrastructure can cost in excess of $1 billion – excluding the cost of developing the gas fields concerned. Such expense is normally covered by borrowing by the producer country. Depending on the ratio between equity and loan capital, such a shortfall in production factors can result in an actual loss of money by the natural gas exporter. This is not calculated to instil confidence in liquefaction and

LNG export trades as an optimal manner of disposing of a nation's gas reserves.

Over and above these disadvantages, many liquefaction plants have run into project delays as a result of commissioning problems. There is no sign, furthermore, that the costs of construction for these plants have diminished as more and more have come on stream. On the contrary, despite some twenty-three years of liquefaction experience, there have been few or no savings from scale plants or from the increasing experience in their design and construction.

The technology of LNG carriers has also encountered difficulties. One particularly popular design of LNG ship, the Gaz Transport Membrane design, has experienced problems in the integrity of its LNG tanks. It is customary, in order to maintain low temperatures, for such carriers to carry a 'heel' of LNG cargo on their return voyage (this avoids the delay of again cooling down the tanks prior to reloading). In practice, the 'sloshing' of this LNG heel has proved violent enough to damage the tanks. Because repairing these tanks is a lengthy and expensive business (all natural gas has to be 'purged' from tank insulation prior to repairs), many of these ships have continued to sail with only partial use of the carrier's total capacity. In mid-1981, ships with a capacity of 1,017,010 m^3 of natural gas – out of a total of 5,380,000 m^3 – were laid up on this account alone (Davis, 1981a). Perhaps the worst case of faulty design was that of the three carriers constructed in the Quincy Massachusetts Avondale shipyards that proved to have defective insulation and failed to pass US certification standards. (These were part of the star-crossed El Paso LNG trade with Algeria and a subject of a $300 million insurance settlement in 1981.)

The sailing record of the LNG carriers has been plagued by the difficulties normally encountered by a new technology such as this. The lack of serious untoward incident was broken by the grounding of the El Paso vessel *Paul Kayser* off Gibraltar in 1979. Although the ship's 90,000 m^3 LNG cargo was successfully transferred to another vessel without incident, the episode is said to have been jokingly referred to in Lloyds' circles as the 'day we almost lost Gibraltar'.

The most costly aspect of LNG carrier transportation has been laid-up tonnage, the result of LNG trades that were mooted for a number of years and then cancelled or suspended. As much as 32 per cent of all LNG tonnage has been mothballed. This is more serious than would be the case for oil or other commodity bulk carriers because, owing to the nature of LNG vessels, they cannot easily be brought out and used for 'spot' trades (because of both the time and cost involved in preparing them for service and the very limited and sporadic market for 'spot' trades), but must sit out the slack period and hope for a surge in activity. This is not inexpensive, as interest charges alone for a 125,000 m^3 carrier can amount to $10 million per month. (Added to this are the costs of berthing and storage – another several millions per month).

Reception terminals and storage facilities have had their problems as well. In practice, few storage tanks have not leaked into their insulation barriers. This has led to underutilization of storage facilities, an expensive business at best. At worst, such leakages have caused fatal accident. An insulation fire in a leaky Staten Island storage tank (possibly exacerbated by a gas explosion) killed forty workers as they were attempting to repair the lining. (This destroyed the chances of a promising LNG import trade into New York.) Leaks at a Das Island storage facility have been cited as a major reason why the Abu Dhabi–Japan trade did not exceed 40 per cent of the amounts contracted during the period 1978–9. Finally, an explosion and fire at the El Paso Cove Point terminal killed one worker and caused extensive damage, halving the LNG received by the terminal for a considerable period of time.

In all these cases, a problem accompanying that of economic loss has been the ultimate security of LNG facilities in general. Stories such as those cited above have not served to calm the fears of local citizenry of the consequences of a mishap in an adjoining LNG facility. Here concern is directed not only at international trades, but also at the more common use of LNG tanks as storage facilities for 'peak-shaving' purposes, especially in the United States and Great Britain. On Canvey Island, in Hirwaun, Wales, on Staten Island, and in Newport, Oregon, citizen groups have emerged to fight emplacement of LNG facilities in movements not unlike those of the anti-nuclear drive. These movements have similarly cost LNG trades. They involve repetitive hearings, delays and considerable accompanying transaction costs for proponents of LNG trade or storage.

One final problem has been the cost of feed gas to liquefaction plants and LNG 'boil-off' from ships and storage tanks. Figures vary for the loss of feed gas, depending on the processes and the technology involved. Typically, however, this runs from 16 to 18 per cent, perhaps at times slightly lower. Boil-off for a typical carrier varies as well, but losses are generally in the range of 0.25 per cent per day. LNG boil-off is somewhat less for storage tanks, but averages 2–3 per cent during re-gasification. Thus it is not uncommon to lose up to 25 per cent or more of the energy transported in an LNG trade. LNG engineers have attempted to reduce these losses, primarily by installing dual-fuel engines and re-liquefaction plants on LNG carriers to reduce boil-off. These losses, important in the days when feed gas could be had for pennies per MMBtu, are generally far more serious today, when the price of feed gas can amount to as much as $3.00/ MMBtu. While not really critical where LNG is the only alternative for the producing nation concerned, it is of vital importance where the producing nation may be considering alternative domestic uses for its natural gas or of putting off a decision until LNG technology becomes more reliable.

LNG marketing: problems of 'fit'

What accounts for 'success' in LNG trade? Perhaps not too surprisingly, one of the critical factors is the cost and reliability of LNG technology. Those trades that have succeeded to the satisfaction of both seller and buyer have been those where the equipment has proved reliable, the markets have been stable and capable of absorbing high-priced LNG, the liquefaction technology has been installed without problems, and good financial terms are secured for financing the project.

Yet LNG marketing success has also been conditioned by the nature of the importing market. In each national market to which LNG is supplied there are different philosophies, historical experiences and corporate preferences. A pattern of LNG trade thus emerges. Each importing nation incorporates LNG into its natural gas market in a particular national manner. Success or failure with LNG has to a very important degree been dependent on this national philosophy. The most successful nations in this regard have been Japan and to a lesser extent France. Elsewhere, in part because of this problem of 'fit', LNG trade has fallen on harder times.

Perhaps the worst experiences with LNG have been in the American market. With one exception – that of Algerian LNG to the Boston firm Distrigas – American trades have experienced difficulties. The fate of two typical trades is indicative of this point.

The least successful deal was the planned importation into two terminals, one at Cove Point, Maryland, and the other at Elba Island, Georgia, of some 10 billion m^3 per year of natural gas from Algeria. This deal, signed by El Paso, Columbia Gas and a series of other American transmission companies in 1971, took advantage of a temporary Algerian surplus. The price paid for the LNG fob in Algeria was just $0.39/MMBtu – clearly a 'knock-down price'. The project was ambitious right from the start; at the time it was the largest LNG deal ever signed. It was also from the beginning plagued with difficulties. For one thing, the enormous liquefaction plant at Arzew developed problems. Then, even before the commencement date, the Algerians were pressing for significant price increases. El Paso, responding to these demands, agreed to the conclusion of a second contract for an additional deal, the so-called El Paso II contract. 'We couldn't change the contract,' one authority is quoted as saying, 'so we revised the price in the second one upwards so that the average of the two more closely reflected the existing market prices' (L. N. Davis, 1979, p. 236). This contract immediately ran into difficulties with US regulatory authorities, who objected to the price indexation linkage of the LNG concerned to oil products. (The first of these two contracts is included in Appendix IV.) When the project finally got underway, the difficulties persisted. The liquefaction plant at Arzew never reached design performance, working at a 58 per cent production factor in 1978 and a 65 per cent production factor in 1979. Three of the nine ships built for trade, as noted earlier, were faultily designed with defective insulation and could not be used. A fourth went aground off Gibraltar with severe damage to the hull. The reception terminal at Cove Point was devastated by an explosion that significantly reduced its capacity. Finally, the Algerians, impatient at the terms of payment in the 1971 contract, halted LNG deliveries in 1979. Numerous attempts by El Paso and the US State Department notwithstanding, the Algerians held firm in their demands for a price that would reflect the price of alternative Algerian crudes in 1980 – a very high price indeed. The El Paso group, caught between the US State Department's desire that LNG prices be restrained and the Energy Regulatory Agency's desire to hold the line on prices of imported natural gas (and particularly on escalation clauses reflecting the price of crude oil), was in a dilemma. It solved this dilemma by closing its reception terminals, selling its vessels – including the three defective carriers from Avondale – and claiming losses on its American tax returns. The El Paso I deal, which had begun with such hopes, was no more.

The fate of a second Algerian deal, the Trunkline project, remains somewhat in the balance. The plan here was to import 4.8 billion m^3 per annum to the Trunkline transmission net in Louisiana. But it has run into regulatory difficulties since its inception. The Trunkline contract was signed for a pre-1973 price, reportedly in the region of $0.56/MMBtu fob Algeria, and was about 40 per cent the size of the El Paso trade. Trunkline (also called Panhandle Eastern, after the second partner in the American end of the deal) was scheduled to start up in 1980. This date was not held. Much as with the El Paso deal, the Algerians notified Trunkline that they wished to renegotiate the contract in light of the higher oil prices of 1980. Trunkline refused. The Algerians then cancelled the deal, but backed down under a Trunkline threat to carry the cancellation to an arbitration tribunal. This trade started unofficially in late 1982, but without final US approval. The FERC, concerned about the high price of the

Algerian gas (roughly \$7.20/MMBtu delivered to the trunklines outside Lake Charles) and the impact of the high Algerian price on other American sources of supply (Mexico and Canada were receiving some \$4.94/MMBtu), delayed its approval until February 1983. Even now, there is sentiment within the US Congress against the Trunkline deal, which could still imperil its fulfilment.

The most successful experience of LNG on the American market was a series of deals concluded between the Boston firm Distrigas and Sonatrach, the Algerian national company. Distrigas lies in the Northeast corridor of the United States and supplies markets that are relatively isolated from the US transmission line network. Further, it has been using its Algerian LNG almost entirely for 'peak-shaving' purposes for the Boston area alone. These two factors have enabled Distrigas to satisfy US regulators about the prices paid for Algerian LNG on the one hand, and to pay for LNG at the increasingly high prices that the Algerians have demanded on the other. The Distrigas-Algerian contracts have in fact been negotiated/renegotiated no fewer than five times. The first Distrigas contract was for a minimal amount of natural gas. The second contract was for 1.1 billion m^3 per year – an amount that has been largely forthcoming. While the El Paso contract price lay around \$1.17/MMBtu cif in 1978, the price that Distrigas paid rose from \$2.59/MMBtu cif in that year to \$5.23/MMBtu cif in 1981.

What were the reasons for these American difficulties? In part (as with El Paso I) these difficulties were technical. More importantly, however, they were of a regulatory nature. LNG imports to the United States were judged by their impact on regulated US natural gas markets. As stressed earlier, 'fit' in the American market means bureaucratic 'wrangling'. Testimony on the pricing of LNG before the FERC (and before it the FPC) runs into hundreds of pages. Algerian (or Indonesian) prices are not judged on whether they are too high or too low, although considerable comment on this point is evoked; rather, at issue is the indexing of such prices to crude oil or crude oil products. Here again, it is the regulatory aspects that colour the picture. Similarly, Algerian (or Indonesian) prices cif are a matter for concern, but primarily in terms of their regulatory aspects: whether or how these prices can or should be 'rolled in' to the overall price structure of the transmission lines concerned. In short, while Algerian militancy has been a problem, this factor has been complicated immensely by the American regulatory process, a process that Algerians, perhaps with considerable justification, have felt was giving American buyers an oligopsony rent.

LNG imports into Europe have been somewhat more successful. Reference has been made to the small (1 billion m^3 per annum)

Algerian LNG trade with Great Britain. Here there was less of a problem of 'fit' than in the United States, Algerian LNG being first used as a supplement to manufactured gas and later used as 'peak-shaving' gas for the Metropolitan London area. With the coming on stream of the North Sea fields (and increased Algerian prices) the British Gas Corporation never acted to expand its LNG trade with Algeria. The amounts currently arriving at Canvey Island generally remain at about the level they have always been. An abundance of British gas purchased at monopsony pricing rates has thus undercut Algerian LNG imports. Domestic marketing considerations once again impinged on the nature of an LNG trade.

The French stand in considerable contrast to the British. Gaz de France has long sought diversification of sources of supply, and has been eager to supplement the dwindling domestic production of natural gas. Algerian gas, which fulfils both needs admirably, has been introduced into French markets where it was most needed: first at Le Havre (1965) with contracted quantities of some 567 million m^3 per annum, and later at Fos-sur-Mer (4 billion m^3 starting in 1971–2) and Montoir (5.7 billion m^3 starting in 1981–2). Each of these terminals is located close to an industrial centre: Le Havre to the Le Havre–Rouen industrial concentration, Montoire to the Nantes industrial area, and Fos-sur-Mer to Marseilles–Toulon. These locations, and the desirable peak-shaving characteristics of LNG, have allowed for the direct introduction of liquefied natural gas into nearby industrial/urban conglomerations with a maximum of flexibility in distribution and end-usage. Fos-sur-Mer had the additional advantages that it supplemented the declining Lacq production and that it was located at such a distance from the Dutch sources of supply that LNG proved highly competitive and desirable in terms of natural gas usage both in the South of France and in the Lyons industrial area.

Yet, despite these positive factors, French imports of LNG have suffered only a marginally better fate than the American. With the negotiation of each new contract, for each new terminal, Gaz de France has had to renegotiate all the preceding contracts as well. (This is probably because of the existence of MFN provisions in Algerian contracts with France.) Algeria has also unilaterally terminated LNG deliveries to support its demand for price increases, insisting that LNG keep apace of price hikes in the world of oil. This interruption of supplies during the oil price increases of 1979–81 was settled in the end only by French government intervention and payment of higher prices ($5.14/MMBtu cif) than Gaz de France was willing to consider. The result was a considerable state-to-state binding of trade on the side of the Algerians in return for government subsidization of the LNG prices paid by Gaz de France.

A worse fate overtook another European natural gas deal – between Spain's Enagas with Italy's ENI and Libya. Here the amounts involved were small (some 1.12 and 2.52 billion m^3, respectively). Deliveries of the LNG began in 1971 after an eighteen-month delay, and subsequent events were similarly inauspicious. The Libyans disliked the terms of the original contract intensely, particularly its inclusion of natural gas liquids in the overall pricing of the cargoes delivered – an inclusion they felt underpriced their methane relative to the Algerians. Libyan LNG prices soon fell behind oil prices. At this point the Libyans acted, terminating their deliveries to Italy and to Spain in 1974 in order to increase the prices paid for LNG from \$0.96/thousand ft^3 to \$1.86. Spain, fully dependent on imported LNG for its markets, gave in and acceded to Libyan demands; Italy's ENI held out and after lengthy negotiations arrived at a compromise settlement whereby prices escalated on an interim basis to the full \$1.86 in 1977. The Libyan deal remained star-crossed, with the Libyans reportedly offering their Marsa el Bragha liquefaction plant first to the Russians and then to ENI (L. N. Davis, 1979, p. 237). In 1982, Esso, the operator of the plant, disgusted at the combination of the slow development of the LNG trade, Libyan and American politics, and the lack of profitability from its LNG investments, saved the Libyans the trouble of nationalizing the liquefaction plant by unilaterally withdrawing. The Libyan trade remains moribund as of this writing.

The European transmission companies en bloc have also, true to form, evinced an interest in importing LNG. Here, their efforts have 'flopped'. An intra-European effort to purchase Algerian LNG, the Société d'Achat de Gaz Algérien pour l'Europe (SAGAPE), signed a contract in 1972 for a large quantity of Algerian natural gas at a price of \$0.47/thousand ft^3. SAGAPE bears much resemblance to the consortia described in Chapter 8, being composed of seven companies from five nations (France, Belgium, Austria, Germany and Switzerland). However, it came 'unhinged' in 1974 when the Algerians suddenly revised their asking price upwards to \$1.50/thousand ft^3. This was too much for the two German and the Austrian partners, who withdrew. The other members each went their separate ways, some signing their own bilateral agreements (Davis, 1979, pp. 234–5). It is interesting to speculate what might have happened if the Europeans had remained united, for, as we shall note in the next chapter, the effective result of these bilateral deals has been to remove the LNG trade into Europe from any consortium or cartel control.

The upshot of importer experiences outside Japan, therefore, can best be described as 'mixed'. By 1981, LNG prices had risen to unimaginable heights, as indicated in Table 9.2. As can additionally be

Table 9.2 *LNG pricing: 1978 and 1981 (US $/MMBtu cif)*

Trade	1978 (average)[a]	1981	1981 oil product prices[c] Gas oil	Fuel oil 3.5% S_2
Abu Dhabi–Japan	2.20[b]	6.60[b](7/81)	6.34(7/81)	4.20(7/81)
Algeria–France	1.81	4.63 (1/81)	7.25(1/81)	5.49(1/81)
Algeria–Spain	1.36	4.85 (1/81)	7.25(1/81)	5.49(1/81)
Algeria–UK	0.99	5.50 (7/80)	7.22(7/80)	3.73(7/80)
Algeria–USA (Distrigas II)	2.59	5.23 (7/81)	6.83(7/81)	4.18(7/81)
Algeria–USA (El Paso I)	1.17	—	—	—
Brunei–Japan	2.09[b]	6.38[b](6/81)	6.45(6/81)	4.45(6/81)
Indonesia–Japan	2.80[b]	5.95[b](6/81)	6.45(6/81)	4.45(6/81)
Libya–Italy	1.17[b]	4.62[b](1/81)	7.25(1/81)	5.49(1/81)
Libya–Spain	0.99[b]	4.62[b](1/81)	7.25(1/81)	5.49(1/81)
USA–Japan	2.30	5.92 (7/81)	6.34(7/81)	4.45(7/81)

Sources: 1978 figures: *The Petroleum Economist*, December 1980, p. 514; 1980/1 LNG prices: *World Gas Report*, 14 September 1981, p. 12.

Notes: a 1978 figures: 1,000 ft^3 = 1,047,000 Btu. Others unspecified.

b Liquid petroleum gases included in price (not strictly comparable).

c Gas oil and fuel oil prices are for various products, various places:
Prices for US trades: New York (fuel oil with 2.8%S_2).
Prices for France, Spain, Italy: Mediterranean (fuel oil with 3.5% S_2).
Prices for Japan: Singapore cargoes (medium fuel oil).
All prices in metric tons converted to MMBtu at conversion rates 43.0458 MMBtu for gas oil and 40.456 MMBtu for fuel oil.

seen, the Algerian and other OPEC member demands for crude oil pricing parity fob had led to the pricing of LNG in virtually every market in excess of competitive residual fuel oil and gas oil. Belgium and France remain significant LNG importers, having signed large contracts with the Algerians in 1981 and 1982, respectively, much to the distress of other interested European purchasers. That they did this, however, owed more to political considerations than commercial calculation.

Liquefied natural gas and the Japanese market

The Japanese account for roughly 74 per cent of all international LNG trade, a market share that they have achieved, moreover, with a minimum of the troubles experienced by other European LNG importers.[2] What accounts for this Japanese dominance?

In part, the Japanese role in LNG trade is the result of its own lack of domestic supplies. Japan is almost entirely dependent on imported natural gas to cover its needs. Japanese onshore reserves are estimated to be in the region of 33 billion m^3. (There are probably considerable additional offshore reserves, as indicated by the discovery of the Aga Oki, a field containing an estimated 12 billion m^3.) At present, domestic natural gas production in Japan accounts for just 12 per cent of domestic consumption. While the Japanese government hopes in time to double this percentage, the balance of the gas consumed must be either imported or manufactured. Given the particular geographical location of Japan, LNG imports tend to be favoured.

That the bulk of natural gas consumed in Japan has historically been manufactured is curiously enough also an advantage to the burgeoning LNG trade. As recently as 1977, some 19 per cent of Japanese gas consumption was coal gas; another 31 per cent consisted of gas produced from oil feedstocks: kerosene, naphtha, and crude oil itself. Furthermore, these percentages do not reflect a relative constancy in these markets: the consumption of coal gas declined by an annual rate of 0.9 per cent between 1968 and 1977, while the production of gas from oil feedstocks actually increased in this period. These are all relatively costly production processes, which has ensured the continuance of high prices for Japanese commercial and residential use of town gases. (Hypothetically, this could mean that the Japanese were thus more willing to purchase natural gas at the higher prices that LNG exporters were likely to demand. This is not wholly the case, however, as we shall note.)

Another factor making for Japanese success was the ability to switch to LNG imports in a significant manner. Thus, in a period when the rest of the world debated shifting away from exporter nations as 'unreliable' sources of supply, the Japanese were switching to LNG exporters for security of supply reasons. There are two reasons for this decision. First, the Japanese can import LNG from nations that are not major sources of crude for the Japanese market – nations such as Malaysia, Australia and perhaps even Canada. Second, the Japanese have 'cottoned on' to the fact that LNG contracts are long-term contracts extending over twenty years. The alternative fuels to oil are not characterized by such contract length. In terms of length of contract, the Japanese are buying (at a price) an additional degree of security.

This added security of supply is complemented by the record to date of the various liquefaction plants supplying the Japanese markets. These plants are not only the largest in the world, they also have the best performance factors. They have even been reported as having exceeded their designed performance. This has led to considerable exporter contentment, in notable contrast to the Algerian case. Nor are the exporters particularly exacting in their demands regarding economic rents in selling to the Japanese. A not infrequent practice in the area is to charge no money for feed gas and to collect all profits through fob deliveries to LNG carrier 'flange'. This is again in notable contrast to the Algerians, and bespeaks a greater reasonableness among the suppliers to the Japanese market. This 'reasonableness' can also be seen in the fact that many Japanese LNG contracts include other natural gas liquids than purely methane. These gas liquids can be sold at a much higher price than methane, and their separation from the LNG is a relatively simple and inexpensive process. The commercial sale of butane and propane can thus reduce the actual price of the methane considerably. For LNG from Abu Dhabi (the higher-priced of the various LNGs listed in Table 9.2), for example, the price of the actual methane content of the mix could be as low as $4.50/MMBtu delivered cif.[3]

The geographical location of the major urban areas of Japan is also ideal for the LNG trade. Both the Kanto and Kansai areas are characterized by good harbours and harbour facilities, so that, although they are cut off by chains of mountains, their natural gas can be furnished directly from the sea without the additional expense of laying transmission lines. Added to this is the availability of markets large enough for the sale of LNG in the quantities imported. First there are the large electrical utilities, to receive the early start-ups of LNG deliveries; then there are the large natural gas utilities in Osaka, Nagoya and Tokyo that can be counted on to sell LNG in the quantities imported after the longer-term process of conversion to natural gas has been completed. It should be emphasized in this context that these two markets are strategically interlinked; the electrical generation market buys time for the distribution companies to 'get their act together'.

The idea of using LNG with a nominal cif value of $6.00 (actually considerably lower in the case of cargoes with LPG liquids) to generate electricity would seem madness. The ideal cheap fuel for power generation and for freeing Japanese electrical utilities from OPEC oil is coal. Why, then, have the Japanese selected LNG, perhaps the most expensive fuel around, to burn off in electrical power plants where thermal efficiency is only of the order of 35 per cent? The key lies in the Japanese concern for pollution, a serious problem for the major Japanese conurbations. This concern has led to a set of tough emission standards for electrical plants in these areas.

If the desire to switch from oil and the tough emission standards are seen as two juridical givens, the economic framework for introducing LNG into the electrical power plant business, far from being an economic loss to the electrical industry, had considerable economic advantages. Under certain circumstances, in fact, one could claim that LNG was the cheapest alternative. To understand this rationale, it is necessary to step into the shoes of an electrical utility. Granted that one must switch from oil, which way should one go? To switch to coal – the cheaper of the two alternatives – involves high conversion costs for the plant, increased operation costs for coal storage, pulverization, unloading and the like, and worst of all, high costs for installing the sulphur scrubbers necessary to comply with the tough emission standards. The cost of converting to natural gas, on the other hand, is a fraction of that for coal, operation costs are lower (lower even than operation costs for oil-fired generators), the life of capital equipment is longer as well, and there are very few, if any, sulphur emissions.

In economic terms, the choice is exemplified in Table 9.3, which shows the cost of LNG with a \$5.50 and a \$4.50 per MMBtu price – the latter to reflect the net price of methane separated from a cargo of LNG where natural gas liquids were present. It also shows the annualized cost of the alternative in terms of capital costs, annual operation costs, and the like. The price of steam coal is that from September 1982. The plant is a 350 MW plant working at a 68 per cent load factor. Conversion assumptions are based on Danish experience.

As can be seen in the table, even given the relatively low-priced Danish assumptions, a 350 MW plant utilizing LNG pays a lower 'premium' than might otherwise be expected: 0.56–1.44¢/kWh. (This could amount to an estimated 3–12 per cent of the overall domestic tariff in Japan.) Moreover, there are several factors not taken into account in the table that reinforce the advantages of LNG usage. In the first place, the assumptions, particularly the conversion assumptions, are low; the cost of land (for coal storage, etc.) in Japan is many times the cost of land in Denmark (from where the data in Table 9.3 are derived), and this cost is significant in both cases. Second, the benefits of operating with gas-fired power plants are not included in the economic data in the table. These include: elimination of fuel stocking facilities, total absence of combustion wastes (and of apparatus and plant for their removal), elimination of soot or other fouling deposits on boiler heating transfer surfaces, elimination of sulphur products with improvement of boiler performance and life, and greater peak-load flexibility and efficiency (Medici, 1974, pp. 193, 295).

With electrical power generation now accounting for some 75 per cent of all LNG usage, there are plans afoot to utilize this gas in other sectors. Such plans may, however, run into a barrier. The organization of the natural gas industry in Japan is reminiscent of that in the United States. There are about 250 natural gas undertakings in Japan

Table 9.3 *Increased costs of conversion and use of fuel. A 350 MW generating plant using LNG and coal as two possible alternatives to oil. (6,000 full load factor equivalents p.a.) (US $ thousands)*

	Steam coal[a]	*LNG ($5.5/MMBtu)*	*LNG ($4.5/MMBtu)*
Cost of conversion:			
Oil to alternative p.a.[b]	5,812	258	258
Annualized cost of UHDE 'scrubber':			
Capital cost p.a.[c]	2,640	—	—
Operating costs	5,471	—	—
Operating costs in excess of oil-firing alternative	2,506	[d])	[d])
Costs of fuel	55,965.8	102,358.4	83,748
Totals	72,394.8	102,616.4	84,005
Cost per kWh (US cents)	3.44	4.88	4.0
LNG 'premium'	—	1.44	0.56

Sources: ELSAM, 1982. Interviews: Vestkraft, Esbjerg; Jysk-fynske elsamarbejde (ELSAM), Fredericia, Denmark. Gas efficiency factor from Medici (1974). The use of Danish data in this table is due to an inability to acquire the relevant material from Japan. The Danish electrical industry has achieved a wide reputation for its efficiency and for ease of information access, which makes it an 'ideal' shadow price case for the Japanese electrical industry. It is likely that these Danish estimates are low in comparison to estimates elsewhere and in Japan.

Notes: a Average cif. September 1982: $64.02/metric ton; 6,500 kcal/kg 3% less efficiency than fuel oil.

b Cost of conversion estimated at $52.9 million for coal, $2.36 million for natural gas. Annualized even payments over 15 years principle and interest. 7% interest.

c Capital costs of UHDE 'scrubber': $24.1 million. Amortized as with cost of conversion.

d There are operating cost savings not entered here. See text for explanation. It is assumed as well that LNG has 10% more boiler efficiency than oil.

of which less than a third are municipally owned, the balance being in private hands. There is no national grid. There are fewer than fourteen types of gas being distributed (Peebles, 1980, pp. 108–9). Organization of this industry and its regulation by government has had as many 'ups and downs' as in the United States. (In fact, for a brief period, the Japanese tried to imitate the FPC system of regulation – Peebles, 1980.) Work on converting manufactured gas distribution networks to natural gas usage in major cities is progressing slowly and will not be finished for another five–six years. The price of manufactured gas, too, is high, which has limited Japanese demand for it as a fuel. The greater use of LNG as a source of energy must therefore overcome the problems of organization inherent in the wider use of natural gas, and it will have to be marketed on a basis that Japanese industrial, commercial and domestic customers find attractive. In the words of the International Energy Agency, 'supply [on the Japanese market] is no problem. Demand is' (IEA, 1981b).

Japanese imports of LNG have largely been an unqualified success. In the period 1973–9 they accounted for 100 per cent of the marginal increase in the country's energy imports. From some 23.6 billion m^3 in 1980, the Japanese hope to expand their LNG trade to some 45 billion m^3 annually by 1990. Things have gone well thus far, but it would appear that the Japanese could face considerable difficulty in selling the amounts of gas involved in their 1990 projections.

Conclusion: LNG patterns of success and failure

LNG has largely passed through its 'teething' stage. Despite the problems encountered in the last twenty years, today's industry seems poised for further growth. It has learned from past disappointments and adjusted its sights accordingly. Many of the equipment failures endemic of the earlier years seem to have been overcome to some degree. An encouraging sign, for example, is industry reports that some liquefaction units in SE Asia have exceeded their design performance criteria. Problems do remain, chiefly those of cost and safety, but attempts are being made to remedy these. Yet, while technical and design problems seem less important, marketing problems still remain.

What is striking about LNG is that so little of it is used for 'normal' markets. This is exemplified in the various national uses of the fuel. In the United States, the use of LNG cannot be separated from the question of price regulation. LNG importers there have without exception used the fuel to alleviate shortages of conventional gas and for 'incremental' or 'rolled-in' pricing. Apart from one case, LNG could not stand on its own during the 1960s and 1970s. The exception, as we saw, was the Distrigas deal with Algeria, where LNG was largely used for peak-shaving purposes. This obviously somewhat specialized use had the effect of putting LNG in the 'premium' price class, thereby justifying the increased costs of its acquisition.

In the UK, the small LNG import trade into Canvey Island, whatever its original intention, became with the discovery of natural gas in the North Sea of primarily experimental interest. Canvey Island was the first area in the UK to undergo conversion from manufactured gas. LNG thus aided the learning process of how to accomplish conversion. Its lessons were then widely applied elsewhere in the UK. LNG further served as a means by which the Gas Council finally tipped the balance in its own favour *vis-à-vis* the various area boards, as described in Chapter 5. With the advent of North Sea gas, Canvey has lapsed in importance. Its prime function now is the same as that for Distrigas in Boston: peak-shaving. The expiration of the LNG contract has led to negotiations between the Algerians and the British over contract renewal, with several provisional trades in the meantime. The ultimate question in these negotiations is not whether Algerian LNG can make a major contribution to the UK British Gas net; rather it is whether or not the prices demanded by Algeria justify the use of Algerian gas for peak-shaving.

Elsewhere in Europe, the costs of LNG for main line grid purposes have exceeded the cost of other varieties of natural gas, with the possible exception of the recent Statfjord gas contract. France is the only nation that to date has depended heavily on LNG for its gas supplies. But even here, as with Italy, payment for LNG has come to mean the payment of a politically determined premium to the exporter nations concerned. The main European concern so far has been to 'hold the line' and coordinate LNG purchases from various sources, an effort that has not notably succeeded.

In LNG, as in many other industries, the Japanese are simultaneously both odd man out and worth learning from. Japanese LNG strategy is a blend of both commercial and political considerations. Commercially, LNG imports have been stimulated by Japanese realization that natural gas has non-polluting characteristics well worth paying for. The result has been that they have utilized 75 per cent of all imports for electrical generation. This is a sector use considered 'wasteful' in Europe and the United States. The Japanese do not consider it wasteful at all. Electrical power plant sulphur emissions are a problem in urban areas so it is logical to apply a premium end-use to the fuel where it is most needed. This appreciation of natural gas as a fuel is matched by a relatively free uninhibited gas market in Japan. This enabled the Japanese to graft yet another form of gas usage onto an existing gas industrial structure with a minimum of problems. Finally, the very nature of LNG deals, which involve project coordination and finance from multiple sources, is ideally suited to the blend of public utility, private corporate, governmental, banking and trading house activities characteristic of the Japanese.

Politically, with its long industrial contracts and its availability from non-OPEC sources, LNG constituted from the outset an ideal fuel from the viewpoint of security of supply. These factors supplemented the non-pollutant aspect of the fuel, making it doubly a premium fuel. This realization has enabled the Japanese to adapt to the rapidly changing natural gas market with an equanimity strange to Occidental eyes.

Who is right in the end: the Japanese, who are willing to pay a significant premium for a non-polluting fuel; the Europeans, who insist that the fuel should 'fit' into existing natural gas (and oil and coal) markets; or the Americans, who think of natural gas in terms of its impact on federal regulation of prices? If the criterion of 'right or wrong' is success, there is little doubt that the Japanese may have something to show the world.

Notes

1 This technology has been used at Compagnie Algérienne du Methane Liquide (CAMEL) Arzew and Kenai by Technip-Pritchard and Phillips, respectively.

2 This is not to say that the Japanese experience has been trouble free. Jilco, a Japanese consortium, ran into considerable disagreements with the Indonesians over the financing of export projects from the Arun and Badak fields. The differences amounted to roughly $500 million on five liquefaction plants – they were eventually covered by loans from the Japanese purchasers.

3 This is based on material presented by Drayton, 1981, pp. 33, 39–40. It might be added that this practice is being terminated by the SE Asian producers (Bintulu).

10

Trading Trends

Pipeline versus LNG – the future of international trade

While much of our focus has been on advances in LNG, similar progress has been made in pipeline technology. The completion in 1980 of the Sicilian segment of the Trans-Mediterranean Pipeline between North Africa and Italy was a landmark in developing pipeline technology. The line had crossed the Mediterranean at depths of up to 1,800 feet, a development thought impossible a few years before. A modern transmission line actually has little in common with the early all-welded lines in the American Southwest. Composed of advanced steel alloys, controlled by computers and the latest microprocessor technology, such a line is almost organic, its computer brain compensating for simultaneous events thousands of miles removed from each other. Thus a sudden drain on natural gas in Western Russia or Eastern United States will lead to increased compressor activity thousands of miles distant. In this manner, pipeline technology provides a serious rival to developments in LNG technology.

The degree to which nations opt for one form of transport over the other will profoundly shape the character of future international natural gas trade. In this respect as well, the Trans-Mediterranean line is of interest: the line is to be utilized by Algeria, an early leader in the LNG trade. Throughout the world – in Alaska, the Canadian West and Arctic, Northern Norway, even the Persian Gulf producers and Nigeria – the natural gas industry is weighing the two alternatives.

At the moment, pipeline trade is by far the more prevalent of the two modes. The projects in the OECD area are illustrated in Table 10.1 (excluding trade between Third World countries, which is insignificant – about 2 per cent of the total in 1980; and intra-Comecon trade, which accounted for another 17.6 per cent). As this table shows, pipeline transportation is a growth industry: the amounts of natural gas to be transported will increase by a factor of 80 per cent

Table 10.1 *Present and potential OECD pipeline gas trade (billion m^3 p.a.)*

Pipeline connection	Status	Throughput 1980	1985	1990	1995
Mexico–USA	Operational	2.9	3.1	6.2	10.0
Canada–USA	Operational	22.9	28.0	28.0	28.0
USA–USA (Alaska)	Possible	—	10.7	21.4	21.4
TOTAL into USA		25.8	41.8	55.6	59.4
Netherlands–Europe	Operational	47.5	40.0	31.0	13.0
Norway–Europe[a]	Operational	24.6	26.0	26.0	26.0
USSR–Europe	Operational	24.3	25.0	25.0	25.0
Algeria–Italy (Trans-Med)	Operational	—	12.5	18.0	18.0
Subtotal		96.4	103.5	100.0	82.0
USSR–Europe	Possible	—	—	40.0	40.0
Norway–Europe[b]	Under construction	—	—	7.0	7.0
Algeria–Europe (Trans-Med II)	Tentative	—	—	12.5	18.0
Algeria–Europe (Segamo)	Tentative	—	—	10.0	15.0
Subtotal		—	—	69.5	80.0
TOTAL into Europe		96.4	103.5	169.5	162.0
TOTAL USA and Europe		122.2	145.3	225.1	221.4

Sources: International Energy Agency, 1982, p. 120; 1980 figures from Cedigaz as reported to *Noroil*, October 1981, pp. 34–5.

Notes: a Includes pipelines to both the United Kingdom and Continental Europe. Estimates of production from fields now being produced indicate a decline to 22 billion m^3 in 1990 and 12 billion m^3 in 1995.

b Does not account for possible pipeline for production north of the 62nd parallel.

between 1980 and 1995. The 1985, 1990 and 1995 figures are characterized by a certain amount of insecurity, however; over half the anticipated increase to 1995 falls into the categories 'possible', 'tentative' or 'under construction'. The figures in Table 10.1 are in strong contrast to those in Table 9.1 for LNG. Thus, in 1980, LNG accounted for less than 20 per cent of all international trade in natural gas – some 32.1 billion m^3, of which 22.6 billion was confined to trades between Southeast Asia and Japan. Despite this, the International Energy Agency is very bullish on the future of LNG, predicting increases of some 340 per cent in the period 1985–95.

In Chapters 2 and 3, the interrelationship between industrial dynamics and intra-industry stability and bargaining relationships was stressed. The same interrelationship should also exist for international trade. Given that LNG and pipeline technologies are quite different, the obvious question becomes one of how LNG will affect international trading relationships. It has often been argued that LNG transport is the more flexible of the two forms. If so, would increased use of LNG end the national segmentation of natural gas markets – a segmentation noted particularly in Europe – and begin to integrate them into a single worldwide natural gas market? This question will provide the first focus of this chapter.

Interrelated with the impact of LNG transport on world markets is the question of technology selection. To what degree will pipeline gas or LNG be the preferred mode of transport in the future? At present, there is only one case where the two options have been considered at length with the result that one of them has actually been implemented: the Trans-Mediterranean line. What can the deliberations behind this choice tell us about the two transportation forms and their future impact on international trading relationships?

Does LNG possess more bargaining possibilities?

Behind the assumption that LNG allows more flexibility than pipelines (on the part of both the seller and the buyer of natural gas supplies) is the perception that LNG carriers can go anywhere, limited only by the number of accessible liquefaction plants and reception terminals: Indonesian LNG can be delivered in Boston (Everett Terminal); Algerian LNG can be delivered to Tokyo. If this were so, the national segmentation that has plagued international natural gas trade would gradually become a thing of the past and there would be a true world market for natural gas. But how valid is this supposition? To what extent is LNG a less constrained form of international trade than pipeline gas?

Table 10.2 illustrates briefly the differences in the nature of the constraints on LNG and pipeline trade. As can be seen, LNG trade may have considerably more limitations than would appear at first glance. To determine whether or not this is in fact the case, let us look more closely at each of these constraints.

Table 10.2 *Behavioural constraints: pipeline versus LNG trade*

Constraints	*Pipeline*	*LNG*
International variation in end-use value	Differences exist, but are small given proximate markets	Variations greater than in pipeline trade – not confined to proximate markets
Distance	Differing distances can potentially exclude buyers from bidding, sellers from selling natural gas (proximate)	As with pipeline trade, only less constrained
Possibility of rerouting supplies	Confined to existing pipeline network and degree of international cooperation	Possible within distance constraints (above)
Spot trades	Confined to use of existing pipelines	Confined to use of operational LNG carriers
Dedication of capital plants	Pipeline capacity can be allocated/reserved to differing customers in an international trade	Liquefaction lines dedicated to specific trades
Contract provisions	Dependable technology results in more binding contracts	Contracts looser. Hard to deliver or pay, owing to technology. Contracts typically cover risks re. liquefaction plants

International variation in end-use values

As noted in Chapter 9, variations in the end-use of imported natural gas by individual nations can involve widely differing national bids for international gas supplies. Cowhey (1982, p. 4) cites a Shell Oil conclusion that differing end-uses meant that Japan was justified in paying $1.00/MMBtu more than Western Europe for LNG, and Western Europe was similarly justified in paying roughly $1.00/MMBtu more than the United States. Following the logic of such marginal cost pricing, there is nothing to prevent widely varying bids for pipeline gas delivered to several national markets. In fact, this is seldom the case. Prices may vary, but the range is nothing like that mentioned by Cowhey for LNG. Various provisions, most probably MFN provisions, serve to keep proximate markets more or less in line regarding bidding for new gas supplies (this was the conclusion of Chapter 8 in the case of Europe). The situation is reversed for LNG trades; here, within the limits imposed by distance, it is possible to take advantage of differing end-use values. As can be seen from Table 9.2 in the previous chapter, Japan pays a higher price for LNG than does Europe, and Europe pays a higher price than the United States.

Distance

Distances work to constrain the effect of end-use values. Table 10.3 gives a rough MMBtu cost of LNG transport to various major ports.

Table 10.3 *Cost of LNG deliveries to major world ports from major production centres (LNG 125,000 m^3 carrier–fifth year) (US \$/MMBtu)*

	Production centre				
Port	*Cape of Good Hope*[a]	*Iran*	*Suez–Suez*[a]	*Arzew*	*Sumatra*
Rotterdam	1.529		1.188	0.353	2.092[b]
Philadelphia	1.757		1.457	0.666	2.32[b]
Yokohama		1.145		—	0.626
Los Angeles		2.005		—	1.443

Source: Office of Technology Assessment, pp. 58 and 97–8.
Notes: a Return voyage through Suez for the Cape of Good Hope route. Suez route through Suez both directions.
b Via Cape of Good Hope.

As can be seen from the table, distances are a critical factor for both potential exporters and importers. In only a few instances is end-use value unaffected by trading routes. For example, the Iran–Yokohama and Iran–Rotterdam (through Suez) routes involve almost identical costs. Here the Japanese have an advantage through their higher end-use value criteria, which enables them to outbid the Europeans. If the Suez option is not open, the Japanese stand in an even better position. They can outbid the Europeans not only through the end-use factor, but they have a 38¢/MMBtu transportation advantage as well. In other words, the Japanese can bid 38 cents under the European end-use value, and appropriate the resulting rent (some \$1.38/MMBtu). The same logic applies to competition between the United States West Coast and Japan for Sumatran gas. Here, the transport differential is 82¢/MMBtu and the difference in end-use values is $2.00/MMBtu. Even between Europe (Rotterdam) and the United States (Philadelphia) the same phenomenon recurs – a phenomenon that perhaps does much to explain why the Algerians were unhappy with the terms of the El Paso and Trunkline deals. Seen in reverse, LNG trade is also constrained by a distance factor that can either exaggerate or minimize end-use value. Algeria is really bound to Europe, Iran and Indonesia to Japan. This form of constraint is not unlike that of a large transmission line or transmission line net; the underlying logic is the same. The parameters, however, are different.

There are limits to this economically rational view of LNG constraints. Clearly, Algeria does trade with the United States, and Indonesia has expressed an interest in California. Over and above policy considerations, there are solid grounds for such alternative trades. The more that end-use value reflects premium uses alone, the quicker that value will fall as more and more non-premium uses are found for the natural gas involved (unless, as in Japan, government environmental restrictions make it relatively attractive for an electrical industry to use LNG as a fuel). In such a case, differentials between various markets disappear quickly and it becomes useful to have another outlet for one's natural gas. (This was particularly true of the Algerians in the late 1960s and early 1970s, when nobody really 'wanted' their gas.) From the European and Japanese point of view, the existence of alternative trades gives them a 'shadow price' to point in their negotiations with the Indonesians and the Algerians, and makes extraction of an oligopoly rent the more sure.

The situation is different when costs of transportation are similar and end-use values do not vary between the different markets. Here, LNG trade provides a degree of bargaining flexibility that can be used to great effect. This is particularly true of the North African–European trades where the existence of multiple national purchasers – Enagas (Spain), ENI (Italy), BGC (UK), Distrigaz (Belgium), and Gaz de France (France) – and of terminals at more or less uniform distances – La Spezia, Fos-sur-Mer, Barcelona in the Mediterranean; Le Havre, Montoir, Zeebrugge and Canvey in the Channel Approaches – has enabled Libya and Algeria to play a sort of natural gas 'ping-pong' in screwing LNG prices up to these markets. This is done in several ways: signing one contract more favourable than others and then insisting that it apply to existing contracts as well (through the use of MFN provisions if available); termination of LNG deliveries unless purchasers agree to a planned price increase; or a combination of the two tactics.

More recent use of these tactics (also mentioned in passing in the context of Chapter 9) has been a 1981 double cross between Belgium and France. In the Belgian case, Distrigaz desperately wanted its own supplies of natural gas. These were to be supplied by Algeria through a terminal at Zeebrugge (although early deliveries were to be made via the French Montoir terminal). Negotiations with the Algerians were escalated to include the Belgian government, which through its representative agreed to a price for Algerian LNG that was far out of line with the other prices then being paid for LNG on the Continent (see Table 10.4), especially with regard to the price escalation formula. By October 1981, the price for Belgian LNG was rumoured to be at $5.60/MMBtu cif level. Having struck this deal with Distrigaz, the Algerians recommenced their negotiations with the French, which they had been pursuing intermittently for two years. Aided by the French Socialist desire to come to terms with Algeria, they were able to sign a contract for $5.10/MMBtu fob, which translates to some $5.80/MMBtu cif at Montoir (*World Gas Report*, 8 February 1982, pp. 1, 11). Particularly interesting in this deal is the role of the French government, which is directly picking up the tab for 13 per cent of the

Table 10.4 *Distrigaz price versus Continental alternatives, 1 January 1980 (US $/MMBtu cif)*

Distrigaz	*Gaz de France*	*Distrigas (USA)*	*Enagas (Spain)*	*ENI (Italy)*
4.80	3.75	3.35	4.575	3.08

Source: World Gas Report, 14 April 1981, pp. 1–11.

fob price, and receiving in return Algerian guarantees for future industrial orders from France. This arrangement means that Gaz de France pays $4.437/MMBtu fob (some $5.137/MMBtu cif). There can be little doubt that this contract has caused some distress on the Continent – despite the fact that France can claim that the Gaz de France price is considerably lower than that of Belgium.

It is the Belgian deal that has particularly drawn fire from the other Continental companies (*World Gas Report*, 8 February 1982, pp. 1, 11). Their distress is undoubtedly due to the fact that this contract takes as its basis the cif price delivered to Belgium, thus depriving the Belgians of any transport rent that would otherwise be accumulated with an fob pricing basis. A further aggravation is that the most favoured nation clause of the contract binds Belgian conditions to conditions in other existing Algerian contracts. This means, in effect, that, for the Belgian conditions to obtain, the Algerians must bring pressure on other contracts. (This is obviously the case with the Gaz de France contract and perhaps with the Italian SNAM–Sonatrach Trans-Mediterranean Pipeline controversy.) A final major source of irritation is that the Belgian and the French contracts are for the first time directly tied to a crude oil MMBtu basis. (All previous contracts have, as an initial contract price, a price that is a percentage of an MMBtu equivalency price, which then is tied to an index – most commonly oil or oil products.)

Rerouting: the possibility of spot trading?

In addition to these constraints, there are some technological limitations to flexibility. There are, for example, very few spot cargoes of LNG, and hence no spot market as with crude oil and oil product prices in Rotterdam. Such trades do exist, but the LNG industry literature is silent about their extent. There is, however, increasing evidence that the trades mentioned in Chapter 9 are not the only frame for LNG commerce. Thus in 1977 two shiploads of Algerian LNG were delivered to Distrigas in Boston; as the loads did not fulfil the contracted amounts, these were 'made up' the following year. In 1980, 350,000 tons of LNG, approximately six shiploads, were delivered from Indonesia to Tokyo Electric on a 'spot' basis (Segal, 1980, p. 515). More interesting were reports of a single load of LNG from Indonesia to Distrigas on the US East Coast in the winter of 1980–1. (This was a contemplated deal to alleviate gas shortages in the American Northeast without drawing on the more convenient (more political) Algerian supplies; the reported cost – ca. $9.00–10.00/MMBtu – was outrageous, and it is doubtful whether the trade was implemented.) There are undoubtedly other deals. For example, Algerian cessation of contractual supplies to France in the period 1979–81 contingent on increased prices was 'made up' by a series of cargoes with independent billing (request for advance payment on prices demanded). There is little doubt that there were hard negotiations over these shipments, but to regard them as spot trades is dubious.

There is a good technological reason why spot deals are so few. A random cargo needs a random vessel. The problem here is that the vessel laid up ('mothballed') in a Norwegian fjord is extremely expensive to prepare for a single voyage or two, and, when ready, a vessel needs more or less constant employment.

There could also be problems with mixing cargoes from different sources. Because LNG is a mixture of a number of different liquefied gases, some loads may be heavier or lighter than others, or colder or warmer. Mixing different cargoes of different weights and different temperatures is liable to create a process called 'roll-over' in which the two volumes exchange places with much rumbling, boiling and escaping gases. ('Roll-over' is not discussed much in industry literature, but in at least one case it is thought to have nearly ruptured a tank, with the resultant danger of a large-scale LNG spillage and fire.) Until this problem is solved, it may be felt best to keep LNG cargoes from different sources separate, or to empty and 'neutralize' storage tanks before interchanging LNG cargoes – an expensive process.

Summa summarum, there is little to indicate that LNG to date has been used that much more for international spot trades than international transmission lines. Spot pipeline sales may take the form of small sales to particular customers not previously contracted for. Natural gas 'swaps' in Europe, for example, are liable to become even more common as the European transmission net nears completion and the various companies become more accustomed to such trades.

Dedication of liquefaction plants

Technological aspects limiting LNG flexibility are not confined to LNG carrier availability. There are indications that technical specifications for the unloading of LNG cargoes in different trades are different (Cowhey, 1982, p. 12). Then, too, the technology of liquefaction has enabled importers to specify contractually the liquefaction 'train' that is dedicated to them. (This specification is not included in the El Paso contract in this book.) Evidence is admittedly sketchy, but such terms are included in the Distrigas (Boston)–Sonatrach deal as noted by the Federal Energy Regulatory Commission (FERC):

> Distrigas was successful in obtaining priorities of production from lines 1, 2, 3, and 4 of Sonatrach's Skikda liquefaction plant so that if production falls below that required for fulfilling all Sonatrach LNG contracts from the plant, Distrigas customers in the United States will have a preferred position on the Skikda production that is available. . . . Distrigas shall have a 12% interest from lines 1, 2, and 3 and a 'first call' on production from line 4. (US Federal Energy Regulatory Commission, 1977b, pp. 3–4)

In this sense, rights to certain liquefaction lines in a liquefaction plant might be compared to transit rights on a common carrier pipeline such as those in Europe. In both instances, the capacity available for the individual trade is limited by technology – a limit that is expressed in terms of contractual obligations.

Contractual constraints

Curiously enough, there seems to be a more general lack of contractual constraints endemic to LNG trade than to pipeline trade. This is no doubt due to the differing transport technologies. A careful analysis of the two contracts reproduced in Appendix IV is revealing in this regard. First, as might be expected, the El Paso–Sonatrach LNG contract is very specific about exactly what kind of liquefaction facilities/berthing facilities/port facilities will be required for LNG handling and delivery; this specificity is not present in the Argentine–Bolivia pipeline contract. Another difference directly attributable to the dissimilarities in technologies is that much is made of the problem of the measurement of the LNG – a substance which, since it is transported and stored as −162°C, is always boiling or sloshing and has a peculiar effect on measuring instruments.

Perhaps most interesting, though, is the difference in language when buyer and seller obligations are specified. In clause 12 of the Argentine–Bolivia contract, Gas del Estado, the Argentine purchaser, has certain rights. To the degree that there is an interruption in supply or a diminution in the quantities supplied of more than 30 per cent for over seventy-two hours (three days), the purchaser can demand to have the missing balance filled either in the year that the supply failure occurred or in the following year. Failing this satisfaction, Gas del Estado can demand a specified cash penalty of 18.5 US cents for each cubic meter not supplied by the Bolivians. If the delivered volumes fall below those specified by the contract by 25 per cent or more for a period of 365 consecutive days, Gas del Estado can terminate all acceptance of deliveries without prejudice to the sums of money outstanding. These provisions give considerable protection to the purchaser. They are also in considerable contrast to the provisions of the El Paso contract.

The El Paso–Sonatrach contract has few if any clauses that protect the purchaser. There are no provisions beyond the general ones stating the amounts to be delivered (perhaps an implicit form of obligation), which penalize Sonatrach for non-delivery. Should El Paso pay for amounts that are not delivered, these amounts can be made up at later points in time. But even this provision is very unspecific. Rather, the contract protects Sonatrach. The purchaser is subject to 'take or pay' provisions: all the LNG offered by Sonatrach must be purchased by El Paso.

One can wonder at this lack of security in an international contract. Although the blame for the absence of penalty clauses covering non-delivery of LNG might be laid at the doors of El Paso, it is more likely that the unpredictable technology of liquefaction renders such non-delivery meaningless. Nor would the 1969 El Paso contract seem to be particularly unique, although evidence is hard to come by here. (Even the Federal Power Commission, the competent US authority in passing on LNG trades, has complained of not being able to acquire copies of the LNG contracts concerned – on one occasion even going on record as meeting a 'blank wall' on this point; US Federal Power Commission, 1972a, p. 17.)[1]

The various declarations by the US regulatory agencies constitute the best available source of evidence. These are hardly grounds for confidence in the purchaser's rights to assured delivery of the contracted amounts of natural gas. In 1977, the FERC noted, regarding an Algerian–Distrigas contract, an absence of obligation on the part of the sellers to 'make up' defaulted deliveries. FERC was particularly concerned about the nature of *force majeure* in the contract:

> *Force majeure* provisions include events commonly found in such clauses, but also other events such as (1) severe accident with respect to operations or facilities from the Algerian gas fields through the exit pipeline from the Everett regasification terminal, and (2) 'any act or failure to act on any public authority of Algeria or any other country entailing suspension of operations'. . . . *Given Sonatrach's relationship with the Algerian government, this amounts to a unilateral right to abrogate.* (US Federal Energy Regulatory Commission, 1977b, p. 4; emphasis added)

Elsewhere, lack of purchaser rights is conspicuous in their absence from US regulatory authorities' reference to penalty/non-delivery clauses; thus for El Paso II:

> [El Paso] is obligated to *take or pay for the annual contract quantity to the extent that such quantity is tendered for sale by Sonatrach.* Any deficiencies in take may be made up at any time during the contract term as long as Atlantic first satisfies its obligation to take the annual contract quantity before taking make-up gas in any given year. (US Federal Energy Regulatory Commission, 1977a, p. 43)

The reasons for the inability of US regulatory authorities to deny the validity of Algerian contracts as a basis for US imports of LNG are in part their lack of legal competency in this particular field (although they do have authority regarding price escalation clauses), and in part a feeling that Algerian good behaviour will be motivated by economic incentives. Thus, in passing on the El Paso agreement, the FPC stated:

> While it [the Department of State] recognized that there can be no absolute guarantee that there will be no interruption of supply, it felt that security of supply was enhanced by Sonatrach's financial

> stake in the project, and by the project being a central part of its economic development effort and a source of foreign exchange. We defer to the Departments of State and Defense in respect of these matters of foreign policy and national defense and accept their findings. (US Federal Power Commission, 1972b, p. 9)

Similarly, in passing the Distrigas agreement, the FERC asserted:

> The fact of the matter is that Sonatrach is building facilities to process and sell gas, that it is not to its advantage deliberately to delay deliveries of the additional volumes, and it is not to its advantage to intentionally withhold deliveries at a price which is now competitive and will remain so due to price adjuster indices tied to other hydrocarbons. (US Federal Energy Regulatory Commission, 1977b, p. 10)

These statements seem ironic in light of the fate of the two largest US LNG import projects from Algeria: El Paso I was terminated and Trunkline has only recently commenced after much controversy.

Although the bulk of the evidence here is American, it is unlikely, given the nature of gas contracts, that the Algerian–European contracts are much better from the buyer's point of view than the ones cited. This suspicion is confirmed generally by Algerian behaviour towards France and other Algerian customers where supply cut-offs and other measures have been used to 'up' prices.

To summarize the analysis of international contracts briefly, one is left with an impression that international pipeline transport involves a tried and true technology, one that is reflected in good contractual custom; trade in LNG, in contrast, is more 'wheeling-dealing', freed from a dependence on pipelines and detailed legal contractual precedents. As we shall note in the next chapter, however, this situation is changing rapidly.

Concluding this discussion, there seems little doubt that constraints do exist in the LNG trade. Exporters are tied to importers through end-use values, distance variations, contractual provisions and problems of technology. There are exceptions to this rule, and the constraints are not identical to those for international pipeline trade. The exceptions, of course, are Algeria and Libya, with their ability to switch cargoes and obtain higher prices on the European Continent. It should be noted in this context that, despite their various 'suspensions' and 'terminations', the Libyans and Algerians none the less did not experience a year in the period 1974–80 when revenues from natural gas and natural gas liquids exports declined by more than 7 per cent. This includes the Algerian cancellation of the El Paso contract and the suspension of the Trunkline deal. In the period 1974–80, Algerian gross receipts rose from $41.3 million to $940.9 million and Libyan gross receipts rose from $56.9 million to $627.9 million. Despite this success (which has effectively destroyed Libyan credibility and hurt Algerian reputations), it is well to note that the bargaining context for these two countries was exceptional, and currently does not exist elsewhere in the world.

LNG versus pipelines – the problem of choice

Industrial literature on the choices between LNG and pipeline transportation technologies is seldom satisfying. Comparisons are made in isolation, well removed from the problems of marketing and the particular advantages that one form of transport may have over the other form given very particular situations. The Bonfiglioli and Cima (1980) analysis, widely quoted and cited in the industry literature, while excellent in many respects, resolves the 'pros and cons' of LNG vs. pipeline in terms of only two factors: the distances that the natural gas is to be transported, and the cost of feed gas for both pipeline and LNG liquefaction plant. This is illustrated in Figure 10.1. Depending on the type of pipeline used, the break-even point for LNG is in the

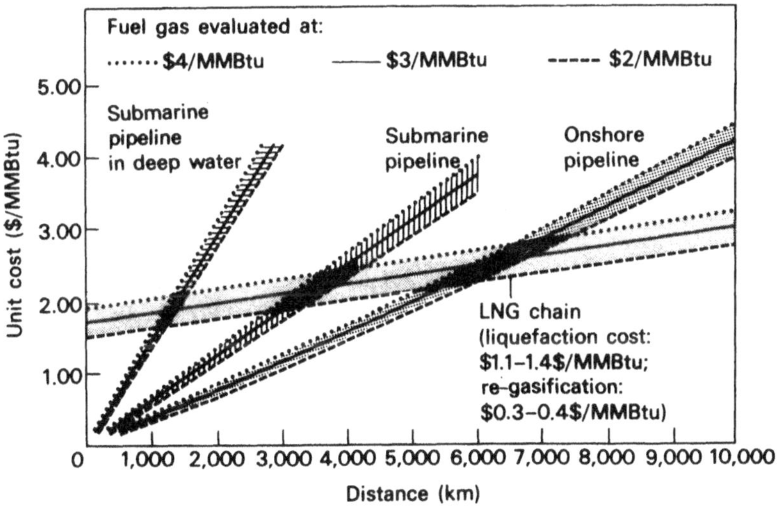

Figure 10.1 *Comparative costs of natural gas transport (annual quantities of 18 billion m^3)*

Source: Bonfiglioli and Cima, 1980

1,000 km range in relation to the deep water submarine pipeline alternative, 3,000–4,000 km in relation to the submarine pipeline alternative or 6,000–7,000 km for onshore lines.

Yet there are problems to such a presentation. To begin with, the quantities assumed are high: 18 billion m^3 per annum. Smaller quantities could very well serve to change the interrelationship considerably. Certain projects, as we have noted, are deemed too small for pipelines. These then must qualify for LNG, and hence LNG must have some advantages for small quantities of natural gas over longer distances not included in the Bonfiglioli and Cima figure. Other factors, on the other hand, argue more against LNG. I have been informed through private correspondence with the authors of the study illustrated in Figure 10.1 that the performance factor is 85 per cent. As noted in Chapter 9, this is considerably in excess of the actual operating average for most projects.[2] LNG costs will also vary more than is indicated in the figure – especially if harbour facilities and other necessary infrastructure are taken into account.

Theory is one thing; operational experience something else again. The problems associated with LNG have been discussed in Chapter 9. How have these affected the actual earnings of the LNG exporting nations? Table 10.5 gives the Libyan and Algerian revenues from

Table 10.5 *Algerian and Libyan export receipts: natural gas liquids and liquefied natural gas, 1966–1980 (US $m.)*

Year	*Algeria*	*Libya*
1966	7.6	—
1967	0.4	—
1968	30.0	—
1969	35.3	—
1970	35.16	0.0
1971	26.9	7.07
1972	30.5	37.7
1973	51.5	56.95
1974	41.29	80.16
1975	149.6	136.53
1976	185.6	165.13
1977	172.4	156.3
1978	275.51	218.07
1979	873.04	351.01
1980	940.85	627.95
Totals	2,855.68[a]	1,836.86[a]

Source: United Nations, 1967–81.
Note: a May differ from the totals in the columns owing to rounding.

both LNG and natural gas liquids for the period 1966–80. Given the costs of liquefaction plants and other facilities, one might say that the experience has not been a very profitable one. In the first place, these are export earnings: the actual costs of capital and operations are not deducted (and these run to hundreds of millions, if not billions, of dollars). Second, the figures also include the export of natural gas liquids and are thus biased upwards. Third, these figures do not reflect time value. Discounting the cash flows for both the Libyans and Algerians by 8 per cent, the net present values of these totals are $714.6 million and $940 million, respectively. Subtracting operation and capital costs, these net present values could well be negative. It is highly likely, therefore, that the Algerians would have earned more, at less risk, by depositing $700 million in 1966 in a bank savings account. This would have had the added advantage of allowing the Algerians to retain their natural gas in the ground for future sales. It is little wonder that the Algerians have been so bitter and so militant.

An actual case study, that of the Trans-Mediterranean pipeline, illustrates the sterility of reducing LNG versus pipeline to a question of relative distances. Table 10.6 illustrates the reasons why the Trans-Med line was substituted for an original LNG agreement between the Algerians and the Italians. The table represents the different routes that LNG and pipeline trade would take, the different capital costs of the two options, the different markets that could be serviced, scheduling of start-up, tariffs for transportation and the nature of additional capacity. It can be seen that, in this case, pipeline factors work together to allow for lower costs of service (due to earlier start-up), better marketing flexibility, and greater, cheaper additional capacity. It should be noted that in this table the assumptions are only approximate. This is particularly true of the LNG option, which assumes an entirely new LNG project with new facilities at each stage – see notes a and b. Although the basis of the *Wärme Gas* figures is not given, they are in fact ENI figures. It should also be noted that the Trans-Mediterranean Pipeline crosses Tunisian territory to Cape Bon; thus, a disadvantage to this mode of transport, one not mentioned in the table, is that it is dependent on Tunisian good will.

What is particularly interesting about the table is that it shows that, in terms of capital costs, the pipeline route is at a disadvantage compared with the LNG alternative. Despite this, it is also clearly to the advantage of both the Algerians and the Italians that the pipeline be constructed. The Algerians save about $2.5 billion in capital costs by not opting for the LNG alternative. The Italians must shoulder the major cost burden instead (paying more than $3.2 billion additionally), but obtain the following advantages: (1) they can open up Southern Italy to new supplies of natural gas, which could be a

Table 10.6 *Advantages and disadvantages of pipelines and LNG as they affected the Trans-Mediterranean Pipeline Decision*

	LNG route	*Pipeline route*
Routes	Algeria–N. Italy	Algeria–Tunisia–Sicilian Straits–Sicily–N. Italy
Capital costs[a]	$4,220m.	$4.860m
Of which:		
Algerians	$3,498m.	$898m.
Italians	$722m.	$3,962m.
Markets served	Existing N. and C. Italian markets	Existing markets plus new areas in Sicily and S. Italy
Commencement of operations	Dependent on all facilities being in place operational	Can commence after completion of Sicilian Channel segment. Sales in S. Italy can help finance 881 miles of line in Italy 'piggy-backing'
Representative tariffs for transportation	$2.874/MMBtu	$1.328/MMBtu
Of which:		
Algerian	$2.308/MMBtu[b]	$0.268/MMBtu[b]
Italian	$0.566/MMBtu[b]	$1.06/MMBtu[c]
Additional capacity	Additional liquefaction trains (at $100m. per copy), and ships (at $100–125 m. per copy)	Can take up to 5.5 billion m^3/year more with additional compressors

Notes: a Author's capital cost assumptions: $1.2 billion for liquefaction facilities, $1.2 billion for Algerian port facilities, $898 million for gathering and pipelines in Algeria, $400 million for carriers (two Algerian, two Italian) and $522 million for Italian reception terminal. Trans-Med figures are from industrial sources.

b LNG figures based on 10 per cent operation costs/$3.50 MMBtu feed, 15 per cent IRR to equity, 1:4 equity/loan gearing, 40 per cent taxes. (Same assumptions apply to field gathering and pipeline in Algeria.)

c Wärme Gas International, June 1982, p. 388.

profitable exercise; (2) operations can commence as soon as the North African and Sicilian Channel segments are completed – this means earlier sales, and has the further advantage that these segments can pay for the 881 miles of line in Italy, thus reducing 'exposure' to below that of LNG and allowing for lower tariffs and a greater range of marketing opportunities; (3) the line is already characterized by additional capacity – it is reputed to be able to take up to 18 billion m^3 per year, while the current contract calls for some 12 billion. To add additional capacity to the LNG trade is more costly, involving the building of additional liquefaction 'trains' (at about $100 million apiece) and new carriers (about $150–200 million per copy).

The upshot of this mutual advantage is that there is a considerable difference between the costs of transport: the pipeline is an estimated $1.50/MMBtu cheaper than LNG transport. The Italians and Algerians could be expected to differ on how this pipeline rent should be divided, and they did. An additional complicating factor was the LNG shadow price. Given the $5.12/MMBtu fob price for LNG destined for the Belgian markets, the price of LNG landed in La Spezia or some other Northern Italian port would be in the neighbourhood of $5.50/MMBtu. Assuming that the price of Algerian natural gas delivered to Northern Italy would be around the same level, this would put the price of natural gas in Southern Italy in a much lower category. This cheaper gas, the Algerians argued, was what the Italians were really interested in. The logical price for natural gas at the Algerian frontier was therefore based on the end-use value in Southern Italy, which could arguably be higher even than $5.50/MMBtu.

These differences, reinforced by mutual suspicion, led to the non-utilization of the Trans-Mediterranean Pipeline for over a year in a celebrated row between ENI and Sonatrach. The conflict was ended only by the intervention of the Italian government and the imposition of a compromise (which essentially gave in to the Algerian demands). The final price at the Algerian border was set at $4.41/MMBtu, which translated to $5.13/MMBtu landed in Italy. ENI has yet to be reconciled to this Solomonic decision on the part of the Italian government. The Algerians have every right to be satisfied; by my calculations, they have gained about $1.43/MMBtu more with this compromise than they would have received under an LNG deal as they struck with the Belgians. This amounts to some $635 million per year over two decades at 12 billion m^3/year. Despite being upset at the loss of an advantageous position in Southern Italy, ENI also has some small reason to be satisfied. It is unlikely that the price of Algerian natural gas arriving in Minerbo will exceed the approximately $5.83–$6.00 that re-gasified LNG might have cost at the same point (allowing for costs of delivery from La Spezia to Minerbo and some $0.40/MMBtu re-gasification charges). This is small recompense for the loss of some $600 million per annum, however, and ENI is already claiming that the deal is only an interim one.

While the conflictual nature of ENI and Sonatrach relations is of interest, the main point to be made about the Trans-Mediterranean line is that it is situation specific. The particular reasons for opting for one form of transportation over the other have little to do with general considerations. The important factors here were the Southern Italian markets, the ability of ENI to finance the line through sales of natural gas from the initial segments after these were laid, the bad experience the Algerians have had with LNG, and so on. Distances in this 'game' really are not worth too much. Clearly, in other areas of the world with other circumstances, the choice could well be the reverse. It is local particularities that will tell in each case.

Conclusion

This chapter commenced with two interrelated questions. First, would the introduction of LNG significantly change the manner in which natural gas is traded in the world today? And second, what criteria lie behind various national decisions for one form of transportation or the other? The answers that emerged have been instructive.

Although LNG opens the possibility for new trades between previously unconnected producers and domestic markets, as is the case with Japan and Southeast Asia and Alaska, this trade is dominated by the same constraints as exist elsewhere in the world of natural gas and is far from accentuating a trend towards free unconstrained trade. The image of LNG in the future is not that of a tanker cruising the Indian Ocean with a spot cargo up for bidding among consuming nations. Rather, it is one of LNG carriers dedicated to specific trades which have established routes and schedules that they will keep throughout their operational lives. LNG is a new mode of gas transport, not a revolution in natural gas bargaining relationships. It does not signify the emergence of one international natural gas market. This market will remain as segmented as it is today.

Why do nations choose LNG over pipeline trade? This choice, we maintained, is too complex a subject to be covered in terms of one or two variables (as in Figure 10.1) that are then applied generally. With regard to the Trans-Mediterranean Pipeline, the choice of technology presented a rare opportunity for both parties to try to maximize their 'take'. The Algerians could maximize their take through escaping much of the investment burden involved in liquefaction plant facilities and LNG carrier ownership. The Italians could do so by letting the various completed segments of the pipeline pay for the portions still under construction. As soon as the line was completed across the Channel, the natural gas could begin to be marketed in areas new to natural gas. Revenues from these additional markets could help to pay for the transmission line through Italy to Minerbo, a transmission line that would also open up additional markets. The dispute between the parties over how the economic rent from the pipeline would be divided appears to have been won by the Algerians, at least preliminarily. The point here is that, without these unique characteristics of the joint Italian–Algerian decision, it would be difficult to understand the rationale behind their choice of the technology to be utilized.

Notes

1 . . . [T]he basic contracts were withheld from the record because they were considered "confidential".'
2 If one assumes liquefaction plants perform as in the past, the actual performance factor can mean a reduction in cash flows measured in net present value terms by as much as 35 per cent (depending on the circumstances).

11

The Future

The coming challenge

The nineteenth-century American natural gas industry was characterized by a 'boom and bust' cycle. Natural gas, when discovered, was used locally. When the reservoir was depleted, the local industry collapsed as there was no way to tap gas resources at a further distance. Today's industry is not threatened in the same manner; its limitations are not technical. As we have seen, the technology exists to unite markets with sources of supply thousands of miles away. Yet today's industry is challenged by the possibility of another type of 'boom–bust' syndrome. The roots of this challenge exist in the establishment of stable international trading regimes.

Industrial countries have been drawing on proximate reserves of natural gas in a pattern not unlike that of the local nineteenth-century American enterprise. In the United States, for example, the East relied on Appalachian natural gas in the beginning; as these supplies began to be depleted, they switched to sources in Texas, Louisiana and Oklahoma. The United Kingdom drew first on proximate natural gas reserves from the British Southern Basin fields. These are now being supplemented by increasingly distant (and more costly) natural gas from the Northern North Sea and the Irish Sea. On the Continent, the Groningen field's location in relation to major potential markets was a gasman's dream: very close to the major centres of European population and industry. This gas is about halfway depleted presently and the search is on for newer, more expensive gas from further away: Ekofisk, Frigg and Statfjord in Norway; Hassi R'Mel in Algeria; Orenburg and Urengoy in the Soviet Union. Even in the Soviet Union the trend is the same. Here the centre of production has 'zigged' from the proximate finds in the Baku and Volga Urals areas to the Central Asian fields, then 'zagged' from this area to the super-giant fields in Western Siberia.

Given the relative concentration of most national markets, this trend away from proximate resources leads logically to more and more reliance on international trade, which is thus growing in importance for the national natural gas markets involved. Curiously, this international trade is nearly as segmented as national markets. Much as in the days when the Pope divided the known world between Spain and Portugal, various countries have carved out sources of supply for themselves: Europe has acquired Soviet Union, Norway and Algeria; the American orbit thus far includes Canada and Mexico (with minimal amounts coming from Algeria); the Japanese are re-creating the natural gas equivalent of the Greater East Asia Coprosperity Sphere in the Pacific basin. The interstices of this segmentation are subject to competitive bidding: the Middle East (Europe versus Japan), and the Canadian Arctic and Nigeria (the United States and Europe).

Despite this segmentation, the stable marketing relations of a self-sufficient national market are nonetheless replaced by the somewhat less stable cooperative relations between two or more sovereign political entities. This trend raises several important issues.

First, how will natural gas be utilized in the future? Copious reserves of natural gas are not enough. There has to be agreement among the parties concerned that the reserves are to be used specifically for export. The alternatives are for the producing country to store the gas for better times, to utilize it domestically, or to utilize it for its industrial value, in terms of either its thermal or petrochemical properties. Should exporting be chosen as a viable option, both exporter and importer must select their respective partners with care.

Second, international relations are complicated by the quickly rising costs of producing, transporting and marketing natural gas. After paying an estimated $5.50 for Statfjord gas in 1984, European transmission companies are confronted with Norwegian demands for a political security premium. While this is designed to extract the last ounce of blood from the Continental transmission companies, it is also necessary to cover the enormous costs of development of the various northern Norwegian natural gas fields, particularly the huge Troll field. The cost of a middling LNG project now runs in the neighbourhood of $4–6 billion – without the cost of producing and gathering facilities. The cost of incremental Siberian gas is also increasing, although the Soviet planners, tied to their notions of average costs, may not fully realize it. This is new natural gas. Old natural gas is also rising in price because of the increased marginal costs of extraction. Coal gasification technology remains an alternative for the future. Capital costs here are growing apace with the cost of natural gas, and costs of coal feed to these plants is increasing perhaps even faster.

Third, although underemphasized in the context of this book, natural gas trade will increasingly become the hostage of geopolitics in general, and of great and superpower rivalries in particular. Several US–USSR LNG deals have foundered on the rocks of mutual distrust and improving/worsening relations between the two superpowers, most notable among these being the North Star Project. Chapters 6 and 8 have touched on the Yamburg pipeline project as seen by Europeans and Americans. There was a day when the active natural gas community involved in the buying and selling of natural gas within Western Europe comprised about thirty men. This group wielded incredible power over an industry that makes up 15–20 per cent of Western European energy consumption. (Nowhere else in the energy field is there a similar phenomenon.) The power of the industrial brokers has been curtailed by an expanding policy community interested in the international political implications of natural gas trade in general and East–West trade in particular. Protestations by the very able Soviet Soyuzgaz personnel that they and their Western counterparts are in the same boat (and, of course, that there should be no fears of termination of Soviet deliveries) are to little avail when the hard-eyed politico-military strategists at the Pentagon and in the Kremlin look at the consequences of such trade.

These trends have definite implications for market stability or instability. Throughout this volume we have reviewed various national and corporate responses to instability. Each national market has its own set of remedies for the problems faced by the gas industry. Even with these single national markets, reliance on the 'word of Princes' has been chancy. The future challenge involves not one national market with a single sovereign entity; rather it will involve multiple sovereign powers, multiple problems, and multiple legal jurisdictions. Nor is the international challenge to be confronted only by governments. A vertically integrated firm faces different problems when it is integrated across national frontiers than when it is integrated within a single country or national market. A national grouping of companies becomes an international group, each member of which is subject to the laws of the country within which it is domiciled, as well as to the differing industrial practices within the country. Contractual relations between the corporations of the group are no longer confined to a single legal jurisdiction but are interpreted by two or more legal jurisdictions. To say that this creates problems is to understate things. In Chapter 9 we noted how the American system of regulation led to misunderstandings with the Algerian authorities. In Chapter 8 we saw how a natural gas contract signed by Deutsch BP with BP Norway was acknowledged in Norway and disavowed by the German antitrust authorities. In summary, the replacement of a single sovereign authority by several sovereign authorities, of one law of contract by multiple laws of contract, of a single national industrial practice by an array of practices, can work to destabilize trade even when the trade occurs with the best intentions of all parties concerned. (And this last condition is something of a rarity in the world of international natural gas trade.)

The size distribution of reserves

Jensen Associates Inc., in a recent study for the US Office of Technology Assessment, estimated that proved gas reserves at the end of 1978 amounted to 72,406 billion m^3 (Office of Technology Assessment, 1980, p. 45). There is little doubt that this is a low figure and that much more natural gas will be proved up in the future. Tiratsoo, for example, reports that there are no fewer than 600 sedimentary basins that hold promise of natural gas. Of these, 200 have been inaccessible to date, 240 have been lightly explored with the discovery of some hydrocarbons, and 160 basins have been found to hold commercial hydrocarbons (Tiratsoo, 1979, p. 330). Furthermore, giant gas fields – size is important for their development – have been discovered at a fairly rapid rate. Only seventy-two of the ninety-six 'giants' in Table 2.1 were known in 1972. The others, containing 33 per cent of the total for giant fields were found between 1972 and 1979. This trend is likely to continue. Furthermore, it is also contended by many geophysicists that natural gas occurs in older and more deeply buried rocks. These sedimentary layers will increasingly be the target of exploration wells as existing hydrocarbons resources become more and more scarce (Tiratsoo, 1979, p. 310). Some of these more expensive drilling targets are already known to have much promise – the Khuff (Permian) sediments in the Middle East, for example, only preliminarily investigated to date.

Yet discovery of reserves should not be confused with their availability. Some natural gas is reinjected to repressurize oil reservoirs. Other natural gas is simply too inaccessible to major markets to be utilized. Still other gas can be held in reserve by the nations in which it is found – deferred to a more appropriate time for marketing. Of the gas used commercially, much might be reserved for internal domestic usage. The gas that is exportable (the residual gas) has in many instances already been committed to an export market (as with Indonesia and Algeria). In only a few instances, despite the plentitude of resources, is natural gas available in sufficient quantities for new export projects to the major markets of the world. This exportable surplus is estimated to be in the region of 23,000 billion m^3 or 31.8 per cent of the total proved reserves (Office of Technology

Assessment, 1980, p. 46). Of this, some 78 per cent is located in three countries: the Soviet Union, Iran and Algeria; and of these nations, Jensen Associates estimate that Algeria has committed all of its 254 billion m^3 to international trade. The other two nations possess exportable gas in quantities close to 18,000 billion m^3. The location of these reserves is shown in Figure 11.1. While the figure ignores the small fry, which account for 1,252 billion m^3, and concentrates on the major exporters, the impression given is that of a future international natural gas market with concentration first and foremost in the Soviet Union (with over half the available resources), followed by the Middle East (Iran and Saudi Arabia), Nigeria, Mexico and the Pacific (Indonesia and Australia). Given Soviet preponderance it seems likely that natural gas trade will increasingly be tied to East–West relations.

This size distribution of reserves leads automatically to what might be termed the 'sourcing problem'. Those nations with well-developed national markets are precisely those with dwindling reserves. As noted in Chapter 4, US reserves, however they are calculated, are limited (some 10–15 years of production from proved reserves) – especially if the Alaskan Natural Gas Transportation System fails. Alternatives to increased reliance on imported gas, whether from Mexico, Canada, Indonesia, Trinidad or Argentina, involve relatively small, highly expensive volumes. The same might be said for the British and Continental markets (some 18 and 19 years, respectively, from proved reserves) – although in this case the sourcing problem is perhaps less serious. As Figure 11.1 makes clear, there are no two ways about it: East–West relations will increasingly be tied to natural gas. This reliance should give both national governments and their respective industries pause while mutually defining national depletion policies, policies that will be hard to define in the best of circumstances.

The problems of cost

International conferences are often more interesting for the corridor 'politicking' than for the substantive materials presented in scholarly or industrial papers. The 1982 World Gas Association conference was no exception to this rule. The issues were those of cost and price; the spokesmen were the Algerians and Ruhrgas. The Algerians came to the conference complaining of the high costs of natural gas field development and export projects. The Algerian Valorization Hydrocarbon Development Plan (VALHYD) envisages the expenditure of $34 billion (in 1976 dollars) in the period 1976–2005. As a result, the

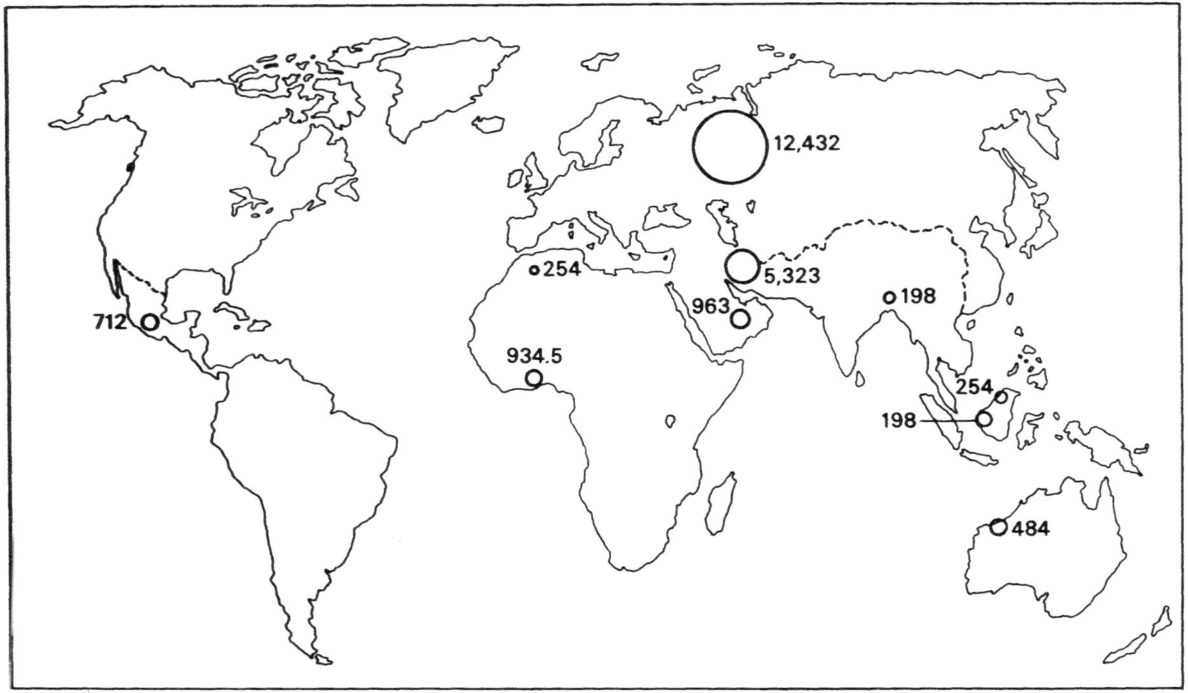

Figure 11.1 *Major uncommitted gas reserves exportable to world markets (billion m^3)*
Source: Office of Technology Assessment, 1980, p. 52.

Algerians said they had no alternative but to charge the prices they did. Journalists were informed by senior Sonatrach executives that the production costs alone of natural gas had risen to over $3.00/MMBtu. The Nigerians echoed these sentiments. Current estimates for Nigerian Bonny LNG were put even higher, at $4.50/MMBtu. In both instances the price excluded financing costs and did not allow for any capture of economic rent. Capital costs of gas export facilities had risen between three and four times since the inception of the VALHYD programme.

In the opposite corner, Ruhrgas's Karl Liesen was arguing for the linkage of natural gas prices to coal and not to fuel oil or crude. Liesen's points were simple and direct:

> First of all natural gas purchased on parity with crude oil will, by definition, stand no chance on the energy market.
>
> Secondly, an importer purchasing gas exclusively on the basis of the price level of fuel oil will lose the power station market and part of the other industrial users.
>
> Thirdly, if natural gas exporters are unwilling or unable to supply natural gas at a price allowing gas to compete against coal, and in the future against nuclear energy, the share of natural gas in overall supplies to the industrialized nations will decline.
>
> Fourthly, natural gas projects where the costs are so high that a reaction to the market situation described becomes impossible will have to be postponed. The same will apply to projects where the security of supplies or the security of contracts is inadequate in the eyes of the users. (*Gas 2000*, I, no. 0, p. 7)

Dr Liesen's comments, curiously enough, broke with a decade of European pricing experience, which had tended to tie prices of natural gas to the prices of oil products. The reasons for this break are clear: oil product prices had risen to such heights that future expansion of natural gas markets resulting in lower end-use values would not be possible unless natural gas prices were tied to those of a cheaper fuel, in this case coal.

The natural gas market, then, is caught between the pincers of higher costs and lower end-use values. Nowhere is this more evident than in the debate over contractual prices. Table 11.1 illustrates the bases of most of the international trade contracts. As can be seen, prices in international trade are extremely high, particularly on a cif basis. The reader should note the increasing practice of linking contractual prices to crude oil prices. This trend, perhaps the most serious threat to the continued expansion of natural gas markets in Europe and the United States, is a result of OPEC gas policies and

Table 11.1 *Base price and escalation factors in international gas trade*

		Present world gas prices ($/MBtu)					Indexation terms		
Importer	Exporter	Vol. (billion m^3/year)	Fob	Cif[a]	As of	Period	Alpha factor	Index	How applied
USA	Canada	28.3	—	4.94[b]	1.82	Irreg.	1.00	Crude imports to Toronto/US alt. fuels price	
USA/Border Gas	Mex/PEMEX	3.1	—	4.94[c]	1.82	Qtrly	1.00	Arith. avg of 5 crudes: Saharan, Arab Lt, Tia Juana Med. UK Forties, Isthmus	% change
UK/British Gas[d]	Alg/Sonatrach	1.1	[b]		Expired				
France/Gaz de France	Alg/Sonatrach	9.1	5.12[be]		2.82	Qtrly		Eight crude oils	Absolute Btu equiv.
Belgium/Distrigaz[f]	Alg/Sonatrach	2.5[g]	5.12[bh]		1.82		.50	Crudes exported by LNG producers	Absolute Btu equiv.
							.50	Crudes imported by Belgium	
Italy/SNAM[d]	Alg/Sonatrach	12.5	Under negotiation						
Spain/Enagas	Alg/Sonatrach	4.5	4.575		1.81				
US/El Paso	Alg/Sonatrach	9.8			Suspended				
US/Distrigas	Alg/Sonatrach	1.4	3.94	5.82[b]	1.82	1/2 yrly	.5/.5	Prices of No.2/No. 6 oil in New York Harbor	% change
US/Trunkline[af]	Alg/Sonatrach	4.5	3.94[b]	7.20	1.82	1/2 yrly	.5/.5	Prices of No. 2/No. 6 oil in New York Harbor	% change
Italy/SNAM	Libya	2.4			Suspended				
Spain/Enagas	Libya	1.2	3.48[b]		1.80				
Switzerland	Neth/Gasunie	0.5		4.45[b]	10.81				
Germany	Neth/Gasunie	23.9		4.45	10.81			Low-sulphur fuel oil	
France/GdF	Neth/Gasunie	11.6		4.45	10.81		.95	(1%) with 5-month lag	
Belgium/Distrigaz	Neth/Gasunie	9.8		4.45	10.81				
Italy/SNAM	Neth/Gasunie	6.7		4.45	10.81				
UK	Norway	9.4		[b]	7.81				
Neth/Gasunie	Norway[i]	2.6		4.25[j]	7.81				
Germany	Norway[i]	9.7		4.25	7.81				

Table 11.1 *Base price and escalation factors in international gas trade — continued*

		Present world gas prices ($/MBtu)				Indexation terms			
Importer	*Exporter*	*Vol. (billion m^3/year)*	*Fob*	*Cif*[a]	*As of*	*Period*	*Alpha Factor*	*Index*	*How applied*
France GdF	*Norway*[i]	*2.2*		*4.25*	*7.81*				
Belgium-Distrigaz	*Norway*[i]	*2.1*		*4.25*	*7.81*				
Germany et al.	*Norway/Statfjord*	*6.7*		*5.50*[j]	*7.81*				
Austria	*USSR*[k]	*3.0–4.1*							
France GdF	*USSR*[k]	*4.0*							
Germany	*USSR*[k]	*11.0*				*Yearly*	*.96*	*.64 low-sulphur fuel oil, .22 heating oil, .10 wage developments*	
Italy/SNAM	*USSR*[k]	*7.0*							
Finland	*USSR*		*c.4.90*		*8.81*			*Heavy fuel oil*	
Germany	*USSR/Yamal*	*10.5*	*4.65*[b]	*4.65*[l]	*7.81*		*1.00*	*20% crude oil, 40% gasoil, 40% low-sulphur fuel oil*	*% change*
France/GdF	*USSR/Yamal*	*8.0*	*4.35*[b]	*c.4.65*[l]	*7.81*				
Switzerland	*Germany*	*0.6*[m]			*1.82*				
Japan	*Abu Dhabi*	*3.0*	*—*[b]	*6.36*	*11.81*				
Japan/Chubu Elec, Kansa Elec, Toho Gas, Osaka Gas	*Indo/Badak*	*4.6*			*4.81*				
Japan/Tokyo Elec, Tohuku Elec	*Indo/Arun*	*4.8*			*4.81*			*Primarily based on crude oils imported by Japan*	
Japan/JILCOGP (Osaku Gas, Kyushu Elec, Chubu Elec, Kansa Elec, Nippon Steel)	*Indo/Badak*	*10.5*		*5.93*[b]	*11.81*				
Japan	*Brunei*	*7.5*		*5.77*[b]	*11.81*				
Japan	*US/Alaska*	*1.5*		*5.86*[b]	*11.81*				
Argentina	*Bolivia*	*2.1*		*4.13*	*1.82*	*Irreg.*			

Argentina	*Chile*	*0.1*	
USSR	*Afghanistan*	*2.5*	
USSR	*Iran*	*5.0*	*Suspended*

Source: International Energy Agency, 1982, pp. 134–5.

Notes:
a For LNG projects, approximately 40¢/MBtu should be added for re-gasification costs, except for Trunkline/Algeria and US Distrigas/Algeria, which include re-gasification.
b Basis of contract.
c To achieve parity with Canada; formula price at 1 January 1982 was $4.63/MBtu.
d Italicised importers indicates LNG.
e 13.5% of the fob price will be paid by the French government.
f Not in operation.
g 1982–5. At that time, price is renegotiable and volume increases to 5 billion m^3.
h Owing to favoured nation clause, actual contract price will depend on outcome of Algerian negotiations with France, Spain and Italy.
i Does not include Statfjord gas.
j Cif Emden.
k Does not include proposed additional gas through the proposed Soviet (formerly 'Yamburg') pipeline.
l Contracts are denominated in importing country currency and contain a floor price of about $5.40/MBtu.
m Annual volumes vary from year to year.

Algerian lobbying. The idea is a simple one: the MMBtu price of crude oil exported from an OPEC country and the MMBtu price of natural gas from that country should be the same. This most extreme view is perhaps held only by Algeria. Other OPEC nations, such as Indonesia and Abu Dhabi, are perfectly willing to settle for crude prices cif to the importing country as a basis for price linkage. This effectively means that the fob price of $5.12 charged to the Belgians in Table 11.1 is around $5.50 in cif terms (perhaps a bit more). The Japanese contracts, as we have noted before, are not strictly comparable owing to the percentage content of natural gas liquids in these cargoes.

While it is true that crude oil fob must also be shipped, refined and distributed, these costs are far lower than the costs of shipping, re-gasifying and distributing LNG: oil transportation costs alone are under half the costs of transporting LNG; while refining charges for crude imports run to some 22¢/MMBtu as opposed to re-gasification charges of roughly 40¢/MMBtu.

More importantly, the markets served by crude oil products and by natural gas are not entirely the same. About 14 per cent of a barrel of refined crude results in gasoline, for which there is no other competitive fuel for a significant part of the transport market. Diesel oil (a form of gasoil) is also important in this market, which is untouched by any form of natural gas if one excludes LPG. Some 32 per cent of a barrel of oil is dedicated to the gasoil market. Gasoil has many qualities that make it ideal for home heating and commercial use, but natural gas has many premium qualities that enable it to compete to advantage. The problem with this competition is that home heating and commercial use of natural gas are costly in storage and load factor terms. Finally, some 34 per cent of a barrel of oil is sold as residual fuel oil. Here too, natural gas has premium qualities that put it at a pricing advantage. But again, residual heating oil is priced to compete with coal; natural gas thus enjoys a premium with regard to a very low-priced oil product. This, of course, reduces any income from natural gas. Furthermore, much of the residual oil substitution is in the form of interruptible supplies, a function of the need to increase load factors. In this case, a good portion of natural gas is actually sold at a discount in relation to the cheapest oil products. Table 11.2 gives an impression of the disadvantages of natural gas *vis-à-vis* various oil products – assuming that the equivalency price of natural gas is the Btu content of crude oil.

As can be seen from the table, only 21.5 per cent of the natural gas as against 45.9 per cent of crude oil products in the European markets was sold at or above a $6.75/MMBtu equivalent of gasoil. Additionally, a disproportionate amount of natural gas is sold in competition with coal or with low-price residual fuel oil grades. Furthermore, there is a refining trend away from residual fuel oil towards producing more of the higher fraction oil products; this is technically possible through use of catalytic cracking technology. It must be stressed in this context that Table 11.2 is an oversimplification. Much depends on the particular end-use market, the relative heat efficiencies of the fuels concerned, and other factors not included in the table.

While it can be argued that European natural gas companies in the major markets should sell more gas in premium markets, problems of distribution expense and load factor make this a difficult process. To the degree that small amounts of natural gas can be sold in this market on an incremental basis, there will be a tendency to overbid for

Table 11.2 *Relationships between oil product markets and natural gas: Europe,[a] 1980*

	Oil market		*Natural gas market*
Oil products	*% product share/bbl*	*Price (US$/MMBtu)[b]*	*% market share competing with oil product in first column*
Gasoline	13.6	8.33 }	21.5
Gasoil	32.3	6.75 }	
Fuel oil	33.7	5.19[c] 4.35[c]	72.7
Other	20.4		5.9

Notes: a Germany, Italy, France only.
b Rotterdam 1.1.81.
c Higher price for 1% S_2, lower for 3.5% s_2 content.

supplies. Otherwise, unless the state subsidizes the purchase of natural gas, as in the case of Belgium, France and Italy, there is a limit to further expansion no matter what the Algerians or Indonesians claim.

What of alternative uses of natural gas in the producing countries? Do these hold much promise for future natural gas production? Do they constitute a significant alternative to natural gas exports? To date, oil-producing countries have mounted impressive efforts to find uses for associated natural gas and reduce costly flaring. Many of the reputed petrochemical plants currently under construction in OPEC lands are aimed at associated gas. So successful have these programmes become that, with the exception of Nigeria, most OPEC associated natural gas is destined for alternative use.

This confines the field to non-associated natural gas. Bonfiglioli and Cima (1980) have discussed the non-export uses of natural gas in a paper presented to the Organization of Arab Petroleum Exporting Countries (OAPEC) in Algeria. The shadow prices of natural gas feedstock for these alternative uses are given in Table 11.3. These prices (or 'netbacks') are generally lower than the prices that producers might hope to get from natural gas exports. The exceptions to this rule are the use of natural gas as a feedstock in refining and in the manufacture of ethylene or methanol. However ethylene is made from ethane, which makes ethylene manufacture an impractical alternative to natural gas exports as these exports are typically based on methane. (Processes based on methane instead of ethane have been discovered and will come into future commercial use.) Another

Table 11.3 *Alternative uses of natural gas: market absorption capacity and shadow price of feedstock*

	Market absorption		
Alternative use	*Million metric tons/year*	*No. of plants*	*Shadow price of feedstock (US$/MMBtu)*
Refinery fuel	8.7	—	3.50
Ammonia	2.56	7	1.00
Methanol (chemical)	1.46	3–4	2.70
Methanol (automobile use)	36.5	15–20	2.50
Sponge iron	0.6–1.2	1–2	1.10
Alumina + aluminum	0.15–0.3	1–2	0.30
Ethylene	3.65	7–10	7.30[a]

Source: Bonfiglioli and Cima (1980), assumptions are included in Appendix V.
Note: a Price of ethane. A process for making ethylene from methane has been developed but is not yet in commercial operation.

point against the petrochemical alternatives is that the markets for these products are highly structured and it could well be difficult for ethylene producers to make much headway in European and American markets. Limits also exist on how much these markets can absorb. The authors estimate in most instances that more than seven plants would face difficulties in selling their products in Europe. (The exceptions to this are ethylene and automobile methanol.)

Another set of factors mitigating against alternative uses is that the capital consumption of such plant and infrastructure when related to the amounts of natural gas absorbed are quite high (see Table 11.4). An alumina–aluminum works, for example, consumes as much capital as an LNG liquefaction plant but allows a price netback to natural gas of $0.30/MMBtu and absorbs only 26.1 million MMBtu per year. Plants for the manufacture of methanol as automobile fuel, while allowing for more favourable netbacks, are also at a disadvantage when it comes to natural gas consumption. To achieve a reasonable level of gas consumption (400 million MMBtu), some $2.5 billion would have to be invested in multiple methanol plants, and the prices here are only $2.50/MMBtu. Ethylene is another matter altogether, however.

According to Bonfiglioli and Cima, there is little attraction in alternatives to natural gas export. This must be a tentative conclusion based on the assumptions of their study. Unfortunately, they do not specify the annual operating costs of the differing plants, an omission

Table 11.4 *Alternative uses of natural gas: capital absorption per plant*

Alternative use	*Investment*[a] *per plant ($m.)*	*Absorption of natural gas (million MMBtu)*
Refinery fuel	n.a	9.3
Ammonia	220	11.2
Methanol (chemical)	140	11.2
Methanol (auto)	375	56.0
Sponge iron	170	9.3
Alumina/aluminum	1,850	26.1
Ethylene	300	11.9

Source: Bonfiglioli and Cima (1980).
Note: a Including 12 per cent interest during construction. These figures include all *but* the following items: pipelines from wells to plants, port facilities, and escalation.

that makes it impossible to replicate their results. (The assumptions are included as Appendix V.)

There might be other reasons for selecting one of the alternatives in Tables 11.3 and 11.4: domestic jobs, self-reliance on domestically produced raw materials, the desire for a diversified export base, and satisfaction of domestic financial and industrial interests. Nevertheless, in non-political economic terms there seems relatively little of interest in alternatives to exporting the natural gas concerned.

A significant challenge to the future growth of natural gas exports lies in the double uncertainty of prices and alternatives. Caught between the inability to earn as much from natural gas as from crude oil and low revenue domestic operations, the potential exporter is liable to set aside the resources for future use. Meanwhile, rising costs of imported gas could significantly dampen demand so that it will be difficult to sell the natural gas contracted for. If natural gas is to live up to its future potential, these problems will have to be solved in a realistic manner.

The geopolitics of supply

In addition to the issues raised by the size distribution of world natural gas reserves and by the issues of cost, natural gas will increasingly become hostage to what might be loosely termed 'the geopolitics of supply'. The constraints of supply distribution of reserves on the one hand and of depleting resources in the United States and Europe on the other will necessitate an increasing reliance on natural gas imports from the Soviet Union or from countries that are either members of or ideologically aligned with OPEC. Both these features go beyond a discussion of natural gas in terms of domestic political economy to geopolitical considerations. These involve constraints imposed on natural gas trade by superpower rivalries, and those imposed by Third World/OPEC militancy.

Constraints of superpower rivalries

What was unique about the Yamburg pipeline controversy was not that it was the first East–West natural gas deal to become hostage to superpower interests. Previously, two major US LNG projects, the Tenneco North Star Project and the El Paso/Texas Eastern/Japanese Yakutsk LNG project, had run foul of hardening US–Soviet relations. What was exceptional about the pipeline was that it was opposed by a superpower not immediately a party to the complex of contractual commitments, acting on behalf of what it perceived as its own interests. As the controversy continued, the pipeline increasingly took on a symbolic significance in a discussion that revolved as much around Atlantic relationships, the necessity of doing business with the Soviet bloc and the problem of technology exports to the East, as it did address the natural gas question itself. That this controversy subsequently lapsed into relative obscurity was due to several factors.

To begin with, both sides in the conflict backed off. Two gas security studies were prepared by the IEA and the OECD for the May 1983 Williamsburg Summit between Canada, the USA, Japan and the European powers. The IEA study recommended that no European country should be dependent on any single supplier of natural gas for more than 30 per cent of its annual need. The recommendation was leaked amidst much controversy before the work on the papers was finished – reportedly the result of American (or French?) pressure (*World Gas Report*, 11 April 1983, p. 13). In the Summit and in later, lower-level meetings, European nations refused to be bound by the IEA recommendations, which were generally regarded as too stiff and simplistic, but would accept them as guidelines. The issue that had begun with a 'bang' ended in an unclear compromise. Neither side was wholly satisfied with this compromise, but nonetheless it would appear that the compromise was face-saving on both sides of the Atlantic.

Another factor was that demand for natural gas in Europe was stagnating. Between 1975 and 1982 industrial consumption of natural gas fell by 15 per cent, and power generation gas by 37 per cent. Only domestic usage rose in this period, by a European average of 8 per cent (*World Gas Report*, 20 June 1983). In 1982 there was a further decline of around 6 per cent overall. Stagnating demand lessened pressure to import natural gas from the Soviet Union generally, while the pattern of stagnation has made European industrial markets less dependent on Soviet gas supplies in particular.

Stagnating demand also led to a decline in Dutch Groningen gas exports. Gasunie can accept such a fall-off with a certain degree of equanimity; a shortfall today can be made up in a couple of years and at a higher price. Yet the Dutch government, however, under recessionary pressure, can ill afford the loss in revenue that this shortfall entails. In 1982, some 19.4 per cent of total government income derived from the sale of natural gas, while 1983 figures promise to represent a 2 per cent drop (*World Gas Report*, 9 May 1983). Studies by the high-power Sociaal Economische Raad have urged resumption of exports to European markets and use of the resulting revenue for a programme of energy conservation. So long as the natural gas market remains 'soft', therefore, there seems every possibility that Dutch exports will continue at present levels and that new contracts may be signed (especially for the Belgian market). It would appear that Dutch depletion policy has been waived not for superpower political reasons but for reasons of continued governmental income in a worldwide economic recession.

Finally, other domestic political considerations may have entered into the question of European security of supply. The Thatcher government has been reportedly open to the notion presented by the IEA to interconnect the Dutch and British markets. If accomplished (within the limits of the physical characteristics of the gases involved), the security of European supplies might be even further enhanced.

While one should not overstate the point, it is clear that the recent confrontation fortuitously coincided with a slump in natural gas demand, which provided a corrective remedy where political persuasion had failed. This does not mean that controversy will not rekindle in the future. The real issues – of East–West trade, relationships with the Eastern bloc, and the shape and nature of a future Western Europe – are too close to the surface to disappear permanently. These issues will be superimposed on any deal with the Soviet Union in the foreseeable future. They will most certainly make for an interesting decade for Western Europe natural gas.

Constraints of Third World/OPEC militancy
The Soviet and Chinese exportable surplus aside, approximately 82 per cent of all other prospective natural gas trades will involve members of OPEC. If one includes the two non-OPEC nations, Mexico and Malaysia, in this number, the share approaches 95 per cent. All these countries have reason to look at natural gas in opportunity cost terms. The rationale is that of joint producers of oil and gas everywhere in the world: achieving parity between their fuels. Whereas Shell and Esso in Europe wanted natural gas prices linked to oil product prices, and the American natural gas producers are perfectly willing to sell their natural gas at above oil equivalent MMBtu prices, OPEC members naturally wish to sell their gas in conjunction with their sales of crude.

This 'oil factor' was manifested directly in a much-vaunted attempt by Algeria to duplicate OPEC's success through the formation of the Organization of Gas Exporting Countries (OGEC). This attempt was accompanied by much politicking by the Algerians, who sent major delegations to most potential gas exporters to offer advice on how to manage their resources. This led to some concern on the part of the US State Department, which attempted to coordinate the policies of the US allies and lobbied against acceptance of Algerian pricing demands. The Algerian effort came to nothing, foundering on the major Soviet gas sale to Western Europe. That the members of OPEC are not willing to form a parallel OGEC, however, does not mean that oil resources are not a major factor influencing attitudes to natural gas trade, attitudes not thus far taken account of in the assertion that natural gas imports from the Third World must increase 'because it is there'.

Oil will obviously influence the gas producers within OPEC. The finds of associated gas with oil, gas that in many instances is the basis of industrial projects, can work to undermine necessary production cutbacks within the organization. Reducing oil production thus means not only loss of oil revenues, but also that associated gas projects, in many cases involving billions of dollars, become white elephants. OPEC has already cautioned its members to prepare non-associated gas reserves to replace associated gas should substitution become necessary. Beyond this, however, it is uncertain how much OPEC as an organization of oil producers will become interested in natural gas. Certainly the organization did not fully support the Algerians in their efforts to form OGEC.

More important will be the presence of oil reserves in the individual producing states. The major effects of the oil factor are listed in Table 11.5, for both oil-producing and non-oil-producing gas exporters. In many ways, oil-exporter interest in natural gas resembles the alleged interest of the multinational oil companies on the European Continent: natural gas should be marketed so that it does

Table 11.5 *The 'oil factor': effects on potential gas exporters*

	Potential gas exporters:	
Effect	*with significant oil reserves*	*without significant oil reserves*
Price effect	Natural gas price tied to crude oil prices, which enables assured income with automaticity and ensures that natural gas is not underpriced *vis-à-vis* crude	Not a significant factor, except as a means of maximizing returns
Financial reserves	Increase in oil prices leads to intransigence in natural gas negotiations aided by increased earnings from oil ('feedback effect')	Increase in oil prices/worsening of national finances leads towards import substitution policies re. natural gas
Market share	Excess of natural gas sold to OECD energy markets further undermines market shares for crude oil	Not a factor

not interfere with more profitable oil markets. Pricing policy is also somewhat the same. But where oil companies have led the way in tying the price of natural gas to prices of other fuels, particularly refined oil products, OPEC countries are naturally interested in tying prices to their products: in this case a blend of crude oils. For virtually all of the OPEC gas exporters, the increased income from the oil price rises in 1979 has not only justified the hiking of prices, at times violating contractual arrangements, but also given countries such as Algeria the means to suspend gas exports without affecting overall export earnings. Thus, the linkage between gas and oil prices here has a double-edged effect, about which all parties are aware.

The problem would perhaps be less if oil exporters could be assured that the netbacks in natural gas would be larger than those of oil. Unfortunately, as we have noted, this is not possible. The difference between oil and gas processing and transport costs is simply too great to be reversed.

It cannot be denied that this presentation of the oil factor is to a degree simplistic. Clearly, countries such as Indonesia and Nigeria can use any foreign exchange they earn, whether through the export of natural gas or through the export of oil. Other oil countries (Qatar, for instance) have large non-associated gas reserves and a relatively small oil production. For them, natural gas exports can serve as a supplement to declining oil revenues. Nevertheless, even for these countries, reference prices have always been crude oil prices, and crude oil exports have always come first.

Consuming nations should find that potential gas exporters in the non-oil-producing lands are more natural partners. Here there is less concern with oil markets in Western Europe and the United States, and perhaps more willingness to agree to OECD pricing and indexation terms. Furthermore, increases in oil prices for these countries lead to further indebtedness, not a flood of new revenues with which to organize potential 'OGECs'.

Yet there are also significant limits to such a 'natural partnership'. The countries concerned are few, especially when confined to those that do not have alternative plans for their natural gas. In some of these countries (Thailand, Cameroons), it is not certain that the amounts of natural gas involved are sufficient to support even a minor LNG project. In other countries (Malaysia and Trinidad, for example), more and more consideration is being given to increased domestic use of natural gas or alternative export projects.

The security factor, then, presents the Western world with yet another squeeze, one reminiscent of that between producer price and end-use value. This squeeze is between the insecurity of reliance on Soviet exports and the potential unreliability of Third World/OPEC exporters. There is no way in which reliance on these two sources of natural gas can be avoided if natural gas use is to increase in the future. Reliance, however, should ideally be balanced between these two sources.

The future challenge: the governmental response

Given the factors of size distribution, cost and geopolitics, how will governments respond to the trend away from market-proximate natural gas resources? Curious though it may sound in this period of natural gas deregulation in the USA and privatization in the UK, it seems unlikely that governmental roles in organizing the natural gas industry will decline consistently over time. This is in part due to the factors that have been discussed in the national chapters of this book. More importantly, however, it will also be due to the increasing degree of international sourcing of supplies – the topic of this last chapter. Commercially, governments will become more active in establishing the rules on which international relations will be conducted and with underwriting the industry, in terms of both domestic high-cost projects and politically uncertain international trades. Politically, the security concerns behind natural gas supplies will lead to an increased natural gas policy community, a community composed not only of industrial experts and a smattering of Foreign Ministry or State Department expertise, but of numerous experts in multiple branches of government that previously had little to do with the natural gas industry itself.

The first of these trends, that of an increased governmental commercially supportive role, is obvious. The recent LNG/pipeline deals with Algeria could not have been signed without government support. In different ways, the Belgian, French and Italian governments have interfered in commercial negotiations between Sonatrach and their national natural gas firms. This intervention has been castigated by the industry as 'misguided'; but the industry itself was not willing to do without the incremental amounts of natural gas under negotiation. The companies, too, would have wanted government to underwrite their LNG terminals and ships and natural gas pipelines should these have incurred losses due to lack of the natural gas for which they were designed. Meanwhile, American officialdom is currently negotiating price reductions of Canadian and Mexican imports. Price decreases may be forthcoming but it is uncertain how long they will last.[1] There is notably less criticism of the American government's role from the American purchasers of Canadian gas.

This need for commercial intervention is underlined by the need for governmental underwriting of much of the risk involved in developing the high-cost resources that characterize safer OECD sources of natural gas. Table 11.6 gives a good idea of the dimensions of the problem. Today, about 53.2 per cent of expectable proved reserves in the OECD area can reasonably be classified as 'high cost' because of location in deep waters offshore, small reservoir size, transportation distance to major markets, or a combination of these circumstances. The situation is even more serious for potential reserves. Here, almost 78 per cent are categorized as high cost – and the real figure is almost certainly higher (owing to lack of information on Australian resources and other factors).

Many of these resources are likely to be dedicated to domestic or alternative uses. As we have seen, one cannot count on either the UK or the Netherlands to increase exports: the UK, far from exporting, has appeared as a major competitor for Norwegian supplies; while the Netherlands is on the way out as the major European gas exporter. The Norwegians, with the potential to become the major OECD trader in natural gas, are less concerned about developing export

Table 11.6 *Major present and potential OECD natural gas exporters: proved and potential gas reserves (billion m^3)*

	Proved reserves	*Potential reserves*
Canada[a]	2,525	10,002
of which high cost[b]	473	6,412
Netherlands[c]	1,578	672
of which high cost[b]	—	672
Norway[d]	1,318	2,800–4,000
of which high cost[b]	918	2,800–4,000
United Kingdom[e]	739	820–1,395
of which high cost[b]	371	820–1,395
Australia[f]	850	n.a.
of which high cost[b]	453	n.a.
Total	7,010	14,294–16,069
of which high cost[b]	2,215	10,704–12,479

Source: International Energy Agency, 1982; data taken from material presented in tables.

Notes: a Proved reserves are those classified as 'marketable natural gas: remaining reserves'. Potential reserves are those classified as 'undiscovered potential'. High-cost reserves are those lying in the Arctic, Beaufort Sea, Hudson Bay, Labrador areas and in the MacKenzie Delta.

b Reserves that are small in size, remote in location, or expensive to extract.

c Proved reserves are those classified as 'proved'. Potential reserves are those classified as 'possible and probable'. High-cost reserves are those estimated to lie in small North Sea structures.

d Proved reserves are those classified as 'proved' by Norwegian authorities. Potential reserves are those classified as 'additional' by the IEA. High-cost reserves are those for which no contracts have been signed at present. These are from the Troll field and north of 62° N.

e Proved reserves are those classified as 'proved'. Potential reserves are those classified as 'probable and possible' plus 'reserves in future discoveries'. High-cost gas includes all but Southern Basin fields and the amounts under contract from them at present.

f Proved reserves are those classified as 'proved' by Australian authorities. High-cost reserves are those estimated on the NW Shelf.

markets than they are about becoming overdependent on North Sea resources for their economic base. There is at present little to indicate that the country's resulting 'go slow' policy will significantly reverse itself in the next decade. Canadian natural gas exports are plagued by dissension between Alberta and Ottawa. The provincial authorities are interested in increasing natural gas sales to the USA and elsewhere and are fretting under the federal government's pricing policy (which aims to reduce US sales). Additionally, much of the Canadian reserves are horrendously expensive, whether delivered by Arctic pipeline or by LNG icebreaking carrier. The capital costs of development of Arctic and Beaufort Sea reserves are high and the costs of transport do not allow for a comfortable netback to the producing well. Australia similarly is utilizing its resources domestically, although talk of LNG exports to Japan may yield results.

Interlinked with the enhanced commercial role of government is the interface between natural gas trade and geopolitical security issues. Throughout the recent transatlantic controversy over Soviet natural gas, a gap between the natural gas and security policy communities was obvious. The natural gas traders, with their long experience with Soviet counterparts, tended to minimize the security consequences of their contracts. The security community, largely ignorant of the *minutiae* of natural gas trade, superimposed their view on an industry with which they were dimly familiar. This gap between the two communities was not coincidental. The European gas traders representing Ruhrgas, Gaz de France, Distrigaz and the like had dealt with their Soviet counterparts for a decade. Within the constraints of mutual economic advantage expressed in contractual form among the parties involved, there had existed a certain community of interests underwritten by the prospects of greater profits and expanded markets. On the other hand, the fundamental characteristics of natural gas trade – the importance of long-term contractual relationships, of economic interdependence of interest – were foreign to men prepared to think of the Soviet Union as 'the enemy'. The result was perhaps predictable. Security specialists examined the nature of back-up measures to counterbalance overdependence of Western economies on Soviet gas and found these either non-existent or unsatisfactory; they then advanced unworkable remedies. These remedies were based less on commercial considerations than on possibilities conceived by the security community. A study was reportedly commissioned on the possibilities of establishing an emergency storage capacity, a remedy that even if practicable would have cost many millions and would take years to implement. Similarly, assumptions were made about early start-up of gas deliveries from the giant Norwegian Troll find, setting dates that were not only impracticable from the technical point of view but undesirable from the commercial angle. American delegations talked of coal and LNG as substitutes for Soviet gas. The natural gas community could point out flaws and did so. But they, for their part, were hard-pressed to answer some of the security-oriented questions relating to dependence on Soviet natural gas. Perhaps the ultimate expression of difference of views was the initial IEA proposal of a flat 30 per cent limitation on natural gas from any one source, and European national protests (perhaps reflecting the interests of the European natural gas policy community) that a 30 per cent limitation was too rigid and inflexible and that there were very real differences among nations both as to the economic role played by natural gas and as to the end-uses for which natural gas was employed. (This point is almost certainly correct, as we have noted in Chapter 7.)

Since, as noted previously, an accent on East–West trade will be hard to avoid given the size distribution of gas reserves and the location of natural gas markets, it should be interesting to note how this natural gas/security ptlicy interface resolves itself. Given the importance of the fuel in direct energy consumption terms, and the significance that it will have for all trade with the Soviet bloc, it seems hardly likely that natural gas imports from that source will be subject to less scrutiny in the future.

The consequences of this scrutiny, of guidelines such as those issued by the IEA, will be to constrain the trade of natural gas still further. Exporters competitive to the Soviet Union should enjoy a relative premium in future energy markets, particularly in Europe. (The Norwegians have already been quick to claim that their gas deserves such premium treatment.) Whether or not this politically two-tiered natural gas market becomes reality is, however, somewhat dubious.

The future challenge: the corporate response

If the coming years bring more and more governmental activity in natural gas markets, they will also be witness to considerable change in the industry itself. The natural gas industry on a global basis is still quite young and dynamic. It will resist change where it is possible to do so, but it can be relied upon to be innovative and dynamic where necessary. This has been characteristic in the past: in the early years of the Gas Council; in the NAM organization of Continental European markets; in the current US industry's adaptation to deregulation, full or partial, rapid or gradual. With this track record, it is unlikely that the industry will be incapable of making the changes needed for further expansion.

In Chapter 3, natural gas market instability was interpreted as being influenced by three factors: the numbers of actors to a given deal, the length and flexibility of commercial relationships, and whether or not there was security against individual parties 'exiting' from the relationships. There are two ways of dealing with instability: internalizing it within vertically integrated or commonly owned corporations; and drafting long-term contracts to specify future commercial relationships (or a combination of both integration and contracts).

This volume has already given some idea of the paths along which future industrial integration will evolve. Chapter 8 examined the manner in which European transmission companies have coordinated their access to new supplies of natural gas. Here, where coordination was possible, it was provided through common ownership of common carrier transmission lines. The Japanese success in LNG is due in part to the prices paid for the fuel; yet it is also due to their LNG consortia's blend of banking, trading house and utility interests. Similarly, joint activities in several stages by multiple national transmission companies may become more important in the future. As prices rise and demand dips, efforts at such coordination will succeed; the issue then will be less one of gaining individual access to natural gas reserves and more one of coordinating activities in a purchasers' cartel.

National firms – British Gas, Société Nationale Elf-Aquitaine, ENI, Transco – have all integrated either forward or backward into several stages of their national operations. Such vertical integration confers benefits that should be considered more widely. First, it distributes profits and risks. Transco, an American transmission company, has been so successful in finding new natural gas resources that its exploration affiliate is pulling the rest of the group to advocacy of deregulated prices. For other firms, success in obtaining higher prices for field gas can offset declining revenues in transmission activities. Secondly, vertical integration increases industrial expertise and knowledge about sectors of the industry that were not immediately familiar. Vertical integration can also increase knowledge about end-use values – a notoriously difficult concept to define precisely, and one that many exporting firms view with justified misgivings. Thirdly, vertical integration enables orderly marketing, a lesson that Esso and Shell/Royal Dutch have put to good use in their multiple activities on the European Continent. Such orderly marketing would enable intercorporate flexibility in the event of termination of Soviet supplies as well as controlled use of natural gas as a fuel.

To date, such vertical integration is an option yet to be fully explored in an international context. The degree to which producing companies simply are unaware of the commercial marketing problems of their customers is detrimental both to the interests of the customers and ultimately to their own interests. One of the problems faced by Sonatrach is that it is not engaged in selling natural gas in either the American or European markets, and, indeed, shows little interest in entering the Continental markets – as a partner of ENI, for example. Norway's Statoil, too, prefers to sell to the European transmission companies rather than join in their transmission facilities as a shareholder either of a corporation such as Ruhrgas or as a joint-owner of a transmission line. The same can be said of other producing companies. It is a pity that this should be the case, since vertical integration is one way of bridging the gap between exporters and importers.

While organization change as discussed here is hampered by many obstacles, contract law is likely to change rapidly in the next decade. Contractual 'wrinkles' are constantly being found and implemented to the satisfaction of buyers and sellers alike – although not necessarily of governments. Most favoured nation (MFN) provisions, now common features in many natural gas contracts, are a good example of how corporate contractual ingenuity can confront a challenge. In this case it was the monopsony powers of the large US transmission lines that gas producers opposed. Through the insertion of MFN provisions in their contracts, they were able to improve their bargaining positions to such an extent that MFN provisions have been blamed for the price rises that resulted in the control of field gas prices in the United States. Similarly, in the 1960s, tying prices of natural gas to oil product prices – while desirable in end-use markets – was not commonly resorted to in producer–transmission line contracts. Today, it is difficult to find a contract in which prices are not at least partially linked to oil prices. LNG contracts as unsophisticated as the El Paso contract in Appendix IV are today as out of date as an Edsel Ford. According to one contractual expert, the modern LNG contract includes:

- a guaranteed return on project investment;
- escalation clauses for cost increases in labour, maintenance, materials, etc.;
- clauses for currency revaluation and devaluation;
- clauses allowing for base prices at the well-head;
- escalation factors tied to competitive liquid fuels. (Khan, 1982)

As can be seen, the modern contract has 'something for everyone'.

Contractual prices practices will continue to evolve. Here, there are precedents to be adopted from American practices (see Chapter 4). 'Off-system sales', use of *force majeure* clauses, 'marketing out' clauses, and price indexation with ceiling prices are increasingly used in response to the decontrol in American gas prices and a parallel (although not wholly consequential) slump in natural gas demand. Such clauses can also be utilized with inter-European or other international contracts. Although producers may jibe at these clauses, the point remains that legal contractual obligations that spell out buyer and seller rights are worth little if there is not more to the relationship. This is the lesson of the current American experience: both seller and buyer have much to lose if they opt for the letter of the contract over a mutually flexible arrangement based on common understanding. Over the longer term, such understanding is in the interest of the producer, whether the producer is a Third World country or the Norwegian state company Statoil. This is the essential point that the Japanese have grasped in their relationships with Indonesia and other Southeast Asian suppliers. As in so many other areas, the world can learn from the Japanese example.

A closing note . . .

Natural gas has without doubt come a long way in a relatively short period of time. From being a nuisance fuel fit only for flaring in the 1930s, it is now appreciated for its premium qualities. Yet although appreciated and better utilized, the continued expanded use of the fuel will involve significant future challenges. The international sourcing problems as described in this chapter, together with cost and geopolitics, threaten to put limitations on its future expansion. This is particularly true when depleting reserves enter the equation. How long will US reserves last? Can Canada and Mexico make up for American shortfalls in production? What of European supplies? Despite the relative abundance of the fuel, there is already considerable cause for concern. (The Yamal pipeline controversy would never have come into being had it not been for a Dutch depletion policy that led to non-renewal of Dutch European contracts.) This book does not purport to answer these questions; such a task is left to others.

There is none the less, room for optimism. Ultimate reserves of natural gas today are anybody's guess. Exploration for the fuel is only in its infancy. Furthermore, alternatives to reservoir gas are plentiful. Gas-from-coal technology has not progressed much beyond the stage it was in during the early 1960s, at least in commercial terms. Coal remains an ample feedstock for future 'substitute natural gas', and various experimental processes aiming at low or high calorific gas products are currently under study. Gas is also obtainable from other sources: from plant biomass, from solid wastes, from semi-frozen hydrates, from underground gasification of coal in seams, from oil shale distillation. There are even theories of a large geo-gas layer – methane trapped beneath the earth's crust – dating from the formation of the planet. There will be no lack of alternatives when available reserves begin to run out mid-way through the next century.

This volume ends where it began, with the astronauts' view of the flares at the Soviet Samotlar and Middle Eastern fields and of the great urban corridors to which natural gas is a major contributor. In Chapter 1, the technological feat of the natural gas industry was compared to that of modern space-age technology. Our analogy does not end there. As with natural gas in the future, the utilization of outer space is not merely a function of technology; it is also dependent on international cooperation and understanding. The limits to commercial exploitation of both resources are bounded only by Man's ability to master his future.

Note

1 The authoritative *World Gas Report* (25 April 1983, p. 7) notes that the Canadian Energy Minister, Jean Chrétien, had announced a drop in the export price of Canadian natural gas from $4.94/MMBtu effective from 11 April 1983. It was doubted that this price reduction would have much effect on Canadian exports.

Appendix I Volumetric and Energy Measures

Given its gaseous qualities – expansion and contraction with temperatures and pressures, the inclusion of vapour, and other factors – the measurement of natural gas can be singularly imprecise at times in this volume. Generally speaking when data have not been available, I have utilized the volumetric measurements 28.31 m^3 per 1,000 ft^3 and 35.15 ft^3 per m^3. This relationship only exists if the two volumes are being measured at equal temperatures and volumes. Similarly, where other evidence is not available, it is assumed that1,000 ft^3 are equivalent to 1 million British thermal units (MMBtu). That these figures are imprecise as well is unavoidable. Unfortunately, sources are generally precise about Btu content of natural gas but other measurement conditions are not specified.

For this reason Table I.1. specifies the more commonly used energy/ volumetric measurements and conversions. The volumetric conversions – cubic meters to cubic feet and vice versa – differ according to the different pressures and heats of each standard utilized: the ft^3 as utilized in the US, the 'normal' Nm^3 as utilized in Europe and the SI m^3(st) as recommended by the International Gas Union (IGU). To relate energy content to volume, it is assumed that 1,000 ft^3 are equivalent to 1 MMBtu and 1 Nm^3 to 9,400 kcal.

Table I.1. *Volumetric and energy conversions: natural gas and other fuels*

From / *To*	*10^6 Btu*	*10^6 kcal*	*ft^{3a}*	*Nm^{3b}*	*$m^3(st)^c$*
10^6 Btu	1	0.252	1,000	26.9	28.3
10^6 kcal	3.967	1	3,967	106.38	112.23
ft^{3a}	0.001	0.000252	1	0.0269	0.0283
Nm^{3b}	0.03722	0.0094	37.22	1	1.0548
$m^3(st)^c$	0.035315	0.0089	35.315	0.948	1
1 barrel oil equivalent	5.8	1.462	5,800	155.5	164.02
1 ton coal equivalent	27.337	6.888	27,337	732.9	773.06
1 ton oil equivalent	43	10.838	43,000	1,152.8	1,215.97

Source: Shell International Gas Ltd, 1969.

Notes: a At 60°F 30 inHg dry (14.73 lbf/in^2). This is a commonly used ft^3 measurement in the English-speaking world.

b At 0°C 760 mmHg dry. This is the unit most widely used in Europe.

c At 15°C/760 mmHg dry. This is the unit recommended by the IGU and is the standard designated by the international system of units (SI).

To convert from the upper row multiply by the relevant factor. Working from this table, the reader is cautioned that all the heating values are gross heating values, not net heating values; i.e. they do not account for the lower temperature combustion due to water/steam byproducts of such combustion. It is particularly important to note this when comparing thermal properties between natural gases and other fuels. This factor should not be overrated, however; fuel efficiency in specific end-use is the most important marketing criterion for natural gas.

Appendix II Glossary of Gas Industry Terms

Absolute pressure – *gauge pressure* plus *atmospheric pressure.*

Annual inflation adjustment factor – a percentage or fractional figure used to adjust price-controlled natural gas for inflation. This is done by multiplying the previous month's price by the monthly equivalent of the annual inflation factor.

Arms' length bargaining – bargaining between two or more unaffiliated firms carried out in a truly competitive manner.

Associated (affiliated) company – a company that is either directly or indirectly controlled and/or owned by another firm or holding company.

Associated natural gas – natural gas originating from underground structures producing both liquid and gaseous hydrocarbons. The gas may be dissolved in crude oil (solution gas), or in contact with gas-saturated crude oil (gas cap gas). In such structures, gas production rates will depend on oil output, with oil usually representing the major part in terms of energy equivalents.

Atmospheric pressure – the pressure of the weight of air and water vapour on the surface of the earth. Approximately 14.7 lbf/in^2 at sea level.

Attribution rule – refers to the *first sale* regulation of natural gas under the US *Natural Gas Policy Act of 1978*, for pipelines and distributors.

Automatic adjustment clause – a clause in a rate schedule providing for automatic adjustment in the amount of a customer's bill, either upward or downward. The adjustment is based on variations in cost from a predetermined level of such factors as fuel, purchased power, taxes, etc.

Bbl – abbreviation for barrel. A barrel is equivalent to 42 US gallons.

Bbl/d – abbreviation for barrels per day.

Btu – abbreviation for British thermal unit. The quantity of heat necessary to raise the temperature of one pound of water one degree Fahrenheit at a specified temperature and pressure (from 59°F to 60°F at atmospheric pressure of 29.92 *inches of mercury*).

Base load – the minimum demand for service of an electric or gas utility over a stated period of time.

Base rate – a fixed amount charged each month for any of the classes of utility service provided to a customer.

Billing demand – the volume of gas for which the customer pays a *demand charge*, typically the *contract demand* or the maximum actual deliveries during a specified time interval (such as a day) within the billing period or a previous period.

Biomass – organic materials (such as trees, plants and crop residues) used as a source of energy.

Blue water gas – also called 'water gas'. Made in a cyclic process in which an incandescent bed of coke or coal is alternately subjected to blasts of air and steam. The gas consists mainly of equal proportions of carbon monoxide and hydrogen and has a *gross calorific value* of about 300 Btu/ft^3.

Boiler fuel – fossil fuel used as the primary energy source to generate steam.

Book cost – the amount at which property or assets are recorded in a company's accounts without deducting depreciation, amortization, or various other items.

Borderline customer – a customer located in the service area of a particular utility, but supplied by a neighbouring utility through an arrangement between the companies.

Bottled gas – liquefied petroleum gas contained under pressure in cylinders (or 'bottles' as they are sometimes called) for convenience of handling and delivery to customers.

Bottom gas – the quantity of gas that is not normally recoverable from storage field operations.

Budget type application – an abbreviated natural gas application. The filing contains all the information and supporting data fully to explain the proposed project, its economic justification, and its effect upon the applicant's present and future operations, as well as the public that would be served. Such an application is submitted where complete data are not required to explain the proposed undertaking (such as changes in sales or service, new construction, etc.).

Burner capacity – the maximum Btu per hour that can be released by a burner while burning with a stable flame.

Butane – a gaseous hydrocarbon usually present in small quantities in most natural gases with a *gross calorific value* of about 3,215 Btu/ft^3. Can be *liquefied* by the application of modest pressure and/or by cooling. Butane can also be produced synthetically. It is commonly used as a household fuel, refrigerant, aerosol propellant, and in the manufacture of synthetic rubber.

Calorific (heating) value – the amount of heat released by the complete combustion of a unit quantity of fuel under specified conditions. For gases, this value may be expressed in Btu/ft^3, kcal/m^3, joule/m^3, etc. Practice varies from one country to another. Values may be quoted on a gross or net basis. Briefly, 'gross' means that the water produced during combustion has been condensed to liquid and has released its latent heat, while 'net' signifies that such water stays as vapour.

Calorimeter – an instrument for determining the heating value of a fuel.

Carburetted water gas (carburetted blue gas) – the gas resulting from the enrichment of *blue water gas* during its manufacture by a simultaneous process of light distillate, gas oil or fuel oil gasification. The gas has a gross heat content of about 500–550 Btu/ft^3.

Charge:

- **commodity charge** – the portion of a natural gas rate based upon the volume actually purchased.
- **connection charge** – an amount levied on a customer in a lump sum, or in instalments, for connecting his facilities to those of his supplier.

• **customer charge** – an amount charged periodically to a customer for such utility costs as billing and meter reading without regard to demand or energy consumption.

• **demand or capacity charge** – the portion of the charge for electric or gas service that reflects a customer's contract requirements or special facilities.

• **termination charge** – a charge levied on a customer when service is terminated at his request.

Circumvention rule – refers to *first sale* regulation of certain sales of natural gas by pipelines and distributors under the US *Natural Gas Policy Act of 1978*. The rule is designed to prevent circumvention of the maximum lawful prices authorized by the Act.

City gate – a measuring station, which may also include pressure regulation, at which a distributing gas utility receives gas from a natural gas pipeline company or the transmission system.

Classes of service – groups of customers with similar characteristics (e.g., residential, commercial, industrial, etc.), which are identified for the purpose of setting a rate for electric or gas service.

Coal gas or coke oven gas – a *manufactured gas* made by destructive distillation ('carbonization') of bituminous coal in a gas retort or byproduct coke oven. Its chief components are methane (20–30 per cent) and hydrogen (about 50 per cent). This gas generally has a *gross calorific value* of 500–550 Btu/ft^3. When the process takes place in a closed oven (with gas as a by-product of coke production) it is generally designated 'coke oven gas'; when produced in retorts it is called 'coal gas'.

Coal gasification – the process of converting coal into gas. The basic process involves crushing coal to a powder, then heating this material in the presence of steam and oxygen. The gas produced is then refined to reduce sulphur and other impurities and to increase the methane content.

Co-generation – the sequential production of electricity and useful thermal energy from the same energy source.

Combination utility – a company that supplies both gas and some other utility-type service, e.g. electricity, water, steam, etc. Quite a common type of utility in the United States, but less common elsewhere.

Combined cycle – the increased thermal efficiency produced by a steam electric generating system when otherwise waste heat is converted into electricity rather than discharged into the atmosphere. One of the technologies of co-generation in which electricity is sequentially produced from two or more generating technologies.

Completion location – any sub-surface location from which natural gas is being, or has been, produced in *commercial quantities*.

Combustion turbine – a machine in which high-pressure gases from the combustion of fuel, usually natural gas or oil, expand through fan-type blades to drive a rotating shaft.

Commingling – the mixing of different streams of gas in a well-bore or a pipeline.

Commercial quantities – production of natural gas from a well or reservoir that is either sold or delivered to another operator or is retained by the owner for subsequent economic benefit. Generally, it excludes natural gas used for testing of wells or for other field uses that are production related.

Commercial service – service to customers engaged primarily in the sale of goods or services, including institutions and government agencies.

Condensate – a liquid/hydrocarbon mixture of 45–64 degrees API gravity, which may be recovered at the surface from some *nonassociated gas* reservoirs.

Contract demand (take-or-pay) – the volume of gas that the supplier agrees to deliver and, in general, the amount that the customer agrees to take or pay for, hence the expression 'take-or-pay'.

Cost of service – a rate-making concept used for the design and development of rate schedules to ensure that the filed rate schedules recover only the cost of providing the gas or electric service at issue. This concept attempts to correlate the utility's costs and revenues with the service provided to each of the various customer *classes*.

Costs:

- **fixed costs** – costs associated with capital investment such as equipment, overheads, property taxes.
- **fixed operating costs** – costs, other than those associated with capital investment, that do not vary with the operation such as maintenance, payroll.

Cubic foot/cubic meter (gas) – common units of measurement of gas volume. They are the amounts of gas required to fill a volume of 1 ft^3 or 1 m^3 under stated conditions of temperature, pressure and water vapour content.

Curtailment plan – a plan that recognizes and implements priorities of service during periods of curtailed gas deliveries by any local distribution company or intra-state or inter-state pipeline. A plan required by an electric utility to accommodate shortages of electric energy that restrict the utility's service to its wholesale customers. Gas and electric curtailment plans are developed to ensure public health and safety and the equitable treatment of customers.

Cushion gas – the volume of gas that is used in a *reservoir* to maintain adequate pressure and deliverability. The quantity of gas that would not be withdrawn during operation of the storage field. With respect to gas pricing, a gas cushion refers to a supply of low-cost gas that can be mixed with higher-priced gas to yield an average price.

Cycling/recycling – the process by which an *nonassociated gas* reservoir is produced only for the recovery of condensate. The gas, after removal of the condensate, is reinjected into the formation.

Daily average send-out – the total volume of gas delivered for a period of time divided by the number of days in that period.

Daily peak – the maximum volume of gas delivered in any one day during a given period, usually a calendar year.

Declining block rates – a rate structure in which the charge for energy decreases as the amount of energy consumed increases.

Degree day – a measure of the extent to which the mean daily temperature falls below an assumed base, say 65°F. Thus each degree by which the mean temperature for any day is less than 65°F would represent one degree day. (In Continental Europe, °C are used instead of °F and the assumed base temperature is generally taken as 16°C, equivalent to 60.8°F).

Dekatherm (Dth) – ten (10) therms, which equal 1 million British thermal units (MMBtu) or 1,000 ft^3 of 1,000 Btu gas.

Demand – the rate at which electric energy or gas is consumed.

Demand charge – the portion of the price, tariff or rate charged to gas customers based on the customer's demand characteristics, usually expressed as a fixed sum for a specified period.

Demand charge credit – a credit applied against the buyer's demand charges when the delivery terms of the contract cannot be met by the seller.

Demand day – daily rate of gas consumption. This term is primarily used in pipeline distribution company agreements.

Dependable capacity – the load-carrying ability of a station or system under adverse conditions for a specified time period.

Distribution company/gas utility – a company that obtains the major portion of its gas operating revenues from the operation of a retail gas *distribution system* and that operates no transportation system other than incidental connections to a transportation system of another company. For purposes of American Gas Association statistics, a distribution company obtains at least 95 per cent of its gas operating revenue from the operation of its retail gas distribution system. (See also *transmission company*.)

Distribution system – feeders, mains, services and equipment that carry or control the supply of gas from the point or points of local supply (usually the *city gate* station) to and including the consumer *meters*.

Diversity factor – the ratio of the sum of the non-coincident maximum demands of two or more loads to their coincident maximum demand for the same period.

End-user – a consumer who uses natural gas, electricity, or other forms of power.

Escalator clause – a clause in a purchase or sale contract that permits, under specified conditions, adjustment of price or profit.

Ethane – a gaseous hydrocarbon invariably present in small quantities in most natural gases with a *gross calorific value* of about 1,730 Btu/ft^3. Can be liquefied by cooling. Apart from the United States and a few isolated cases elsewhere, ethane is almost never extracted from natural gas and thus is a constituent of natural gas (predominantly methane) as distributed to customers.

Federal Energy Regulatory Commission (FERC) – the US federal agency that has jurisdiction over natural gas pricing, wholesale electric rates, hydroelectric licensing, oil pipeline rates and gas pipeline certification.

FERC guidelines – a compilation of the FERC's enabling statutes, procedural and programme regulations, and orders, opinions and decisions.

FERC priority scheme – the order in which natural gas customers must be curtailed by an inter-state pipeline if there are insufficient gas supplies to meet the pipeline's obligations.

Federal Power Act – enacted in 1920, the Act, as amended in 1935, consists of three parts. Part I incorporated the Federal Water Power Act administered by the former US *Federal Power Commission*. It confined FPC activities almost entirely to licensing non-federal hydroelectric projects. With the passage of the Public Utility Act, which added parts II and III, the Commission's jurisdiction was extended to include regulating the inter-state transmission of electric energy and rates for its sale at wholesale in inter-state commerce.

Federal Power Commission (FPC) – predecessor of the US *Federal Energy Regulatory Commission*. The FPC was created by an Act of Congress under the US Federal Water Power Act on 10 June 1920. It was originally charged with regulating the nation's water power resources, but later was given responsibility for regulating the electric power and natural gas industries. It was abolished on 30 September 1977, when the new Department of Energy was created and its functions were embraced by the Federal Energy Regulatory Commission, an independent regulatory agency.

Feeder (main) – a gas main that delivers gas from a *city gate* station or other source of supply to the distribution point.

Feedstock gas – natural gas used as a raw material for its chemical properties in creating an end product (e.g., plastics, etc.).

Field price – the price paid for natural gas, usually at the point of first exchange or sale in or near a producing area. It is essentially the prevailing price that a buyer must offer to get a contract for additional gas. It may also be the average weighted contract price at which gas is being delivered to (a) inter-state pipelines or (b) all buyers in the field, including intra-state pipelines, local industrial users and other producers.

FIFO – first-in, first-out method of inventory valuation by which the earliest-acquired natural gas layer in storage is assumed to be sold first; the most recently acquired is assumed to be still on hand.

Firm service – the highest-quality generation and/or transmission service offered to customers under a filed rate schedule that anticipates no planned interruption.

First sale – any sale of any volume of natural gas in the production–marketing chain under the US *Natural Gas Policy Act of 1978*.

Flat rate schedule – a rate schedule that provides for a specified charge irrespective of the quantity of gas used or the *demand*.

Force majeure – an occurrence generally beyond the control of the buyer or seller. A *force majeure* event that results in non-performance will typically relieve a buyer or seller of liability for damages and may result in some financial relief for the buyer.

Fuel cost adjustment clause – a clause in a rate schedule that adjusts the amount of the customer's electric bill as the cost of fuel varies from a specified base amount.

Gas field – a specific geographical and geological area from which natural gas is produced.

Gas, distributors (local) – those who sell gas directly to local consumers.

Gas purchase facility – a facility that connects independent producers or other similar sellers of natural gas to a purchaser's gathering facilities or main line facilities.

Gas works – a plant where a combustible gas (a secondary energy) is manufactured, usually from coal (a primary energy). (See also *manufactured gas*.)

Gauge pressure – the pressure generally shown by measuring devices. This is the pressure in excess of that exerted by the atmosphere. (See *absolute pressure*.)

Geothermal energy – natural heat contained in the rocks, hot water and steam of the earth's sub-surface. Geothermal energy can be used to generate electric power, to heat residences and for industrial needs. Such energy is nearest the earth's surface and most accessible in the western United States.

Gigawatt (GW) – 1 million kilowatts (kW).

Gigawatt-hour (GWh) – 1 million kilowatts (kWh).

Grid – the layout of a gas or electric *distribution system*.

Gross calorific (heating) value – the gross calorific value at constant pressure of a gaseous fuel is the number of heat units produced when unit volume of the fuel, measured under standard conditions, is burned in excess air in such a way that the materials after combustion consist of the gases carbon dioxide, sulphur dioxide, oxygen and nitrogen, water vapour equal in quantity to that in the gaseous fuel and the air before combustion, and liquid water equal in quantity to that produced during combustion, and that the pressure and temperature of the gaseous fuel, the air and the materials after combustion are one standard atmosphere at 25°C. (British Standard definition).

Heat capacity (specific heat) – the quantity of heat in *Btu*'s required to raise the temperature of one pound of a given substance by one degree Fahrenheit.

High priority use – the use of gas in a residence, commercial establishment using less than 50,000 ft^3/day, school, hospital, or similar institution. Any use that, if curtailed, would endanger life, health or maintenance of physical property.

Holding company – a corporation (parent company) in the USA that directly or indirectly owns a majority of all of the voting securities of one or more companies that are located in the same or contiguous states. As most states do not permit a utility company that operates in another state to compete within their own boundaries, the holding company-type of organization is used to bring into one family, without jeopardy to state control, companies that can best be operated as part of an integrated utility system. The Securities and Exchange Commission, as administrator of the Public Utility Holding Company Act of 1935, defines a holding company as 'any company which . . . owns, controls . . . 10 per cent or more of the outstanding voting securities of a public utility company'.

Inch of mercury (inHg) – a unit of pressure (equivalent to 0.491154 lbf/in^2) by which one pound per square inch equals 2.036009 inches of mercury column at 0°C.

Inch of water – a unit of pressure (equivalent to 0.03613 lbf/in^2) by which one pound per square inch equals 27.68049 inches of water column at 4°C.

Incremental gas acquisition costs – *first sale* purchased gas acquisition costs that are in excess of the threshold prices as published by the US *Federal Energy Regulatory Commission*.

Incremental pricing – a cost reallocation measure contained in Title II of the US *Natural Gas Policy Act of 1978* (NGPA). Its objectives are to shield residential and other high-priority users of natural gas from the full impact of higher prices authorized by the NGPA and to create a market-ordering device. It is a pricing mechanism whereby inter-state pipelines, and distribution companies served by inter-state pipelines, must pass on the higher costs of certain new, high-cost, and imported natural gas to their large industrial customers who use gas for boiler fuel to generate steam or electricity.

Incremental pricing surcharge – a charge above the contract price, based on the current price of high-sulphur No. 6 fuel oil, published by the US Energy Information Administration, assessed on industrial boiler fuel facilities.

Industrial customer of market – consumers who use gas primarily for industrial-type applications; in many countries this class of market or customer embraces gas used as a feedstock for the manufacture of chemicals and/or gas used for power generation purposes.

Interconnection – a transmission line permitting a flow of energy between two electric or natural gas systems. The physical connection of natural gas or electric power transmission facilities to allow for the sale or exchange of energy.

Interlocking directorates – the holding of a significant position in one corporation simultaneously with the holding of a similar position with another corporation, or a firm doing business with the corporation.

Interruptible gas – gas made available under agreements that permit the curtailment or cessation of delivery by the supplier of specified times, ambient temperatures, etc.

Intra-state pipeline – a natural gas pipeline company (USA) engaged in the transportation, by pipeline, of natural gas not subject to the jurisdiction of the *Federal Energy Regulatory Commission* under the *Natural Gas Act*.

Inverted rate design – a rate design for a customer class for which the unit charge for energy increases as usage increases.

Kilowatt (kW) – a unit of power equal to 1,000 watts or 3.3414 horsepower. It is a measure of electrical power or heat flow rate and equals 3,413 Btu/hour. An electric motor rated at 1 horsepower uses electric energy at a rate of about 3/4 KW.

Kilowatt-hour (kWh) – the quantity of electrical energy, 1,000 watts, operating for one hour. Electrical energy is commonly sold by the kilowatt-hour.

Lifeline rates – a concept by which the customer is guaranteed a certain subsistence amount of electricity or of natural gas at a relatively low price. The lost revenue may be recovered by raising the rates of residential customers who consume beyond the lifeline level or by increasing rates for non-residential *classes*.

LIFO – last-in, first-out method of inventory valuation in which the earliest-acquired layer of natural gas in storage is assumed to be still on hand; the most recently acquired is assumed to be sold first.

Line pack – a method of *peak-shaving* by withdrawing gas from a section of a pipeline system in excess of the input into the system.

Liquefaction – the conversion of natural gas to a liquid by cold or pressure.

Liquefied natural gas (LNG) – natural gas that has been cooled to about −160°C for shipment or storage as a liquid. Liquefaction greatly reduces the volume of the gas; thus, it reduces the cost of shipment and storage.

Liquefied petroleum gas (LPG) – also known as *bottled gas*, liquefied petroluem gas consists primarily of propanes and butanes recovered from natural gas and in the refining of petroleum. LPG is used as a fuel for internal combustion engines in applications where pollution must be minimized, such as in buildings and mines. The largest use is as a substitute for natural gas in areas not served by pipelines.

Load – the amount of electric power or gas delivered or required at any point on a system. Load originates primarily at the energy-consuming equipment of the customers.

Load curve – a graph in which the *send-out* of a gas system, or segment of a system, is plotted against intervals of time.

Load density – the concentration of gas load for a given area expressed as gas volume per unit of time and per unit of area.

Load duration curve – a curve of loads, plotted in descending order of magnitude against time intervals for a specified period. The co-ordinates may be absolute quantities or percentages.

Load factor – the ratio of the average load over a designated period to the peak load occurring in that period. Usually expressed as a percentage.

Low pressure gas distribution system – a gas *distribution system*, or the mains of a segment of a distribution system, operated at low pressures of less than 15 inches water column.

LPG–air mixtures – mixtures of liquefied petroleum gas and air to obtain a desired *calorific value* and capable of being distributed through a *distribution system* also used for *standby* and *peak-shaving* purposes by gas utilities.

Mcf – an abbreviation for 1,000 ft^3 of natural gas. Used as a measurement of natural gas volume. Most gas involved in producer–pipeline transactions is measured by volume although it it priced by energy content (*Btu*).

MMBtu – an abbreviation for 1 million British thermal units for (*Btu*'s) for petroleum and gas applications which reflects the energy or *calorific value*. In electrical usage, MBtu or 10^6 Btu means 1 million Btu's. (MM is used to denote million in the gas area; M is used to denote million in the electric area.)

Maintenance expenses – the portion of operating expenses consisting of labour, materials and other direct and indirect expenses incurred in preserving the operating efficiency or physical condition of utility plants that are used for power production, transmission and distribution of energy.

Manufactured gas – combustible gases derived from primary energy sources by processes involving chemical reaction. For instance, gas produced from coal, coke or oil products.

Marginal cost – the change in total production costs associated with a unit change in quantity (i.e., demand or energy).

Marginal cost pricing – pricing at the cost of producing the next unit of the particular good.

Marker wells – any US well from which natural gas was produced in commercial quantities after 1 January 1970 and before 20 April 1977.

Master metering – use of one central meter for several apartments so that residents have no means of judging their individual energy consumption.

Megawatt (MW) – 1,000 kilowatts or 1 million watts.

Meter (gas) – a mechanical device for automatically measuring quantities of gas.

Methane – the principal hydrocarbon constituent of most natural gases, with a *gross calorific value* of about 995 Btu/ft^3.

Minimum bill clause (minimum charge) – a clause in a rate schedule that provides that the charge for a prescribed period shall not be less than a specified amount.

Most favoured nations (MFN) clause – a clause in gas supply contracts providing for automatic increases in the purchase price if the price other parties pay or receive is increased.

Moving average cost – the cost change that occurs when each purchase of natural gas injected into storage is added to the inventory balance so that a new average price is computed for the balance in storage. This is used in pricing of subsequent withdrawals.

Natural Energy Act of 1978 – an Act passed by the US Congress on 15 October 1978 and signed into law on 9 November 1978 contained five major bills: *Natural Gas Policy Act of 1978*; Public Utility Regulatory Policies Act of 1978; Powerplant and Industrial Fuel Use Act of 1978; National Energy Conservation Policy Act of 1978; and Energy Tax Act of 1978. Its primary purpose was to encourage more efficient and equitable use of energy, increased domestic production, and the development of renewable energy sources. The *Federal Energy Regulatory Commission* was given primary authority for implementing the Natural Gas Policy Act and several important programmes of the Public Utility Regulatory Policies Act.

Natural gas – a natural hydrocarbon gas composed of a variety of gases including methane, ethane, butane and propane. It comes from the earth, with or without crude oil, and is used to heat the home, as a boiler fuel for utilities, and as a raw material in the petrochemical industry for the manufacture of fertilizer, cellophane, and other products.

Natural Gas Act – enacted on 21 June 1938, the act gave the former US *Federal Power Commission* jurisdiction over the inter-state transportation and sale of natural gas for resale. The *Natural Gas Policy Act of 1978* later superseded, limited or replaced parts of the Natural Gas Act.

Natural gas liquids (NGL) – hydrocarbons that are separated from natural gas through condensation at the surface or by other methods. Such liquids generally consist of propane and heavier hydrocarbons. They are commonly referred to as *condensate*, distillate, *natural gasoline* and *liquefied petroleum gases*.

Natural Gas Policy Act of 1978 – one of five parts of the US *National Energy Act*, it unified the intra-state and inter-state markets into a single national market for natural gas and provided for the phased deregulation of new natural gas. The Act is divided into six titles as follows: Well-head Pricing; Incremental Pricing; Additional Authorities and Requirements; Natural Gas Curtailment Policies; Administration Enforcement and Review; and Coordination with the Natural Gas Act–Effect on State Laws. The *Federal Energy Regulatory Commission* was given primary authority for implementing the Natural Gas Policy Act.

Natural gas reserves – the quantities of natural gas that geological and engineering data demonstrate with reasonable certainty to be recoverable from a known gas or oil *reservoir*. They represent strictly technical judgements, and are not knowingly influenced by attitudes of conservation or optimism. However, the actual methods of evaluating and quantifying reserves do vary somewhat from one country/company to another.

Natural gasoline – those liquid hydrocarbon mixtures containing essentially pentanes and heavier hydrocarbons that have been extracted from natural gas.

Net calorific (heating) value – the net calorific value at constant pressure of a gaseous fuel is the number of heat units produced when unit volume of the fuel, measured under standard conditions, is burned in excess air in such a way that the materials after combustion consists of the gases carbon dioxide, sulphur dioxide, oxygen, nitrogen and water vapour, and that the pressure and temperature of the gaseous fuel, the air and materials after combustion are one standard atmosphere at 25°C. This value is derived from the measured gross figure.

Off-peak – the period during which the service being delivered is not at or near the system's maximum capability.

Off-peak service – service made available during low-demand periods.

Off-system sale – a sale of natural gas, by a pipeline company, to a party that is neither connected to nor formally served by the pipeline company.

Oil shale – refers to a range of materials containing organic matter (kerogen) that can be converted into crude shale oil, gas and carbonaceous residue by heating.

OPEC (Organization of Petroleum Exporting Countries) – an organization formed in 1960 whose objectives are to coordinate and promote the interests of the larger petroleum exporting countries, such as Saudi Arabia, Iraq, Nigeria, etc. OPEC evolved from widespread dissatisfaction among the exporting countries over pricing policies. OPEC nations contribute over 50 per cent of world output and account for 80 per cent of the oil used by importing countries.

Outer continental shelf – the submerged lands extending from the outer limit of the historic territorial sea (typically, 3 miles offshore) to some undefined outer limit, usually the edge of the continental shelf. In the United States, this is the portion of the shelf under federal jurisdiction.

Particulate matter – solid particles, such as ash, released in exhaust gases at fossil-fuel plants during the combustion process.

Peak-shaving – the practice of augmenting the normal supply of gas during peak or emergency periods from another source where gas may have been either stored during periods of low demand, or manufactured specifically to meet the peak demand.

Pressure maintenance (repressing) – a process in which natural gas is injected into a formation capable of producing crude petroleum to aid in maintaining pressure in an underground *reservoir* for the purpose of assisting in the recovery of crude.

Process gas – natural gas used in industrial applications requiring precise temperature controls and precise flame characteristics for which alternative fuels cannot be substituted.

Producer – a company or entity concerned primarily or exclusively with the production of natural gas. In some countries, producers may have complementary interests in the transmission of gas or indeed in other related aspects of the gas business.

Producer gas – a gas manufactured by burning coal or coke with a regulated deficiency of air, normally saturated with steam. The principal combustible component is carbon monoxide (about 30 per cent) and the *gross calorific value* is 120–160 Btu/ft^3.

Proved reserves – the current estimated quantity of crude oil and natural gas that may be recovered from known oil and gas *reservoirs* under existing economic and operating conditions. Reservoirs are considered proved when they have demonstrated the ability to produce, either by actual production or testing.

PSIA – pounds per square inch absolute.

Rate base – the value, established by a regulatory authority, upon which a utility is permitted to earn a specified rate of return. Generally, this represents the amount of property used and useful in public service and may be based on such criteria as fair value, prudent investment, reproduction cost or original cost.

Reforming – the process of thermal or catalytic cracking of *natural gas, liquefied petroleum gas, natural gas liquids, refinery gas*, or various oil products, resulting in the production of a gas having a different chemical compostion.

Refinery gas – a gas resulting from oil refinery operations consisting mainly of hydrogen, methane, ethylene, propylene and the butylenes. Other gases such as nitrogen and carbon dioxide may be present. The composition can be highly variable and the calorific value can range from 1,000 to 2,000 Btu/ft^3.

Resale – the sale of natural gas, all or a portion of which was purchased and resold in transactions that are *first sales* as defined in the US *Natural Gas Policy Act*. Sales to parties other than the ultimate consumer.

Reservoir – a sub-surface, porous, permeable rock formation containing an accumulation of natural gas, crude oil, or both, that is confined by impermeable rock or water barriers.

Retained earnings – the accumulated net income of the utility, less distribution to stockholders and transfers to other capital accounts.

Seasonal rates – the rates charged by an electric or gas utility for providing service to consumers at different seasons of the year, taking into account demand based on weather and other factors.

Send-out – the quantity of gas delivered by a plant or system during a specified period of time.

Service area – territory in which a utility system or distributor provides service to consumers.

Specific gravity of gas – the ratio of the density of gas to the density of dry air at the same temperature and pressure.

Stabilization – the process of adding a gas to the gas normally supplied in order to raise or lower the *calorific value*. Air is often used to reduce calorific value and liquid petroleum gases are used to enrich or raise the calorific value.

Standby charge – fixed monthly charge for the potential use of a utility service.

Standby service – service that is not regularly used but is available as needed to supplement the usual supply.

Storage gas – natural gas that is purchased or produced by a natural gas company and stored in depleted or partially depleted gas or oil fields, or other underground *reservoirs*, for sale or use at another time.

Stripper well natural gas – natural gas produced at an average rate no greater than 60,000 ft^3/day. Commonly, stripper wells are old wells that draw on *reservoirs* that are already largely depleted.

Summer valley – the decrease that occurs in summer months (in appropriate temperature zones) in the volume of the daily load of a gas *distribution system*.

Synthetic/substitute natural gas (SNG) – a gas made from coal or oil products that has a *calorific value* and burning characteristics compatible with natural gas.

Tariff – a published volume of rate schedules and general terms and conditions under which a product or service will be supplied.

Therm – a heat quantity equivalent to 100,000 *Btu*.

Tight formation gas – natural gas produced from geological formations of low permeability that cannot be produced in economic quantities without the utilization of enhanced production techniques, such as fracturing.

Top gas – the part of the total reserve inventory that can be withdrawn, sold and replaced each year. (See also *bottom gas*.)

Town gas – gas piped to consumers from a gas plant. The gas can comprise both *manufactured gas* (secondary energy) and *natural gas* (primary energy) used for enrichment.

Transmission company – a company that operates a natural gas *transmis-*

sion system and that either operates no retail *distribution system*, or, as defined by the American Gas Association, receives less than 5 per cent of its gas operating revenues from such retail distribution system.

Transmission system – the pipeline system of a transmission company or companies.

Trunkline – a large-diameter, high-pressure natural gas pipeline for transporting large volumes of gas over relatively long distances.

Unaccounted-for gas – the difference between the total gas available from all sources and the total gas accounted for as sales, net interchange and company use. The difference includes leakage or other actual losses, discrepancies due to meter inaccuracies, variations of temperature, pressure and other variants.

Unassociated gas – natural gas unaccompanied by crude oil. Also called non-associated gas, dry gas and gas cap gas.

Utilization factor – the ratio of the maximum demand on a system or part of a system to the rate capacity of the system or part of the system under consideration.

Vintaging – originally, a term used for a pricing scheme in which gas is priced at different levels on the basis of age (date of its initial production). More recently, the term refers to a pricing scheme in which prices are based on categories that are only loosely related to age or date of initial production.

Waste heat – heat contained in exhaust gases or liquids that is usually discharged to the environment.

Water gas – see *blue water gas*.

Well-head – the assemblage of controls and other equipment located at the surface of a well and connected to the well's flow lines, tubing and casing.

Well-head price – the price received by the oil or gas producer for sales at the well.

Wobbe Index – the *gross calorific value* of the gas divided by the square root of the density of the gas as compared with air.

Sources: Federal Energy Regulatory Commission, 1982, pp. 45–52; Peebles, 1980, pp. 211–17; Shell International Gas Ltd, 1969.

Appendix III Instability in Natural Gas Markets

The term 'instability' as it is used in Chapter 3 is singularly imprecise. It gives rise to justified questions: Are natural gas markets unstable because of the natural monopoly character (and size) of the actors? Are they unstable because of the political controversies surrounding natural gas? Are they unstable because of the changes that occur in any commercial relationship over time? This appendix is intended to clarify natural gas market instability as that term is utilized throughout this analysis.

Instability in game theoretic terms

As mentioned in Chapter 3 instability is conceived in game theoretical terms. To adhere to this analogy it is necessary to define the characteristics of the game. These are as follows: the number of actors in the game, the payoffs for which these actors play, and the nature of the game. Following this discussion, attention will be focused on the consequences of these first three points.

Number of actors
Natural gas markets, as stressed in the text, are characterized by the participation of three actors: a producer, a transmission company and a distribution company. (A reduction in the number of these actors will be considered subsquently.)

The 'payoffs'
The nature of the payoffs to each actor in the natural gas chain is a major reason for market instability. Each actor will demand a payoff that at a minimum will cover costs. The problem is that each actor can have reasonable expectations of a payoff considerably in excess of costs: the producer will seek economic rent, which he accrues through producing the natural gas; the transmission line will seek to obtain a monopsony rent *vis-à-vis* the producer and a monopoly rent in its sales to the distribution company (or alternative major purchaser); the distribution company in turn will seek a monopsony rent through its purchases and a monopoly rent in its sales. In practice, regulation affects this behaviour. Distribution companies are not allowed monopoly rents in their sales; this does not prevent them from attempting to obtain natural gas as cheaply as possible and passing the monopsony rent thus acquired to their customers. This allows greatly enhanced security, market expansion, generally at the cost of other fuels, and some of the 'frills' that regulated firms often acquire (the Averich–Johnson effect).

The nature of the payoffs may be even more complicated than is indicated above. There may be total disagreement about the end-use value of the natural gas or the pricing principles that will be utilized in customer sales. Thus, perfect information is only theoretically possible. This would have an effect on actor behaviour inducing recalcitrance, overbidding, suspicion and the like.

The nature of the game

The game as discussed thus far is a cooperative game. Assuming that the producer is identified by 1, the transmission company by 2, and the distribution company by 3, the characteristic functions of the game are:

(1) $$v(1) = v(2) = v(3) = 0$$

(2) $$v(\overline{12}) = v(\overline{23}) = v(\overline{13}) = 0$$

(3) $$v(\overline{123}) > 0$$

Expression (1) merely states that the value of the game to each palyer is 0 if there is no cooperation; expression (2) that any two of the players cooperating will not improve their collective payoff (which is still 0); expression (3) defines the nature of the payoff to a grand coalition of all three actors, the payoff of >0 being a definition.

Given the nature of the payoffs, it is clear that the actors will sooner or later form a 'trade'. The problem is that there is no specified division of payoffs among the parties involved. As with cooperative games everywhere, there is an '*n*-tuple' of possible payoffs. Without knowing more of the nature of the game, the number of solution sets within the characteristic functions given are infinite. In this particular game, however, there are some limitations on the number of solution sets.[1]

Limiting factors

The limitations that are worth examining are: the costs of participation in the game, the sequence of negotiations between the players, and the presence of competitive coalitions and players.

The costs of participation Any solution set must cover the costs incurred by the natural gas trade. No player will accede to the grand coalition without covering the costs of participation. In some instances this limiting factor may be sufficient. Thus, if >0 in expression (3) covers the costs of participation of the three actors and no more, then it is a sufficient solution to the game, and will constitute a stable solution. It should be noted, however, that to date this has been an extremely unusual phenomenon, particularly given the constant rise in natural gas prices.

The sequence of negotiations While the nature of the game in expressions (1)–(3) may seem logical, it is also inconsistent. If the grand coalition $(\overline{123})$ yields >0, then certainly the values of any two players cooperating together must be higher than the value of the single players without cooperation. Given a total payoff of 4 to $(\overline{123})$, clearly in a negotiating sense the value of

($\overline{12}$) must be higher than 0, in that they need only buy off player 3 to form a coalition with the value of 4. Thus while expression (2) is true, it also represents a static state. In the absence of the grand coalition, the value of cooperation of any two members is 0. But this is a statement of fact, not a statement of negotiating values once negotiations begin.

Given this inconsistency, it is obvious that the sequence of negotiation between players is of extreme importance. If 1 initiates negotiations with 2 and they arrive at a distribution between themselves that they find satisfying, and if the amount that they can offer 3 exceeds 3's cost of participating, then they should be able to procure 3's agreement and participation. Similarly, if 2 and 3 arrive at an amicable settlement of payoff distributions, they can then approach player 1 and gain his cooperation so long as they cover 1's cost of participation. In that it is difficult if not impossible for 1 and 3 to initiate proceedings with each other, owing to geographical distance and other factors, it is clear that the transmission company (2) is in a favoured position. It can command a high price for participation from either 1 or 3 and can use its position *vis-à-vis* the non-participating member to procure a maximum payoff for any coalition of which it is a part. (This in fact is exactly what happened in the early 1950s in the USA, where transmission lines, in absence of competition from other transmission lines, enforced monopsony equilibrium prices on producers of natural gas.)

In taking sequence of negotiations seriously, the nature of the game is changed to a degree. A three-person game becomes two-person games with 2 as the indispensable party. The need for both 1 and 3 to acquire 2 thereby gives 2 increased bargaining leverage. Additionally, any agreement between 1 and 2 or between 2 and 3 that entails bargaining as a unit *vis-à-vis* 3 and 1, respectively, means that this second negotiation becomes a two-person cooperative game. Given the indispensable nature of 2's position in this game limitation, it should not be surprising that 2 is often integrated either with 3 or with 1 in the real world of natural gas. (This is a common pattern in Europe, for example.)

The conversion of this three-person game into two-person games leaves the game as expressed in (1)–(3) susceptible to the range of solutions available in two-person game theory. Thus there is nothing to prevent 1 and $\overline{23}$ from arriving at a solution bounded by other limitations. Thus, to take the Nash solution, if we assume that there is a set of solutions for 1 and $\overline{23}$ in which solutions for these players are better collectively than alternative solutions, that there is a solution that is Pareto optimal, that irrelevant solutions do not apply, and that the game is characterized by symmetry, then there will be only one solution that satisfies all these conditions simultaneously and there is an element of stability in the game.

The problem in reality, of course, is much more difficult. Generally, any contractual arrangement dividing payoffs is limited in time and also requires considerable commitment of resources. There have been instances where two parties have reached agreement among themselves about a natural gas deal; the transmission company then went to the distributors (negotiating *en bloc*), needing their participation for the grand coalition. (Otherwise, as indicated in expression (2), the contract is worthless.) The distributors, knowing that there were no feasible alternatives to themselves, that the longer they delayed the better terms they would receive, and well aware that natural gas would start flowing according to the contract terms between 1 and 2, refused to sign any contracts or participate in the deal. (This tactic resulted in the destruction of the natural gas trade, particularly as the distributors, according to law, did not need to make profits in their activities but would have their losses covered by the transmission company.) This example not only illustrates the importance of timing, it also illustrates the importance of competition at the various stages of a natural gas deal – the presence of an alternative.

The existence of an alternative Does the existence of alternative producers, transmission company and distributor – N_1, N_2 and N_3 corresponding to 1, 2 and 3 – limit the range of possible outcomes? We shall assume that, owing to costs, end-use value and other factors, the deal represented by $\overline{123}$ is the preferred outcome.

$$(4) \quad v(1) = v(2) = v(3) = v(N_1) = v(N_2) = v(N_3) = 0$$

$$(5) \quad v(\overline{12}) = v(\overline{23}) = v(\overline{13}) = v(\overline{1N_2}) = v(\overline{1N_3}) = v(\overline{2N_1}) = v(\overline{2N_3}) = v(\overline{3N_1}) = v(\overline{3N_2}) = v(\overline{N_1N_2}) = v(\overline{N_1N_3}) = v(\overline{N_2N_3}) = 0$$

Skipping the nature of the rewards to the grand coalitions involved for the moment, let us look at the alternatives N_1, N_2 and N_3. Clearly, these can place restraints on payoffs among the primary actors. Let us assume that the producer (1) wishes to do trade with transmission company (2); he must now consider the following combinations for himself:

$$(6) \quad \begin{array}{c} \overline{123} \\ \overline{1N_23} \\ \overline{1N_2N_3} \end{array}$$

On the one hand, therefore, he has to be cognizant of the varying opportunities in terms of negotiation (although he only communicates with 2 and N_2). On the other hand, 1 must be aware of the alternatives available to 2 and to N_2 (the alternative). For 2, these are:

$$(7) \quad \begin{array}{c} \overline{123} \\ \overline{N_12N_3} \\ \overline{N_123} \end{array}$$

For N_2 the alternatives are:

$$(8) \quad \begin{array}{c} \overline{1N_23} \\ \overline{1N_2N_3} \\ \overline{N_1N_2N_3} \end{array}$$

Therefore, in any negotiation, 1 and 2 must each be aware of the alternative bid to his potential partner – the bids by N_1 and N_2 respectively. Furthermore, the total payoff to all members in the coalition will be determined by the inclusion of either 3 or N_3 in the trade. Here, 2 (or N_2) enjoy an advantage. They can bargain directly with 3 or N_3, whereas 1 (or N_1) can only assume that certain payoffs are going to be arranged between the transmission companies and the final marketers.

The logic of this situation is reversed in the case of 3 (and N_3). In terms of bargaining information and communication, there is little doubt that 2 and N_2 possess advantages (although it is not certain that these can lead to better payoffs to these two actors).

The following coalitions, then, are the ones in which 1, 2 and 3 have an interest:

$$
\begin{aligned}
&J_1: \overline{123}, N_1, N_2, N_3\\
&J_2: \overline{1N_23}, N_1, 2, N_3\\
&J_3: \overline{N_123}, 1, N_2, N_3\\
&J_4: \overline{12N_3}, N_1, N_2, 3\\
&J_5: \overline{1N_2N_3}, N_1, 2, 3\\
&J_6: \overline{N_12N_3}, 1, N_2, 3\\
&J_7: \overline{N_1N_23}, 1, 2, N_3\\
&J_8: \overline{N_1N_2N_3}, 1, 2, 3
\end{aligned} \tag{9}
$$

It should be noticed here that this is a particular form of a six-person game. The coalitions indicated do not necessarily prevent the non-associated parties from forming coalitions themselves. (Indeed, given the payoffs there will be an incentive to do so, but this need not concern us here.)

What rules can we establish for payoffs to the coalitions in (9)? Clearly the combination $\overline{123}$, by definition, is the preferred alternative, the role of N_1, N_2 and N_3 being to underbid the preferred partners. Logically, although not necessarily, let us assume:

$$v(S_{J_2}) = v(S_{J_3}) = v(S_{J_4}) = v(S_j) \tag{10}$$

$$v(S_{J_5}) = v(S_{J_6}) = v(S_{J_7}) = v(S_n) \tag{11}$$

where S and so forth represent the value of the coalition to the coalition formed in the particular set concerned, and $v(S_j)$ and $v(S_n)$ represent the values generalized for the sets. For each expression (10) and (11), differing payoff imputations will create differing dominant sets within the coalitions concerned. By identification, however, the favoured coalition is simultaneously contained in J_1, and the values of two of the preferred partners – 1, 2 or 2, 3 – should be higher than those of the alternatives: N_1, N_2 and N_3. This means that the following relationship should hold:

$$v(S_{J_1}) > v(S_j) > v(S_n) \tag{12}$$

Sincc, individually, N_1, N_2 and N_3 contribute less to the possible coalitions than the other members, $v(S_{J_8})$ is less than other subset values:

$$v(S_{J_1}) > v(S_j) > v(S_n) > (S_{J_K}) \tag{13}$$

Expression (13) defines both the nature of the game and the limits of payoffs within the winning coalition.

The curious aspect of this game is that there are two deals being formed, each of which delimits the possibilities of the other and in some cases provides a reasonably stable solution. The problem with this stability is that it is almost entirely dependent on the values of the subsets in expression (12). Some examples suffice to prove this. Assuming the characteristic functions as expressed in (4), (5) and (13), and giving these numerical values, we can see the problem.

For example, let

$$\begin{aligned} v(S_{J_1}) &= 12 \\ v(S_j) &= 9 \\ v(S_n) &= 5 \\ \text{and} \quad v(S_{J_K}) &= 3 \end{aligned} \tag{14}$$

The problem with solutions to this game is that they are mutually interdependent. Within the boundaries defined by each other they possess solution sets akin to those proposed by Von Neumann and Morgenstern. Let us assume that the two possibilities start out from the following imputations:

$$\begin{aligned} x_1 &= (4,4,4) \text{ for } (\overline{123}) \text{ and} \\ x_2 &= (1,1,1) \text{ for } (\overline{N_1N_2N_3}) \end{aligned} \tag{15}$$

These payoff imputations satisfy the conditions specified earlier. At the commencement of negotiations, let us assume that no two parties in any coalition of three will settle for less than they receive from imputations $\overrightarrow{x_1}$ and $\overrightarrow{x_2}$. The following possibilities present themselves:

$$\overrightarrow{x_3} = (4,4,1), \overrightarrow{x_4} = (1,4,4) \text{ and } \overrightarrow{x_5} = (4,1,4) \text{ for all } S_j \; ((\overline{12N_3}) \text{ etc.})$$

$$\overrightarrow{x_6} = (1,1,3), \overrightarrow{x_7} = (3,1,1) \text{ and } \overrightarrow{x_8} = (1,3,1) \text{ for all } S_n \; ((\overline{1N_2N_3}) \text{ etc.})$$

In each of these situations, two parties are either receiving more in payoff terms or as much as they would receive in imputations $\overrightarrow{x_1}$ and $\overrightarrow{x_2}$. This does not exhaust the possibilities, however, as the 1, 2 or 3 when excluded by the other two in the group can receive no more than 3, whereas the two remaining can receive no more than 4 in alliance with N_1, N_2 or N_3. Since $v(S_j) = 12$, it would pay the excluded member to award the difference between 3 (received with N_1, N_2 and N_3) and the 4 that it would receive in $(\overline{123})$ to encourage the other two favoured parties to defect from any agreement with N_1, N_2 or N_3.

The maximum that the two favoured parties could demand of the third (divided equally) is represented below:

$$\vec{x_9} = (4\tfrac{1}{2},4\tfrac{1}{2},3), \vec{x_{10}} = (4\tfrac{1}{2},3,4\tfrac{1}{2}) \text{ and } \vec{x_{11}} = (3, 4\tfrac{1}{2}, 4\tfrac{1}{2}) \text{ for all } S_{J_1}.$$

The parameters of a solution set for $(\overline{123})$ in terms of individual payoffs are delineated by imputations $\vec{x_9}$–$\vec{x_{11}}$. Under the conditions specified there are two coalitions $(\overline{123})$ and $(\overline{N_1N_2N_3})$. The rewards to the individual members of the latter coalition will probably be 1, 1, 1, in that no member of $\overline{123}$ could consider giving more than 1 to the non-preferred parties.

This form of stability is notable for three reasons: first, it combines the aspects of a Von Neumann–Morgenstern solution set for the one coalition $(\overline{123})$ and preserves the status quo for the other. Secondly, it is critically dependent on the initial payoff imputations to set the terms for interdependent bargaining behaviour. (For example, the situation would be different if $\vec{x_2} = (\tfrac{1}{2},\tfrac{1}{2},2)$ rather than (1,1,1).) Finally, the entire nature of the situation can be changed through changing the values in any of the subsets. (Here, for example, altering $v(S_n)$ in (14) from 5 to 6 would have opened up the possibility of any number of coalitions containing preferred and non-preferred parties almost indiscriminately.)

Stability as a function of external constraint

Particularly striking about the way in which game outcomes in the previous section are limited is that they are limited because of conditions imposed from outside the game. There is nothing intrinsic about a solution to these games in the games themselves. Thus, in each case, the process of sequential negotiations or competitive bidding is limited by rules that are imposed on the games from outside. In this sense, although it is useful, the nature of a six-person game with two possible grand coalitions and uncertainty about who will be in each coalition appears contrived. There is nothing reprehensible about this, however. To widen the limits of the game further would merely add to confusion. Furthermore, there is an element of outside contrivance in virtually every solution to three-person or two-person cooperative games. To take the Nash solution to a two-person game (an analogy discussed briefly in the section on sequential bargaining): the solution is possible only because two conditions – symmetrical outcomes and independence from irrelevant outcomes – are imposed on the players because they are 'reasonable' (Riker and Ordeshook, 1973, p. 235). They are not contingent on utility-maximizing behaviour and are *ad hoc* in nature. This characterization can be generalized to all solutions to cooperative games.

Yet it is this very element of contrivance on which this book is based. Given an '*n*-tuple' of possible solutions, limited only by the number and geographical position of the actors involved, how is the '*n*-tuple' reduced to one or two solutions? Clearly this is the result of imposition on bargaining outcomes by the actors themselves (through themselves structuring negotiation behaviour), by national governments (through setting negotiation rules, regulating actor behaviour, or participating in the negotiations), or by national governments and corporate actors working together. It is in this respect that the various forms of bargaining stability discussed in Chapter 3 become interesting. They are the political–economic contrivances that impose order and stability on national and international natural gas markets.

Note

1 Much in this analysis might suggest to a game theoretician that a "Shapley Value solution" might be applied. This particular solution is eschewed on the following grounds: (1) there is a strong equity basis to the Shapley Value solution; (2) the solution depends entirely on the characteristic function of the game, *not* on the motives which underlie the characteristic function. As it is precisely the absence of an equity principle, the differing motives behind any characteristic function, and the dynamics of the bargaining process which we emphasize in this context, discussion of the Shapley Value solution seems inappropriate to our purposes.

Appendix IV Two Natural Gas Contracts

The two contracts appended here for reader reference – the El Paso–Sonatrach LNG-I contract of 1969 and the Gas del Estado–YPF contract of 1968 – are taken from Centre de Recherche sur le Droit des Marchés et des Investissements Internationaux (1973, pp. 394–427). Natural gas contract texts are notoriously difficult to come by and it is hoped that these will be a useful supplement to this book.

El Paso–Sonatrach LNG-I contract

CONTRACT FOR THE SALE AND PURCHASE
OF LIQUEFIED NATURAL GAS

The Société Nationale SONATRACH, with registered offices at Algiers, Immeuble Maurétania (hereinafter called the "Seller"), represented by its Président Directeur Général. Mr. Sid Ahmed GHOZALI,

on the one hand,

and

EL PASO NATURAL GAS COMPANY, a Delaware corporation, with registered offices at Wilmington, Delaware, United States of America (hereinafter called the "Purchaser"), represented by its Chairman of the Board and Chief Executive Officer, Mr. Howard BOYD,

on the other hand,

have agreed to establish the following conditions governing the sale by the Seller and the purchaser by the Purchaser of liquefied natural gas FOB Arzew, Algeria.

ARTICLE 1 - DEFINITIONS

The terms and expressions used herein shall have the following meanings for purposes of this Agreement :

1 *Natural Gas (NG)*
Any hydrocarbon or mixture of hydrocarbons and non-combustible gases, in a gaseous state, consisting essentially of methane and which are produced from underground structures in a natural state, separately or together with liquid hydrocarbons.

2 *Liquefied Natural Gas (LNG)*
Natural gas at or below its boiling point at a pressure of approximately one (1) atmosphere.

3 *Contractual Cubic Meter (cM^3)*
The quantity of natural gas occupying a volume of one (1) cubic meter at a temperature of fifteen (15) degrees centigrade and at an absolute pressure of one (1) bar.

4 *Standard Cubic Meter (sM^3)*
The quantity of natural gas occupying a volume of one (1) cubic meter at a temperature of sixty (60) degrees Fahrenheit [fifteen and six-tenths (15.6) degrees centigrade] and at an absolute pressure of seven hundred sixty (760) mm of mercury.

5 *Normal Cubic Meter (nM^3)*
The quantity of natural gas occupying a volume of one (1) cubic meter at a temperature of zero (0) degrees centigrade and at an absolute pressure of one thousand thirteen and twenty five hundredths (1,013.25) millibars.

6 *Gross Heating Value (GHV)*
The quantity of heat produced by combustion in air of one (1) cubic meter of anhydrous gas at a constant pressure, the air being at the same temperature and pressure as the gas, after cooling the products of the combustion to the initial temperature of the gas and air, and after condensation of the water formed by combustion.

7 *Standard Cubic Foot (SCF)*
The quantity of natural gas occupying a volume of one (1) cubic foot at a temperature of sixty (60) degrees Fahrenheit [fifteen and six-tenths (15.6) degrees centigrade] and at an absolute pressure of fourteen and six hundred ninety six thousandths (14,696) pounds per square inch.

8 *Thermie (th)*
The calorie (cal) being the quantity of heat necessary to raise, by one (1) degree centigrade, the temperature of one (1) gram of a substance having a specific heat equal to that of water at fifteen (15) degrees centigrade, and at an absolute pressure of one thousand thirteen and twenty five hundredths (1,013.25) millibars, one (1) thermie is equal to one thousand (1,000) kilocalories, one (1) kilocalorie being equal to one thousand (1,000) calories; two hundred fifty two (252) thermies are equal to one million (1,000,000) BTUs.

All references to BTUs, calories, kilocalories, and thermies will be deemed references to gross heating values at constant pressure (GHV) expressed in BTUs, calories, kilocalories, or thermies.

9 *British Thermal Unit (BTU)*
One (1) BTU shall mean one (1) British Thermal Unit, and is defined as the amount of heat required to raise the temperature of one (1) pound of water from fifty nine (59) to sixty (60) degrees Fahrenheit at a constant pressure of fourteen and six hundred ninety six thousandths (14,696) pounds per square inch absolute.

10 *Barrel*
One (1) barrel is equal to forty two (42) U.S. gallons or five and six thousand one hundred forty six ten thousands (5.6146) cubic feet.

11 *Bar*
One (1) bar is equal to fourteen and five hundred four thousandths (14.504) pounds per square inch; one (1) millibar (mb) being equal to one thousandth of one (1) bar.

12 *Day*
A period of twenty four (24) consecutive hours commencing at zero (0.00) hour A.M. Greenwich Mean Time (GMT) and ending at zero (0.00) A.M. GMT on the following calendar day.

13 *Month*
A period commencing at zero (0.00) hour A.M. GMT of the first day of the calendar month and ending at the same time of the first day of the following calendar month.

14 *Contract Years*
Periods of three hundred sixty five (365) consecutive days commencing on the date of the First Regular Delivery and on each anniversary of this date. Any period which contains the date February 29 will have three hundred sixty six (366) days.

15 *Delivery Point*
The point at which the flange coupling at the outlet of the Seller's loading operations joins the flange coupling located at the entry of the receiving apparatus of the LNG tanker of the Purchaser.

ARTICLE 2 – TERM

This Agreement shall become effective on the date of its signature by the parties hereto and shall continue in effect for a period of twenty five (25) Contract Years beginning on the date of the First Regular Delivery.

ARTICLE 3 – INITIAL DELIVERIES

The parties will make every effort to cause the first initial delivery to take place by October 1, 1973. In any event, the initial deliveries shall take place at the latest forty-two (42) months after the effective date of the construction contract for the liquefaction plant.

Until the date of the First Regular Delivery hereunder, the Seller will deliver and the Purchaser will lift as much LNG as is reasonably possible to deliver and to lift during such period.

The quantities thus delivered during such period will be paid for at the base sales price defined in Article 7, paragraph 1).

The initial deliveries will be scheduled by agreement of the parties in order to permit the start-up of the Seller's and the Purchaser's facilities.

ARTICLE 4 – DATE OF THE FIRST REGULAR DELIVERY

The date of the First Regular Delivery shall be the first day of the first month during which the quantity of LNG delivered corresponds to at least one twelfth (1/12th) of the annual contractual quantity provided in Article 6, paragraph 1) of this Agreement.

In any event, the First Regular Delivery shall take place at the latest sixty (60) months after the effective date of the construction contract for the liquefaction plant.

ARTICLE 5 – QUALITY

The LNG delivered by the Seller to the Purchaser shall, in a gaseous state:

— have a gross heating value (GHV) between ten thousand four hundred (10,400) and ten thousand nine hundred (10,900) kilocalories per normal cubic meter [equivalent to between one thousand one hundred six (1,106) and one thousand one hundred fifty nine (1,159) BTUs per SCF]; and contain components varying within the following limits (in MOL percent):

N_2	between	0.60 and 1.40
C_1	between	84.00 and 92.00
C_2	between	6.00 and 8.50
C_3	between	2.20 and 3.00
iC_4	between	0.30 and 0.50
nC_4	between	0.30 and 0.70
C_5+	between	0.00 and 0.02

— contain not more than five tenths (0.5) parts per million by volume of H_2S

— contain not more than ninety five thousandths (0.095) grains/100 SCF [two and three tenths (2.3) mg/nM^3 of mercaptan sulfur

— have a total sulfur content of not more than one and twenty four hundredths (1.24) grains/100 SCF [thirty (30) mg/nM^3].

The gross heating value shall be determined using the physical constants set forth in Annex A.

Verification of the foregoing composition and of conformity with the foregoing specifications and the procedures for such verifications shall be in accordance with the provisions of Article 11.

In any event, the LNG shall not contain any substance capable of harming the Purchaser's facilities.

ARTICLE 6 – QUANTITIES

1) During each Contract Year the Seller will sell and the Purchaser will lift and pay for an annual contractual quantity of four hundred ten trillion six hundred twenty five billion (410,625,000,000,000) BTUs of LNG [equivalent to one hundred three billion four hundred seventy seven million five hundred thousand (103,477,500,000) thermies].

2) If during any Contract Year, the Purchaser fails to lift the annual contractual quantity defined in paragraph 1) of this Article, the Purchaser will be obliged to pay the Seller, at the contractual sales price as defined in Article 7, for the quantities not lifted reduced by, if applicable, the total amount of:

a) the quantities not delivered by reason of an act or failure to act of the Seller,
b) the quantities not delivered or not lifted because of a case of *Force Majeure* or an event assimilated thereto, as defined in Article 9, asserted by one of the parties, and,
c) the quantities delivered and lifted, starting with the date of the First Regular Delivery, in excess of the annual quantities which the Purchaser was obliged to lift pursuant to this Agreement and which have not been utilized during preceding years to compensate any deficiencies in lifting which may have occurred.

Payment shall be made within ninety (90) days following the receipt of the General Final Statement as provided in Article 8.

If the Purchaser asserts a case of *Force Majeure* in accordance with the provisions of Article 9 the effect of which is to delay payments due to the Seller, and if the arbitration tribunal fails to find a case of *Force Majeure*, the Purchaser shall pay to the Seller, in addition to the principal amount due, as liquidated damages, an amount calculated at the rate of fifteen (15) percent per year of the said principal amount.

If, on the other hand, the arbitration tribunal finds the assertion of a case of *Force Majeure* to be justified, the Seller shall pay to the Purchaser, as liquidated damages, an amount calculated at the rate of fifteen (15) percent per year of the contested principal amount.

3) The quantities paid for but not lifted by the Purchaser pursuant to paragraph 2) of this Article will be credited to the Purchaser, who may request delivery thereof at any time during the term of this Agreement, after having taken delivery of the contractual quantity applicable to the Contract Year during which such a request shall have been made.

The Seller shall be required to supply the amounts so requested, either from the liquefaction plant referred to in this Agreement or from other liquefaction plants located in Algeria, only to the extent that the said facilities permit.

For purposes of the foregoing, the Seller agrees:

a) to give priority to production in its own facilities intended for production of the quantities of LNG which are the subject of this Agreement, necessary to make up the quantities referred to in this Article, and then to making up those referred to in the last paragraph of Article 9.
b) to use its best efforts to supply such make-up quantities from other then existing liquefaction plants in Algeria.

ARTICLE 7 – CONTRACTUAL SALES PRICE

1) *Base Sales Price*

The base sales price FOB Arzew of the quantities of LNG which are the subject of this Agreement shall be:

Thirty and one half (30.5) cents, in U.S. dollars, per million BTUs.

2) *Contractual Sales Price*

The contractual sales price applicable to the quantities of LNG delivered pursuant to this Agreement will be calculated on the first day of each Month for the deliveries to take place during said Month, by applying to the base sales price provided in the first paragraph of this Article, the following escalation formula:

$$P = P_0\,(0.80 + 0.04\,T/T_0 + 0.16\,S/S_0)$$

where the symbols are defined as:

P = The contractual sales price.

P_0 = Base sales price provided in the first paragraph of this Article.

S = The latest value of "Average hourly Earnings Excluding overtime of Production Workers on Manufacturing Payrolls" for the "Petroleum and Coal Products" industry, code number C-4, as published by the U.S. Department of Labor, Bureau of Labor Statistics, prior to the first day of the month during which the delivery took place.

T = The latest value of "Steel Mill Products – Price Index", code number 1013, published by the U.S. Department of Labor, Bureau of Labor Statistics, prior to the first day of the month during which the delivery took place.

S_0 and T_0 = The latest published values on the date of the First Regular Delivery defined in Article 4 of this Agreement, of the items in the publications referred to in S and T above.

If the above values and indices should cease to be available, the parties will choose other values and indices by mutual agreement.

ARTICLE 8 – INVOICING AND PAYMENT

Immediately after the completion of each loading, the Seller will prepare and remit to the Purchaser documents concerning the cargo delivered, including in particular: bill of lading, manifest D1, notice of Ready to Receive registration, time sheet, certificates of quantity, quality, authenticity and gauge certificate.

After each lifting, the Seller shall send the Purchaser an invoice, the amount of which shall be stated in U.S. dollars. The Seller shall furnish sufficient information to permit verification of the amount by the Purchaser.

The Purchaser shall pay the Seller in U.S. dollars, to the bank account indicated on the invoice, within fourteen (14) days of the date of receipt of the invoice; if the last such day is not a banking day, the period shall terminate on the next banking day.

In the event of late payment of any invoice, the sums due from the Purchaser shall bear interest at the rate of eight (8) percent per year.

In the event of disagreement concerning an invoice, the Purchaser shall be obliged to make provisional payment of the amount thereof, except in the event of an obvious and substantial arithmetical error. The parties shall have ninety (90) days in which an invoice may be contested. After this period invoices shall be deemed accepted, even in the case of an error.

No later than thirty (30) days after the expiration of each Contract Year. Seller will deliver to the Purchaser a General Final Statement covering the quantities invoiced with respect to such Contract Year.

The General Final Statement referred to shall also cover:

a) The quantities not delivered by reason of an act or failure to act of the Seller,
b) the quantities not delivered or not lifted because of a case of *Force Majeure* or an event assimilated thereto, as defined in Article 9, asserted by one of the parties, and,
c) the quantities delivered and lifted, starting with the date of the First Regular Delivery, in excess of the annual quantities which the Purchaser was obligated to lift pursuant to this Agreement and which have not been utilized during preceding years to compensate any deficiencies in lifting which may have occurred.

ARTICLE 9 – *FORCE MAJEURE*

Any obligation of one of the parties and the corresponding obligation of the other party shall be temporarily suspended for the period during which one of the parties shall be unable to perform the said obligation:

— in the event of a case of *Force Majeure* or an act of God such as, in particular:

- fire, flood, atmosphere disturbances, storm, tornado, earthquake, soil erosion, landslide, lightning, epidemic, etc.
- war, riot, civil war, insurrection, acts of a public enemy, an act of government, administrative decision, etc.
- strike, lock out, etc.

provided that the party asserting a case of *Force Majeure* must prove that the circumstances constitute a case of *Force Majeure*;

— and in the following circumstances that the parties hereby agree to assimilate, for purposes of this Agreement, to cases of *Force Majeure*:

- serious accidental damage to operations or equipment affecting, in Algeria, the natural gas production facilities in the field, transportation by the principal pipeline system, treatment, liquefaction, storage and loading operations; transportation by LNG tankers; and in the United States of America, unloading, storage, regasification, transportation in the principal piepeline or pipelines leading from regasification facilities and in the principal pipeline systems transporting the gas intended for the Purchaser's Principal Customers, of such a nature that the consequences of said damage cannot be overcome by the use of reasonable means at a reasonable cost. The term "Principal Customer" shall mean any customer whose annual purchases of the gas which is the subject of this Agreement shall equal at least ten (10) percent of the annual contractual quantity set forth in paragraph 1) of Article 6. The Purchaser agrees to consult and collaborate with the Seller during the negotiation and the drafting of the *Force Majeure* clauses which are to be included in the agreements between the Purchaser and its Principal Customers.

- an act of a third party affecting the facilities and operations enumerated above, such that said act or its consequences cannot be overcome by the use of reasonable means at a reasonable cost.

The party affected shall as soon as possible following the occurrence of one of the events enumerated above, give notification to the other party by letter, or by telephone or telex confirmed by letter in both cases.

It is agreed that no event shall release the Seller or the Purchaser from its obligations existing on the date of notification, including, in particular, the obligation of the Purchaser to pay the amounts due at that time as payment for the quantities of LNG theretofore delivered.

In any case, the parties shall take all useful measures to ensure resumption of the normal performance of this Agreement within the shortest possible time.

Prior to restoration of the normal situation, the undertakings of the parties shall continue to exist to the extent that their performance is physically possible.

As soon as any cause of *Force Majeure* or event assimilated thereto shall cease to exist, each of the parties may request that the quantities of LNG not delivered by reason of one of the above-mentioned events, be sold or purchased, as applicable, as make-up quantities, within the shortest possible time and to the extent permitted by the capacity of the respective facilities of the parties and the possibilities of marketing such quantities of LNG. The said quantities will be paid for by the Purchaser at the contractual sales price provided in Article 7, which price shall be deemed to be FOB loading port.

ARTICLE 10 – DELIVERY

1) *Schedule of Liftings*

The Purchaser and the Seller shall, before the first of October of each year, establish a schedule of projected liftings for the following year and, for purposes of information only, a projected breakdown by months.

The deliveries will be scheduled for periods of three (3) months, brought up to date on the 20th of each month for the following three (3) months. The Purchaser and the Seller will meet regularly to coordinate the maintenance schedules for their respective equipment and LNG tankers, with the necessary shutdowns taking place, insofar as possible, during the summer.

The parties will fix in the schedules the exact dates and durations of such shutdowns and will attempt to limit the periods of total shutdown to the extent possible.

Except during shutdowns, deliveries will be made as regularly as possible, subject to the provisions of paragraph 5B, of this Article.

2) *Delivery*

The LNG sold and purchased pursuant to this Agreement will be delivered FOB tanker alongside the pier at Arzew by Seller's facilities. The Purchaser shall, however, return to the Seller the natural gas evaporated from the tanker during loading operations.

3) *Receiving Facilities*

A. *Port Facilities*

The Seller has obtained from the appropriate authorities the necessary guarantees permitting it to have use of port facilities capable of receiving LNG tankers with the following maximum dimensions:

length	–	950 feet (290 meters)
width	–	135 feet (42 meters)
draft	–	38 feet (11.60 meters)

The docks and loading facilities referred to above will be available for maneuvers within a reasonable period of time, at any time of the day or night, and under any sea conditions.

These facilities must permit the completely safe execution of all receiving the Purchaser's LNG tankers upon their arrival.

B. *Dock and Loading Facilities*

The Seller will place at the disposal of the Purchaser dock facilities including:

— mooring facilities
— sufficient lighting to permit safe berthing maneuvers, both day and night
— facilities permitting the loading of normal supplies (water and fuel).

The cost of the services referred to above shall be borne by the Seller. The cost of the normal supplies and the telephone connection for the LNG tanker shall be borne by the Purchaser.

The Seller will, in addition, provide at its expense facilities appropriate for loading LNG. The loading will be carried out at an average rate of sixty two thousand eight hundred ninety eight (62,898) barrels per hour [ten thousand (10,000) cubic meters per hour], at a pressure of one hundred (100) psia measured at the Delivery Point.

The loading of an LNG tanker with a capacity of approximately seven hundred fifty thousand (750,000) barrels [approximately one hundred twenty thousand (120,000) cubic meters], will be completed in approximately twelve (12) hours, provided the temperature of the tanks of the LNG tanker is not higher than that provided in paragraph 5 A. of this Article.

The said facilities will include:

— loading arms
— a loading line for LNG
— a line for returning the gas from the LNG tanker to maintain pressure in the storage reservoirs. This line will have a diameter sufficient to prevent the gas pressure in the tanker tanks and storage reservoirs from

exceeding the maximum design working pressure of the tanker tanks and the reservoirs

— storage reservoirs having a capacity of one million eight hundred eighty seven thousand (1,887,000) barrels [three hundred thousand (300,000) cubic meters].

If a pressure greater than one hundred (100) psia (7 kg/cm^2) is necessary to feed the tanks of the LNG tanker at the Delivery Point, such pressure will be provided by the Purchaser.

4) *Safety*

The Seller and the Purchaser will see to it that their respective facilities are equipped with fixed and movable fire-fighting equipment, in accordance with the regulations in force in Algeria, such equipment to be maintained in good condition.

Deliveries of LNG made by the Seller and the receipt of such LNG by the Purchaser will be in strict conformity with all safety regulations and other relevant practices applicable in Algeria.

5) *Delivery Conditions*

A. *Notification of LNG Tanker Arrival*

The Purchaser shall notify the Seller of the dates and hours of arrival at the loading port of the LNG tankers, and of the quantities of LNG to be loaded according to the schedules agreed upon.

The Purchaser will send the Seller the following arrival notices:

— a first notice to be received ninety six (96) hours before the anticipated arrival

— a second notice to be received forty eight (48) hours before the anticipated arrival

— a third notice to be received twenty four (24) hours before the anticipated arrival; and

— a final notice to be received five (5) hours before the anticipated arrival.

In addition to the foregoing, the Purchaser shall notify the Seller by telex of the date and hour which each LNG tanker will leave the unloading port and of its estimated date and hour of arrival in Algeria.

The Purchaser shall insure that the temperature at the bottom of each LNG tanker tank does not exceed minus two hundred twenty (− 220) degrees Fahrenheit [minus one hundred forty (− 140) degrees centigrade] at the time when the "Ready to Receive" notice is given.

The quantity of LNG necessary to cool the bottom of each LNG tanker tank to minus two hundred twenty (− 220) degrees Fahrenheit [minus one hundred forty (− 140) degrees centigrade], if the temperature thereof is between minus one hundred forty eight (− 148) degrees Fahrenheit [minus one hundred (− 100) degrees centigrade] and minus two hundred twenty (− 220) degrees Fahrenheit [minus one hundred forty (− 140) degrees centigrade], and the quantity of LNG necessary for cooling the tanks after drydock of each LNG tanker, which shall not occur more than once per year, shall be invoiced by the Seller and paid by the Purchaser at a price equal to fifty (50) per cent of the contractual sales price provided in Article 7.

Except in the case of normal drydocking of the LNG tankers, if the temperature of the bottom of any tank is higher than minus one hundred forty eight (− 148) degrees Fahrenheit [minus one hundred (− 100) degrees centigrade], shall be equal to the contractual sales price defined in Article 7, of the bottom of such tank to a temperature of minus two hundred twenty (− 220) degrees Fahrenheit [minus one hundred forty (− 140) degrees centigrade], shall be equal to the contractual sales price defined in Article 7.

The quantities of LNG which have evaporated in cooling the tanks shall be returned without cost to the Seller.

Those quantities of LNG which the Seller agrees to supply to the Purchaser for cooling of LNG tanker tanks shall not be taken into account for the purpose of calculation of the contractual quantities.

B. *Notice of "Ready to Receive"*

As soon as the LNG tanker is tied at the dock and ready to receive its cargo, the tanker's Captain or the representative of the Purchaser shall so notify the Seller or its representative at any hour of the day or night.

The Seller will then take all measures useful to permit the LNG tanker to be loaded as quickly as possible.

The Seller shall not be required to load an LNG tanker which arrives for loading sooner than anticipated except to the extent that the LNG is available in sufficient quantities in the storage reservoirs.

C. *Lay Time*

The allotted lay time shall be the time necessary for the loading of the quantity of LNG delivered at an average rate of sixty two thousand eight hundred ninety eight (62,898) barrels per hour [ten thousand (10,000) cubic meters per hour] plus eight (8) hours.

The actual lay time shall begin at the moment when the notice of "Ready to Receive", as defined in paragraph 5 B. of this Article, is given by the tanker's Captain or the Purchaser's representative, or at the start of loading, whichever first occurs, and shall end when the last loading arm of the LNG tanker is disconnected from the Seller's loading line.

D. *Demurrage*

If the allotted lay time is exceeded, the Seller shall pay the Purchaser for demurrage at the rate of twenty four thousand (24,000) U.S. dollars per day, or a prorata portion thereof for any part of a day.

In computing demurrage, the following shall be added to the allotted lay time:

— any period during which the loading was delayed, hindered or suspended for reasons attributable to the LNG tanker
— the time necessary to cool the tanks of the LNG tanker to the temperature specified in paragraph 5 A. of this Article
— any period during which the loading of the LNG tanker was delayed, hindered or suspended through the fault of the preceding vessel
— any period during which the loading of the LNG tanker was delayed, hindered or suspended by reason of a case of *Force Majeure* or event assimilated thereto.

During the period of initial deliveries, the provisions of this paragraph D shall not, however, apply.

ARTICLE 11 – MEASUREMENT AND TESTS

1) *Gauging*

The quantities of LNG delivered persuant to this Agreement will be measured in metric units by gauging the liquid in the tanks of the LNG tanker.

The first gauging shall be carried out by the Seller immediately after the tanker's Captain has given his notice of "Ready to Receive", the LNG tanker being alongside the pier.

The second gauging shall take place immediately after the completion of loading.

These gauging operations shall be carried out by the Seller in the presence of the Purchaser's representatives. The absence of the Purchaser's representatives shall not prevent these operations from taking place.

The Purchaser will give the Seller a certified copy of the gauge tables for each reservoir of the tanker, in metric units, approved by the Bureau of Instruments and Measurements, Algiers–Paris, or by the U.S. Bureau of Standards at Washington D.C., as well as a copy of the correction tables (list, pitch, tank contraction, etc.).

These gauge and correction tables shall be used during the term of this Agreement in the absence of physical change of the tanks, in which case new gauge and correction tables will be used.

The LNG liquid level measuring devices will be chosen by agreement between the Purchaser and the Seller. Each tank will be equipped with two liquid level measuring devices of different types.

2) *Determination of Density*

The density of the LNG will ordinarily be determined by measurement, but, if necessary, by computation. To the extent possible, the density measuring device shall be one of those used for the liquid level measurement of LNG in the tanks of the LNG tankers.

The instruments and the method of computation shall be agreed upon by the Purchaser and the Seller.

3) *Determination of Temperature*

The temperature of the liquid contained in the tanks shall be measured by means of special thermocouples extending from top to bottom of the tank having an accuracy of plus or minus thirty six hundredths (± 0.36) degrees Fahrenheit [plus or minus two tenths (± 0.2) of one degree centigrade], provided that the instruments can give this degree of accuracy.

These temperatures will be either logged or printed.

4) *Sampling*

One or more representative samples of LNG will be taken downstream of and as close as possible to the loading flange of the LNG tankers. The sampling device will be such as to permit the total and continuous vaporization of a quantity of LNG sufficient to permit the taking of a representative gaseous sample.

The device will be chosen by agreement of the Purchaser and the Seller. The samples will be analyzed by means of a chromatograph or by any other instrument chosen by agreement between the Seller and the Purchaser. The analysis, or the average of such analyses, will determine the molecular composition of the LNG.

A calibration of the chromatograph or other analytical instrument used will be performed every seven (7) days in the presence of a representative of the Purchaser by using a known gaseous mixture having a composition very similar to that of the vaporized LNG.

5) *Determination of the Heating Value*

The gross heating value of the vaporized LNG will be computed on the basis of its molecular composition, its molecular weight and the GHV of each of its components expressed in kilocalories per kilogram or in BTUs per avoirdupois pound.

The values of the physical constants to be used for the computation will be taken from the tables of the American Petroleum Institute Research Project 44, as set forth in Annex A hereto; all other values published more recently by this organization will be substituted therefor but not retroactively.

6) *Determination of the BTUs or Thermies Delivered*

The quantity of BTUs or thermies delivered into the LNG tankers shall be calculated by means of the following formula:

$Q = V$ times M times P_c

in which in the case of BTUs:
Q represents the quantity of BTUs delivered expressed in BTUs
V represents the volume of LNG in cubic feet
M represents the density of the LNG measured in the tanks of the LNG tanker in pounds per cubic foot
P_c represents the GHV of the LNG per unit of mass in BTUs per pound

or in case of thermies in which:
Q represents the quantity of thermies (th) delivered expressed in thermies
V represents the volume of LNG loaded in cubic meters (m^3)
M represents the density of the LNG measured in the tanks of the LNG tanker in kilograms per cubic meter (kg/m^3)
P_c represents the PCS of the LNG per unit of mass in thermies per kilogram (th/kg).

7) *Operating Procedures*

The gauging equipment of the tanks of the LNG tankers, as well as the devices for measurements of the density of the LNG, will be supplied, operated and maintained by the Purchaser at its own expense. All of the equipment, devices and apparatus used for determination and testing of the quality and composition of the products shall be supplied, operated and maintained by the Seller at its own expense.

All measurements and all computations relating to the gauging and the determination of LNG density, and all measurements and computations relating to the determination and testing of product quality and composition will be made by the Seller in the presence of a representative of the Purchaser.

The absence of the representatives of the Purchaser having been duly notified, will not prevent either the measurements or the computations from being carried out.

The Seller and the Purchaser shall have the right to inspect at any time the measurement and testing instrument supplied by the other party, provided that such other party be notified in advance.

Calibration and the changing of diagrams shall be carried out solely by the party responsible for the use of the instruments, but the other party shall be given prior notice thereof and shall have the right to be present during such operations. All of the test data, diagrams, computations and other similar information shall, however, be put at the disposition of both parties and kept for a period of at least three (3) years.

8) *Accuracy and Measurement*

The accuracy of the instruments used may be verified upon the request of the Seller or the Purchaser. Such verification can only be made when the two parties are present, utilizing methods recommended by the instrument manufacturers, or any other method agreed upon between the Purchaser and the Seller.

If, at the time of a verification, it is ascertained that a measuring instrument is registering with an error of less than one (1) percent, the previous recordings of such instrument will be considered to be accurate insofar as they relate to the computation of deliveries and such instrument shall be immediately adjusted as necessary.

If, at the time of a verification, it is ascertained that a measuring instrument is registering with an error greater than one (1) percent, the previous recordings of such instrument will be adjusted to zero error with respect to the calibration results for any period which is definitely known or agreed upon and the corresponding corrections will be made with respect to deliveries which occurred during the said period; but in the event that the period during which such error has occurred is not definitely known or agreed upon, such corrections will be made with respect to a period covering one half (1/2) of all deliveries subsequent to the date of the last calibration.

The devices for the measuring of the level of LNG and its density, the temperature of the tanks of the LNG tanker, and the chromatographic analysis of the natural gas, will be those devices having the greatest reliability and the greatest known accuracy at the moment the choice is made.

These instruments will be installed in accordance with recommended practices. The Purchaser and the Seller will attempt to obtain the services of the Measures and Instruments Bureau of Algiers–Paris or the approval of the U.S. Bureau of Standards at Washington D.C. of the devices and measuring instruments used.

The preceding provisions shall not, however, have the effect of modifying any General Final Statement provided for in Article 8, which has been sent to the Purchaser and accepted by the parties.

9) *Disputes*

In the event of a dispute over the choice of the type or the accuracy of measuring devices, their calibration, the result of a measurement, the taking of a sample, an analysis, any computation or a method of computation, the question will be submitted to the Federal Polytechnical School of Zurich (Technische Hoch Schule de Zurich).

All decisions of this organization will be binding on the Seller and the Purchaser.

Expenses incurred in connection with the services of this organization will be shared equally by the Seller and the Purchaser.

ARTICLE 12 – TRANSFER OF TITLE

The LNG shall become the property of the Purchaser at the Delivery Point. At such point all attendant risks shall pass to the Purchaser. The Purchaser and the Seller shall, however, each be responsible vis à vis the other or vis à vis third parties for damages caused by the negligence or acts or failure to act of their employees, agents, or suppliers of goods or services.

ARTICLE 13 – ARBITRATION

Any dispute arising between the Seller and the Purchaser relating to the interpretation or application of this Agreement shall be finally settled by arbitration, at Geneva, in accordance with the Rules of Conciliation and Arbitration of the International Chamber of Commerce (I.C.C.) by one or more arbitrators appointed in accordance with the said Rules; such arbitrators shall be empowered to act as "amiables compositeurs".

Notwithstanding the foregoing, disputes of a technical nature, other than those for which a settlement procedure is provided in Article 11, shall be previously submitted to an expert or organization chosen by both parties; failing agreement, such choice shall be made by the President of the International Chamber of Commerce.

The conclusions of the expert or organization shall be considered as accepted by both parties unless at least one of the parties hereto has expressly rejected them within one (1) month from the date on which they were communicated to such party.

ARTICLE 14 – SUBSTITUTION AND ASSIGNMENT

Either of the parties shall be entitled to substitute for itself any successor or assignee provided that such party has obtained the prior written authorization of the other party.

The prior written authorization referred to above shall not be required in the event that the successor or assignee is a subsidiary of the assignor. For purposes of this Agreement, a "subsidiary" shall mean any company in which the assignor owns more than fifty (50) percent of the capital stock.

In any event, the instrument of substitution or assignment shall include an express undertaking by the successor or assignee to comply with all the provisions of this Agreement and an undertaking by the assignor to continue to guarantee to the other party hereto the due performance by its successor or assignee of the contractual obligations provided herein.

ARTICLE 15 – EXCHANGE OF INFORMATION

In order to facilitate co-ordination of the construction schedules for their facilities, the parties hereby agree to exchange information every three (3) months up to the 30th month after the effective date hereof, and thereafter on a monthly basis.

ARTICLE 16 – SECRECY

Any confidential information or documents which become known to the parties in connection with the performance of this Agreement shall not be used or disclosed to third parties unless so agreed by the parties hereto.

ARTICLE 17 – NOTIFICATIONS

Unless otherwise agreed by the parties, any notification by one of the parties to the other for the purposes of this Agreement, shall be made in writing and airmailed:

by the Seller to the Purchaser, at the Purchaser's offices:

EL PASO NATURAL GAS COMPANY
P.O. BOX 1492
EL PASO, TEXAS 79999, U.S.A.

by the Purchaser to the Seller, at the Seller's offices:

Société Nationale SONATRACH
Immeuble "Maurétania"
ALGIERS, ALGERIA

ARTICLE 18 – TAXES, CHARGES AND DUTIES

Any taxes, charges or duties payable in Algeria in connection with performance of the operations as provided in this Agreement shall be borne by the Seller with the exception of the port taxes, charges and duties affecting the LNG tankers, which shall be borne by the Purchaser. Any increase in any of the port taxes, charges and duties referred to above on or after July 15, 1969 shall be borne by the Seller.

ARTICLE 19 – CONDITIONS

Except for the provisions of Articles 13 to 17 inclusive, 20, 22 and 23, which shall be applicable upon signature of this Agreement, this Agreement is entered into subject to the following four (4) conditions precedent:

a) That the appropriate Algerian authorities shall, on conditions considered acceptable by the Seller, have delivered the required authorizations permitting the sale and exportation to the United States of America of the LNG which is the subject of this Agreement.
b) That the financing of the investments to be made by the Seller to perform its obligations shall have been obtained on conditions considered acceptable by the Seller.
c) That the appropriate United States authorities shall, on conditions considered acceptable by the Purchaser, have delivered the required authorizations permitting the importation and sale in the United States of America of the LNG which is the subject of this Agreement.
d) That the financing of the investments to be made by the Purchaser to perform its obligations shall have been obtained on conditions considered acceptable by the Purchaser.

The parties shall acknowledge in writing the fulfilment of all the foregoing conditions precedent.

ARTICLE 20 – ADMINISTRATIVE AUTHORIZATIONS

Each of the parties shall use its best efforts to obtain the authorizations referred to in Article 19 as soon as possible. For this purpose, the parties shall cooperate closely and will not undertake any action likely to reduce the chances of successfully obtaining the said authorizations.

ARTICLE 21 – GENERAL PROVISION

The Seller hereby confirms that it has at its disposal the quantities of natural gas required for performance of its obligations under this Agreement.

The Seller agrees to reserve the quantities required for the deliveries provided for in this Agreement and to have constructed and operated a gas delivery pipeline system and a gas liquefaction plant, the production of which is to be allocated, on a priority basis, to the performance of this Agreement, having a capacity permitting the production of the LNG.

The Purchaser hereby likewise confirms that it agrees to resell in the United States market the quantities of LNG which are the subject of this Agreement and to have constructed and operated LNG tankers and regasification plants, the use and production of which are allocated, on a priority basis, to the performance of this Agreement, having a capacity permitting the handling of the LNG.

ARTICLE 22 – SCOPE

This Agreement has been signed in accordance with Article 7 of a Memorandum of Agreement executed by the parties on July 15, 1969 relating to the sale of LNG by the Seller and its purchase by the Purchaser. This Agreement constitutes the entire agreement between the parties relating to the sale and purchase of LNG and supersedes and replaces, to the extent necessary, any provisions dealing with the same subject contained in the said Memorandum of Agreement.

ARTICLE 23 – LANGUAGES OF THE AGREEMENT

This Agreement has been executed by the parties in two original counterparts, one in French and the other in the English language. In the event that one of the counterparts should differ from the other and the parties cannot in good faith agree on a common interpretation, the matter shall be settled by arbitration in accordance with Article 13 hereof.

Signed at New York City, New York, U.S.A.
in 4 originals
on October 9, 1969

EL PASO NATURAL GAS COMPANY
By Howard BOYD
Chairman of the Board

Société Nationale SONATRACH
By Sid Ahmed GHOZALI
Président Directeur Général

Gas del Estado – YPF contract

ARGENTINA–BOLIVIA

AGREEMENT FOR THE SALE OF NATURAL GAS*
(Signed 23 July 1968; modified 9 April 1970)

GAS SALES AGREEMENT

In the city of Buenos Aires, capital of the Argentine Republic, on the 23rd day of the month of July of 1968 between GAS DEL ESTADO, a company of the Argentine State, with domicile at Calle Alsina No. 1169, Federal Capital, represented in this act by its interim General Administrator, Ing. Julio T. Carrizo Rueda, hereinafter called "G.E." as one party, and the other party "YACIMIENTOS PETROLIFEROS FISCALES BOLIVIANOS", autarchical entity of the Bolivian State, created by Decree-Law of 21 December 1936, represented by its President of the Board of Directors and its General Manager, Ing. Jose Patino Ayoroa and Lic. Jose Candia Navarro, respectively, with domicile in the city of La Paz, Republic of Bolivia, at Calle Bueno (w/o number) and "BOLIVIAN GULF OIL COMPANY", a com-

*Translation from Spanish.

pany organized under the laws of the State of Delaware, United States of America, whose juridical capacity was recognized by Supreme Resolution of the Republic of Bolivia, No. 72,298, of 26 December 1956, with domicile in the city of La Paz, at Avenida Mariscal Andres de Santa Cruz No. 1322, 6th floor, represented by its Manager in Bolivia and Attorney-in-Fact, Mr. George W. Hall, whose authority is certified by the Power of Attorney which is attached as an integral part of this contract hereinafter called "THE SUPPLIERS", agree to execute this contract for the supply of natural gas of Bolivian origin, which shall be governed by the following clauses and conditions:

CLAUSE 1 – DEFINITIONS

The parties agree to give the following interpretation to the terms expressed below:

I. The term "day" shall mean a period of twenty-four (24) consecutive hours beginning and ending at six (6:00) o'clock a.m. Bolivian time.

II. The term "month" shall mean a period beginning at six (6:00) o'clock a.m. on the first day of a calendar month and ending at six (6:00) o'clock a.m. on the first day of the next succeeding calendar month.

III. The term "year" shall mean (a) a period of twelve (12) months beginning on the first day of the month after the first delivery of gas hereunder and (b) each twelve (12) month period thereafter.

IV. The term "gas" shall mean natural gas including gas produced from gas wells and oil wells, and the residue gas therefrom and/or gas enriched with hydrocarbon mixtures.

V The term "residual gas" means the balance of the gas after excluding 80% of the propane and all the heavier hydrocarbons heavier than propane.

VI The term "liquid gas" means 80 % of the propane and the 100% of the butanes contained in the gas.

VII The term "gasoline" means the isopentane, pentane and heavier hydrocarbons contained in the gas.

VIII The term "heating value" means the higher heating value (Gross Heat of Combustion) or that is the total number of calories which may be produced by combustion at a constant pressure of a quantity of gas saturated with water vapor which occupies a cubic meter, at a temperature of 15 degrees centigrades (15°) and an absolute pressure of 1.033 kilograms, per square centimeter, with condensation of water vapor of combustion (ASTMD 900-55). Its equivalent using British Thermal Units (BTU), cubic feet, 60 degrees F. and 14.73 pounds per square inch absolute shall be defined by : 1 BTU per cubic foot – 8,899 per cubic meter.

For the purposes of this Contract the determination of calorific value shall be done by calculating on the basis of composition of the gas determined by chromatography.

IX The values of the physical constants corresponding to the hydrocarbons which are components of the gas, which may be necessary to adopt for purposes of the calculation, shall be those that appear in the ASTM SPECIAL TECHNICAL PUBLICATION No 109-A "PHYSICAL CONSTANTS OF HYDROCARBONS C_1-C_{10}" translated to metric units.

CLAUSE 2 – OBJECT

The SUPPLIERS, YACIMIENTOS PETROLIFEROS FISCALES BOLIVIANOS and BOLIVIAN GULF OIL COMPANY, jointly and severally responsible for purposes of this contract, obligate themselves to furnish and deliver to "G.E." of the Argentine Republic and this company agrees to receive and pay to said SUPPLIERS under the conditions established in this instrument, the contractual quantities of gas stipulated.

CLAUSE 3 – MEASUREMENTS

For purposes of this contract, the parties agree to use the units of measurements of the decimal metric system, setting out in each case the equivalent in measurements of the United States of North America purely for purposes of simple illustration. The volume of gas to be delivered shall be determined as follows:

a) The volume for the purpose of measurement shall be one (1) cubic meter of gas at a base temperature of fifteen degrees centigrade (15°) and an absolute pressure of 1,033 kilograms per square centimeter.
b) Appropriate correction for deviation from Boyle's Law shall be made in accordance with the American Gas Association Manual for Determination of Supercompressibility Factors, dated December 1962, and as such may be revised in the future, except that experimental determination, duly proved by both parties would indicate an appreciable deviation of the values tabulated. In this last instance, the procedure to be followed to determine this factor shall be agreed upon.
c) The average atmospheric pressure shall be established by mutual agreement taking into account the actual elevation of the point of delivery above sea level and shall be fixed for the entire period of this contract.

CLAUSE 4 – QUANTITIES

The average daily volumes of gas to be supplied at the delivery point by the SUPPLIERS to "G.E." shall contain the following quantities of residual gas:

	Daily Contract Quantity	
	Cubic Meters	*MCF*
a) For the first seven (7) years, commencing with the first delivery of gas and computed in the manner indicated in subparagraph III of Clause 1.	4,000,000 equivalent to	141,261
b) For each of the thirteen (13) years following.	4,500,000 equivalent to	158,919

The daily averages established in this Clause may diminish in the period included between the first of December and 31st March of each year, up to 20% of the daily contractual quantities and "G.E." shall make up this diminution during the rest of the period without exceeding in any one day twenty (20%) per cent over the fixed average.

CLAUSE 5 – COMMENCEMENT OF OPERATIONS

I The deliveries of gas referred to in this contract shall commence after the date in which completion of the facilities of the SUPPLIERS necessary to transport and deliver the product. However, the SUPPLIERS obligate themselves to place into operation said facilities in the maximum period of twenty-four (24) months after the effective date of this contract.

Ninety (90) days in advance, the SUPPLIERS shall notify "G.E." in advance of the date of commencement of the operation. For the compliance of this Clause "G.E." obligates itself to construct, modify and/or adapt the installations to receive the gas in the Argentine Republic, in such a way to be able to receive the contractual quantities after the date in which the facilities of the SUPPLIERS begin their operation.

II During the time in which the construction, modification and/or adaptations of the means of transportation take place, the Parties agree to notify each other mutually every three (3) months concerning the status of progress of the work for the purpose of establishing with the greatest exactness possible, the date of commencement of the operations.

III Any delay which may occur in the commencement of operations shall not amplify any of the terms established in this contract.

CLAUSE 6 – POINT OF DELIVERY

The gas to be supplied hereunder shall be delivered to "G.E." by the SUPPLIERS at the point to be fixed by mutual agreement of the Parties in the vicinity of the town of Yacuiba (Bolivia).

For purposes of this contract the connection of the two transportation systems shall be considered as located in the Argentine–Bolivian border and as the point of delivery of gas.

The SUPPLIERS as well as "G.E." shall install their measuring installations at a distance not greater than one (1) kilometer from said border.

CLAUSE 7 – PRESSURE

The gas shall be delivered at the point of delivery specified in the preceding clauses and the system to receive the gas shall not exceed a pressure of forty-nine (49) kilograms per square centimeter gauge equal to seven hundred (700) pounds per square inch. In case of diminution of the volume of gas which passes through the measuring station lower than the volumes promised, a shortage shall be considered as existing if the pressure is less than forty-nine (49) kg/cm^2 and shortage at the point of reception if the pressure is equal or greater to this value.

CLAUSE 8 – MEASURING STATIONS

I The SUPPLIERS shall install, maintain and operate, at their own expense, at or near the point of delivery, a measuring station designed to measure the gas whose operation and design shall be in accordance with the standards recommended by the American Gas Association, upon which records shall be prepared the respective invoices. Said station shall be equipped with the appropriate measuring devices, such as, volume gauges, gravitometers, gas temperature recorders, chromatographs and all other equipment necessary for the correct measures of the gas volumes to be delivered.

Orifice meters shall be installed, maintained and operated in accordance with the volumes to be computed in conformity with the specifications contained in Report No. 3 of the Gas Measuring Committee of the American Gas Association (Gas Measurement Committee Report No. 3) in the Manual to Determine the factors of Supercompressibility for Natural Gas of December 1962 and to the revisions which in the future may be approved for the aforementioned Report and Manual.

The technical annex entitled letter "A" which the contractors sign considering same a part of this contract contains the specifications and supplementary norms which the Parties agree to adopt for the installation and operation of the measuring equipment and analysis*.

II The right of access to the recording and measuring equipment is recognized in favor of "G.E.", and reciprocally, in favor of the SUPPLIERS the power of access to the control equipment which shall be eventually installed by "G.E.". The calibration and arrangement of the installations mentioned above, as well as the change of charts, shall be exclusively carried out by the party who owns said equipment.

The right of access recognized above, shall include the power to attend and witness the installations, reading, repairs, maintenance, changes, inspections, tests and calibrations which may be carried out in the measuring equipment used by the parties within the terms of this contract.

III The records which may be obtained in the measuring equipment shall be owned by the party which installs same, but, upon written request of the other party the charts and calculations shall be delivered to the other party for its inspection and control. The devolution of said documents must be carried out in the period not to exceed twenty (20) days. The SUPPLIERS agree to deliver daily to "G.E." a copy of the results obtained in the chromatographic analysis.

IV In the event that any of the equipment for measuring and/or analysis does not operate or registers incorrectly the volumes and/or characteristics of the gas delivered the measuring and/or analysis shall be calculated in the following manner:

By use of the record of any testing equipment if same is installed and is recording with precision. If it is impossible to use the system, the party shall proceed to:

a) Correct the error, if the percentage of the latter is determinable by

*Technical Annex "A" has not been reproduced.

calibration, testing or mathematical calculation. Failing this:

b) Calculation of the volume and/or characteristics by comparison with the deliveries which may have been made during the period and under similar conditions in which the equipment has recorded deliveries with exactness.

V If "G.E." should give notice indicating its desire to test the precision of any equipment for measuring and/or analysis, the SUPPLIERS obligate themselves to obtain the prompt verification of the precision of said equipment.

VI If upon verification of any equipment, it is found that an error exists of magnitude which does not exceed the limits admitted in the technical annex "A", the previous records for said equipment shall be considered exact in the computation of deliveries, without prejudice to the obligation of correcting it immediately in order to guarantee a correct reception of the data. If the error found in a recorder, exceeds the limits tolerated in technical annex "A" referred to above, it shall be considered that any previous record of said equipment must be calibrated to error 0 for any period known or upon which the parties may enter into by mutual agreement. However, in the event that such period cannot be established correctly or the parties do not agree on same, the correction shall be carried out covering a period greater than half the time which has transpired since the date of the last test and provided that it is not greater than a period of correction of sixteen (16) days.

VII The parties shall keep all testing data, charts and other similar records during a minimum period of five (5) years.

VIII The SUPPLIERS must install and operate a system of radio communication which will permit "G.E." to connect with same for purposes of interchange of information concerning details of operation.

CLAUSE 9 – SPECIFICATIONS

I The SUPPLIERS agree that the residual gas delivered hereunder shall have a total heating value of not less than 9,300 calories per cubic meter, equivalent to 1,045 British Thermal Units (BTU) per cubic foot.

II The SUPPLIERS also agree that the gas to be delivered shall meet the following specifications:

a) Shall not have a water content in excess of 113 miligrams per each cubic meter of gas measured at 1.033 kilograms per square centimeter absolute of 15°C. (7 pounds per million cubic feet) as determined by continuous use of a recording water vapor content apparatus which the SUPPLIERS shall install, operate and maintain.

b) Shall not contain more than 23 miligrams of hydrogen sulfide per cubic meter of gas (1 gram per 100 cubic feet of gas) as determined by quantitative test after the presence of hydrogen sulfide has been indicated by qualitative test of a type agreed upon by the parties hereto.

c) Shall not contain any free sulphur or mercaptan sulphur nor more than

460 miligrams of sulphur compounds per cubic meter of gas (20 grains per 100 cubic feet).

d) Shall not contain CO^2 in excess of two (2) per cent.

III Except as otherwise specifically provided to the contrary, all measurements of gas mentioned shall have a reference temperature of 15°C, and an absolute pressure of 1.033 kilograms per square centimeter (59°F and 14.696 pounds per square inch).

IV In addition to meeting the above specifications, the gas delivered hereunder shall be commercially free from dust, gums, gum forming constituents, or other solid matter which might become separated from the gas in the course of its transportation through pipelines.

V Taking into consideration, the total duration of the contract, the gas to be supplied shall contain an average quantity of propane and butane not less than 4.15% MOL.

It is recognized in favor of the SUPPLIERS the right to deliver quantities of propane and butane greater than the average established in the foregoing paragraph, but not greater than six per cent (6%) MOL in order to compensate for eventual shortages in the future.

CLAUSE 10 – PRICES

"G.E." shall pay the SUPPLIERS for the gas subject of this contract, the following unitary prices:

a) For residual gas:
US $0.225 for each 28.31635 cubic meters at 9.300 calories/m^3, (US $0.225 per one thousand cubic feet at 1,045 BTU).
In case the residual gas has a heating value greater to the one indicated, the price shall increase proportionately to the greater quantity of calories delivered. Gas with a heating value less than the one indicated shall not be accepted.

b) For liquid gas:
US $19.90 per ton (US $0.04 per gallon at 3.785 liters).

c) For gasoline:
US $0.0475 per gallon at 3.785 liters.

These prices shall be calculated based on volumes delivered and on the result of the analyses which may be registered daily in accordance with the terms of this contract and the procedures established in annexes "A" and "B".

It is clarified that annex "B" referred to, which also forms an integral part of this instrument, has been composed with the sole purpose of realizing the calculation of the prices for billing which the SUPPLIERS shall prepare monthly.

CLAUSE 11 – BILLING AND FORM OF PAYMENT

I The SUPPLIERS shall present an invoice each month, to "G.E." in triplicate, indicating thereon the deliveries made during the preceding month, setting out separately the different items that are contained therein.

Within sixty (60) days following the presentation of each invoice, "G.E." shall deposit in a bank in the city of New York (U.S.A.), which will be indicated, in writing, by the SUPPLIERS and a joint account which they shall open for this purpose, the net amount without any deduction whatsoever, in dollars of the United States of America.

For purposes of facilitating payment of the invoices in the United States of America, the SUPPLIERS obligate themselves to designate a bank which is a correspondent bank of the Bank of the Argentine nation.

II In the event that any one invoice is not paid within sixty (60) days following its presentation the amount shall accrue eight per cent (8%) interest annually from its due date, placing "G.E." in default automatically without necessity of notice, notification of judicial or extrajudicial interposition and without prejudice to the right of the SUPPLIERS to seek recourse through legal channels and suspend the deliveries of gas under the conditions set forth in Clause 12.

III In case of discovery of an error or a difference in the quantities invoiced by the SUPPLIERS, the correction shall be effectuated within thirty (30) days following the claim. In order to comply with the foregoing paragraph, the parties shall have a maximum term of twelve (12) months, to be computed from the date of receiving the invoice, during which they shall formulate the eventual claims for readjustment.

IV When the invoice is questioned by "G.E." in accordance with that provision of the first paragraph of the foregoing subparagraph, said company is obligated to pay according to practice, ninety per cent (90%) of the amount invoiced paying the balance once the parties have agreed on the disputed difference.

V If, in conformity with the provisions of subparagraph II, of Clause 4 "Quantities" of this contract, "G.E." shall be obligated to pay for quantities of gas that have not been effectively received during any of the years in which this contract is in force, the SUPPLIERS, within thirty (30) days following the end of each year, shall present to "G.E." an invoice in dollars, of the United States of America, corresponding to the value of the value of the volumes of gas placed at the disposal of "G.E." but not received.

Within sixty (60) days after presenting the invoice mentioned in the preceding paragraph, "G.E." shall make its payment in the form provided for in subparagraph one (I) of this Clause and under the conditions established in subparagraph II above.

CLAUSE 12 – GUARANTEES

For the purpose of assuring the normal and correct supply of gas and to guarantee the regular payment of such supply, the parties agree on the following:

I The total interruptions and/or the diminutions in excess of thirty (30) per cent in the supply of gas, shall not exceed seventy-two (72) consecutive hours. On the other hand, without prejudice to the provisions of the following subparagraph II, said interruptions and/or diminutions may not exceed seventy-two (72) hours in the period of thirty (30) consecutive days.

While the maximums established above are not exceeded, the interruptions and/or diminutions in the supply of gas shall not incur a corresponding penalty whatsoever to the SUPPLIERS.

II The interruptions which exceed the margins indicated in the proceding paragraph, as well as the supply of gas which may be less than 95% of the total annual amount agreed upon, shall obligate the SUPPLIERS to pay to "G.E." a sum equal to eighteen and a half (18.5) cents dollar for each 28.31635 cubic meters not supplied, provided that the shortages have not been compensated with additional injections of gas from the deposits of the Northern part of Argentina, over the volumes programmed before for these deposits.

III In the event that the supply of gas may have suffered interruptions or diminutions provided for in the preceding paragraph II, the SUPPLIERS shall have the right to deliver in the same year or in the following year to that year in which deficits shall have occurred, volumes equivalent to the additional ones extracted from the aforementioned Argentine deposits, and "G.E." shall be obligated to receive same and pay for them at the prices established in Clause 10 of this contract. For such purposes, "G.E." shall provide the SUPPLIERS with the production programs of said deposits once the same have been approved.

IV In the event that the gas placed at the disposal of "G.E." in the conditions stipulated above during any period of 365 consecutive days, is less than 75% of the total volume undertaken for such period, "G.E." may suspend the taking of gas and its correlative obligation to pay, without prejudice to the sums which it shall owe on that date. In this case, the parties obligate themselves to meet to consider the basis and standards upon which the renewal of the supply shall be based and in the event that the parties do not reach an agreement in the maximum period of three months computed from the date in which the taking is suspended, the rights and obligations of both parties provided for in this contract shall cease. For purposes of this subparagraph IV the parties agree and accept that fortuitous cause or *force majeure* cannot be invoked as a determining factor for the interruption or diminution of the supply of gas. For the calculation which may be necessary in the application of this subparagraph, the parties agree to accept the deficits over the average volume established in Clause 4 during the months of December, January and February with coeficients 0.7 and the months of June, July and August with coeficients 1.3.

V "G.E.", during each one of the years in which this contract shall be in effect, obligates itself to receive and pay for a volume of gas equivalent to one hundred (100) per cent of the average daily contractual quantities, multiplied by the number of days of the year referred to. The obligation imposed in the foregoing paragraph shall be exclusive of the effective takings of gas, and "G.E." shall be obligated to pay all those quantities which the SUPPLIERS shall place at its disposal in the conditions agreed to and in accordance with the provisions of Clause 7 of this contract.

VI In the event that during any year, "G.E." has not been able to receive the volumes of gas agreed upon and same have been placed at its disposal by the SUPPLIERS, from the time that "G.E." has paid for them in the manner provided in the corresponding Clause, "G.E." shall have the right to require and receive the volumes not delivered without obligation to pay for such volumes again, and "G.E." shall lose this right if it fails to request and receive such volumes in the same year or in the year following that in which its takings or diminutions cease. In order to comply with the right recognized in favor of "G.E." in the foregoing paragraph, the SUPPLIERS agree to deliver all of the gas compatible with the existing capacity of its transportation system and with the possibilities of its gas deposits.

VII In the event that the payments by "G.E." shall be delayed more than ninety (90) days in relations to the due date of any invoice, without prejudice to the interests stipulated in the second subparagraph of Clause 11, the SUPPLIERS shall have the option to suspend the supply of gas and "G.E." shall have the obligation to pay the sums owing up to that date. For purposes of this subparagraph VII, "G.E." shall not have the right to invoke fortuitous cause or *force majeure* to determine the delay or failure to comply with payments.

VIII The stipulations of this present Clause shall become effective after one (1) year computed from the date in which the period of twenty-four months established in Clause 5 expires for the termination of the work which both parties must carry out but not before "G.E." shall have concluded its receiving facilities. Consequently, the norms of this clause shall operate by mere expiration of the terms and without the conditions which Clause 20 stipulates for the duration of the contract.

CLAUSE 13 – ARBITRATION

I All doubts or controversies between the parties concerning the interpretation of any of the clauses of this contract shall be submitted to the judgment of friendly arbitrators, one to be designated by each party respectively, and a third arbitrator which shall be designated by mutual agreement by those appointed by the parties or if the parties shall fail to agree, the third arbitrator shall be appointed by:

a) The President of PEMEX (Petroleos Mexicanos).
b) Failing the above, by the President of the French Institute of Petroleum.
c) Failing the two methods indicated above, by the American Arbitration Association.

The third arbitrator shall act only in the event that the arbitrators appointed by the parties differ in their conclusion and must settle the question presented.

II The place and procedure for arbitration shall be determined by both parties through the respective arbitration agreement which shall contain, in detail, the point or points which are the objects of the arbitration. In the event of a difference concerning the clauses in the arbitration agreement, the parties shall submit the decision to the third arbitrator.

III The arbitration award shall be final and cannot be appealed by the parties unless it is issued after the term established in the arbitration agreement or unless it is issued concerning points not agreed upon, it shall thereby be considered null and void. The Courts of the Republic of Argentina and Bolivia shall be competent, as the case may be, to enforce compliance with the arbitration award issued in accordance with this clause and to determine judicially any controversy with respect to the validity or invalidity of such awards. (Translator's note: Apparently there are two lines missing from the Spanish copy.)

CLAUSE 14 – NOTICES AND DOMICILE

The notices which must be given by one of the parties to the other must be addressed in the following manner:

For GAS DEL ESTADO
Calle Alsina 1169
Federal Capital
Argentine Republic

For THE SUPPLIERS
A) YACIMIENTOS PETROLIFEROS FISCALES BOLIVIANOS
Calle Bueno Esquina Camacho
La Paz (Bolivia)
and
Avenida Corrientes 545 – 2nd Floor
Federal Capital
(Argentine Republic)

B) BOLIVIAN GULF OIL COMPANY
Ave. Mariscal Andres de Santa Cruz No. 1322
La Paz (BOLIVIA)

For purposes of any notice to the SUPPLIERS, "G.E." must send the correspondence to the addresses specified above. Any change of address by any of the parties must be given to the other in writing.

CLAUSE 15 – CONTRACTUAL NONCOMPLIANCE

If any one of the parties should fail to exercise its right of claiming or commencing the pertinent legal procedures due to noncompliance with any one of the obligations established in this contract, such fact shall not be invoked as precedent nor shall it imply renouncement to the power of claiming future noncompliance.

CLAUSE 16 – OBLIGATIONS OF THE PARTIES AND TRANSFERS

This contract obligates the parties, their successors and eventual assignees who shall be subrogated to the rights and obligations which are stipulated herein.

The total or partial assignment of this contract may only be made in the event that the assignee can meet the same technical and economic conditions of solvency of the assignor and for such purpose a notification of the transfer shall be given at least ninety days in advance of such transfer. Without prejudice to the stipulations above, the right of the SUPPLIERS to assign the operation and maintenance of the transportation system and delivery to a third company which may be a subsidiary of the latter, shall not be considered as an assignment or transfer.

Notwithstanding the provisions of the foregoing paragraphs, once the total or partial transfer has been effectuated, the assignee shall constitute itself as full and complete guarantor of all of the obligations transferred up to their total and definitive compliance in accordance with the stipulations of this contract.

The parties agree that this contract can be offered as collateral guarantee and/or security of the investments which may be required for the construction, amplification and/or modification of the transportation systems.

CLAUSE 17 – *FORCE MAJEURE*

With the exception of the provisions in paragraph IV and VII of Clause 12, according to which the parties may not invoke *force majeure* and/or fortuitous cause, nor the SUPPLIERS nor "G.E." may be liable for failure or noncompliance of the terms of this contract, if its execution has been prevented, delayed or hindered by fortuitous cause or *force majeure* including within same all events or circumstances which are outside of the control of the parties.

Once the cause of *force majeure* and/or fortuitous cause is produced, the party affected by same shall notify the other in writing and both parties shall exercise the necessary efforts to overcome the situation.

Upon termination of the event or circumstances which serves as a basis for the notification, the incidence of *force majeure* and/or fortuitous cause shall also cease and said incidence shall serve only to suspend, totally or partially, the contractual obligations of the parties only during the period of its duration and same shall not be construed as an extension of the duration of this contract.

CLAUSE 18 – TAXATION

In accordance to the nature of this contractual operation of exportation and importation of gas, and having stipulated that the prices shall be received by the SUPPLIERS without any deduction whatsoever, all taxes, or levies of any kind whatsoever, present or future which may be established by the Argentine Authorities and which could affect them, shall be for the exclusive account of "G.E.", which shall also bear, in total and exclusive manner any other Argentine tax, or fiscal charge which may be derived from the signing and/or execution of the contract. Co-extensively, the SUPPLIERS assume for their exclusive account, all of the taxes and fiscal charges of any kind whatsoever which the Bolivian Authorities have established or may establish in the future and which could be applicable because of the signing and/or execution of the contract.

CLAUSE 19 – COMPLETENESS OF THE CONTRACT

The parties agree that this contract and its Annexes constitute the only and final expression of their understanding to which they have arrived, and all other previous instruments up to the present which are related with the purpose of this agreement shall now be without force and effect.

CLAUSE 20 – TERM

The present contract shall enter into force and effect on the date in which the Governments of the Republics of Bolivia and Argentina execute the corresponding document or documents for the exportation and importation and purchase of Natural Gas between the two nations and shall have a duration of twenty (20) years after the first delivery of gas, or alternatively, the delivery to "G.E." of 31,572,500,000 cubic meters of gas considering the contract as terminated at the moment in which any one of the two events first occurs.

CLAUSE 21 – TRANSITORY PROVISION

To the extent possible and taking into account the technical and economic interests of the construction of the gas pipeline Santa Cruz–Border, the SUPPLIERS shall give preference to the use of materials and equipment of Argentine origin.

The present contract, except for express agreements to the contrary, is irrevocable and irretractable.

In Witness Whereof, two copies are signed which shall be of equal force and effect, at the place and date indicated above.

GAS DEL ESTADO
(Signature not discernible)
Ing. Julio T. Carrizo Rueda
Int. General Administrator

YACIMIENTOS PETROLIFEROS FISCALES BOLIVIANOS
(Signature not discernible)
Ing. Tenl. Jose Patino Ayoroa
President
(Signature not discernible)
Lic. Jose Candia Navarro
General Manager

BOLIVIAN GULF OIL COMPANY
(Signature not discernible)
Sr. George W. Hall
Manager and Attorney-in-Fact

AMPLIFICATION AND MODIFICATION OF GAS SALES AGREEMENT

In the city of Buenos Aires, Capital of the Argentine Republic, on the 9th of April of 1970, between GAS DEL ESTADO, Company of the Argentine State, domiciled in 1169 Alsina Street, Federal Capital, represented in this act by its Administrator General, Ing. Julio T. Carrizo Rueda, hereinafter called "GE" as one of the parties and as the other YACIMIENTOS PETROLIFEROS FISCALES BOLIVIANOS, an autarchical entity of the Bolivian State, created by Decree-Law of 21 December 1936, represented by the President of the Board of Directors and by its General Manager, Col. Ing. Jose Patino Ayoroa and Ing. Rolanda Prada Mendez, respectively domiciled in the City of La Paz, Republic of Bolivia, in Calle Bueno, hereinafter called "THE SUPPLIER", both with Powers and Authorizations granted in legal form. *Whereas* the Government of Bolivia by means of Supreme Decree No. 8956 of 17 October 1969, provided for the reversion of the petroleum concessions granted to BOLIVIAN GULF OIL COMPANY and the nationalization of all of its properties and rights as well as the intervention of YPFB in the technical and administrative control of the installations and properties of BOLIVIAN GULF OIL COMPANY, agree to subscribe the present Ampliatory and Modifying Contract of that signed the 23 of July 1968 for the supply of natural gas from Bolivian source, which is designated Principal Contract. The modifications shall be regulated by the following clauses:

FIRST: On 23 July 1968, the contract for the supply of natural gas of Bolivian origin was signed between "GAS DEL ESTADO" an Argentine State Company, on the one part and on the other "YACIMIENTOS PETROLIFEROS FISCALES BOLIVIANOS" an autarchical entity of the Bolivian State and the private company "BOLIVIAN GULF OIL COMPANY", in the terms and conditions which are stipulated in said contract.

SECOND: "GE" and "THE SUPPLIER", ratify in all respects the Principal Contract subscribed on 23 July 1968, whose stipulations continue to be valid, with the single exception of the modifications which are set forth in the following clauses of the present Contract.

THIRD: All of the rights and obligations, which by virtue of the Principal Contract of 23 July 1968 correspond to YPFB and to BOLIVIAN GULF OIL COMPANY, as a consequence of the situation derived from D.S. No. 8956 of 17 October 1969, dictated between Government of Bolivia, shall be exercised and fulfilled only by YPFB in its capacity as a party originally obligated in joint and several manner. In such sense, it is established that all of the clauses of the Principal Contract of 23 July 1968, insofar as YPFB and BOLIVIAN GULF OIL COMPANY are mentioned as obligated parties or holders of rights under the common designation of "THE SUPPLIERS", arc modified by excluding BOLIVIAN GULF OIL COMPANY and retaining YPFB which shall be called "THE SUPPLIER".

FOURTH: "GE" and "THE SUPPLIER" agree that the Fifth clause "Commencement of Operations (1)", of the Principal Contract of 23 July 1968, shall be modified in accordance with the following text:

"The deliveries of gas, of which the present Contract deals, shall commence with the date on which the installations necessary for the transport and delivery of the product are completed. Nevertheless, "THE SUPPLIER" shall be obligated to put said installations into operation in the maximum period of "Thirty six months" (36), which period is computed from the 5th of August of 1968, the date on which the diplomatic representatives of Bolivia and Argentina exchanged Reversal Notes authorizing the operations of exportation and importation of gas. With not less than ninety (90) days advance notice, "THE SUPPLIER" shall notify "GE" in writing of the date of commencement of the operations".

For the fulfilment of the present clause "GE" is obligated to construct, modify and/or adapt the gas receiving installations in the Argentine Republic, in such manner as may enable it to receive the contractual quantities, as from the date in which the installations of "THE SUPPLIER" begin to function. The other terms and conditions stipulated in said Fifth clause shall continue in force.

FIFTH: "GE" and "THE SUPPLIER" clarify that in the Eleventh clause, section V of the Principal Contract, by mistake is indicated: "the II incise of the Fourth Clause "Quantities"," when it should have stated: "incise V of the Twelfth Clause "Guaranties"."

SIXTH: For the purposes of curing the error out in the foregoing clause and in order to adapt payment conditions to the situation stipulated in the Third Clause of the Present Contract of Amplification and Modification, the parties agree upon the following wording of the Eleventh Clause of the Principal Contract "Invoicing and Form of Payment":

I) "THE SUPPLIER" shall present each month to "GE" an invoice in Triplicate, stating the deliveries made during the prior month detailing the different items covered thereby. Within sixty (60) days following presentation of each invoice, "GE" shall pay its amount by means of a deposit in U.S. dollars of free convertibility with a banking institution correspondent of official banks of the Argentine Republic, which "THE SUPPLIER" designates.

The referred to deposits shall be made with the intervention of the Central Bank of the Argentine Republic and in the net amount of the invoices without any deduction.

"THE SUPPLIER" by means of a written communication from its legal representatives, shall indicate the Banking Institution which shall act as trustee, and in which "GE" must deposit the payments corresponding to the invoices for the supply of gas, in accordance with the stipulations of the Principal Contract and of the Present Contract of Amplification and Modification.

At the time of subscribing the Guaranty Contract with the World Bank, the modalities and conditions, under which eventual withholdings may be requested from the trustee Bank, shall be agreed upon.

II) In the eventuality that any invoice is not paid within sixty (60) days following its presentation the amount thereof shall earn interest at the annual rate of eight percent (8%) counting from its due date, "GE" being in automatic default without necessity of advice, notification or judicial or extrajudicial summons and without prejudice to the right of "THE SUPPLIER" to resort to legal measures and to suspend the deliveries of gas on the conditions foreseen in Clause Twelve.

III) In case an error or difference in the quantities invoiced is discovered by "THE SUPPLIER", a correction must be made within thirty (30) days following the claim.

In order to comply with the foregoing paragraph, the parties shall enjoy a maximum period of twelve (12) months counting from the date of receipt of the invoices, within which to formulate the eventual claims of readjustment.

IV) When the invoice is protested by "GE", in accordance with the provisions in the first paragraph of the preceding incise, said company shall be obliged to pay in usual form ninety percent (90%) of the amount invoiced, paying the rest once the point object of the protest is settled by the parties.

V) If in conformity with that provided for in incise V in the Twelfth Clause "Guaranties" of this Contract, "GE" should be obliged to pay for quantities of gas which have not been effectively received during any of the years of the validity of the Present Contract, "THE SUPPLIER" within thirty (30) days following the end of each year shall present to "GE" an invoice in U.S. dollars, corresponding to the value of the volumes of gas made available but not received.

Within sixty (60) days after presentation of the invoice of which mention was made in the foregoing paragraph "GE" must make payment in the form provided in incise I of the present clause and under the detailed conditions established in incise II above.

SEVENTH: "GE" and "THE SUPPLIER" agree that the Twelfth Clause part VIII "Guaranties" of the Principal Contract of 23 July 1968, shall be modified in accordance with the following text:

"The stipulation of the Present Clause shall enter into force one (1) year from the date of expiration of the period of thirty six (36) months, as established in Clause Five for the termination of the works which must be carried out by both parties, but not before "GE" has completed its receiving installations. Consequently, its standards shall be effective by the mere passage of the periods. That which is provided for in Clause Twenty in relation to the duration of the Contract shall not be applicable".

EIGHTH: Both parties declare that the other Clauses of the Principal Contract of 23 July 1968, which have not been modified and which are not contrary to the Present Contract of Amplification and Modification shall maintain their validity.

NINTH: Both parties give their conformity to the entirety of the Present Contract, obligating themselves to its true and strict fulfilment and in proof they do sign two counterparts and with one single effect.

Appendix V Assumptions: Alternative Uses to Natural Gas Export

The figures in Tables 11.3 and 11.4, as well as the arguments that these figures underpin, are utilized with a degree of reservation. The reason for this is that the assumptions behind the figures are not entirely explained. They are not capable of being replicated fully and changes in parameters of the various options will have a significant effect on the resulting shadow prices for the natural gas feed. I have assumed that the 12 per cent interest in Table V.1 is in actuality a discount rate, that operation costs are around 10 per cent of capital investment per annum, and that project life is fifteen years. With these assumptions, we get close to the resulting price for natural gas feed figures in Table V.1. It should be pointed out in this connection, however, that the 10 per cent operation costs assumption is probably too high.

On the other hand, the product prices for some of the petrochemical products in Europe could well be too high. They are considerably higher than some alternative figures available to me. For instance, the price of ethylene landed in Europe in 1981 is given as $232 per ton (versus $710 per ton in Bonfiglioli and Cima); for ammonia, $134 per ton (versus $170 per ton); methanol $92 per ton (versus $210 and $160 per ton). The exact period of reference for these alternative figures and conditions of sale is not given in the source available to me (*World Gas Report*, 22 June 1981, pp. 7–8). In the absence of further specifications, it is difficult to know how to interpret the figures in Bonfiglioli and Cima. Suffice it to say, the two authors are probably being overgenerous to the petrochemical/industrial alternatives to export as feasible options, a factor that should be borne in mind while reading Chapter 11.

Table V.1 *Assumptions behind alternative uses of natural gas (Bonfiglioli and Cima)*

	Alternative uses						
			Methanol				
Assumptions	*Refinery fuel*	*Ammonia*	*Chemical*	*Automotive*	*Sponge iron*	*Alumina/ aluminium*	*Ethylene*
Investment[a]	n.a.	$270m.	$140m.	$375m.	$170m.	$1,850m.	$300m.
Interest rate[b]	12%	12%	12%	12%	12%	12%	12%
Equity/ capital ratio[b]	n.a.	n.a.	n.a.	n.a.	n.a.	n.a.	n.a.
Repayment terms[b]	n.a.	n.a.	n.a.	n.a.	n.a.	n.a.	n.a.
Plant capacity	200,000 bbls/day	1,000 tons/day	1,000 tons/day	5,000 tons/day	600,000 tons/year	130,000 tons/year	300,000 tons/year
Product price (Europe)	n.a.	$170/ton	$210/ton	$160/ton	$120/ton	$2,100/ton	$4,710/ton
Period of investment	n.a.	n.a.	n.a.	n.a.	n.a.	n.a.	n.a.
Length of project life	n.a.	n.a.	n.a.	n.a.	n.a.	n.a.	n.a.
Resulting price for natural 'feed'	$3.50/MMBtu	$1.00/MMBtu	$2.70/MMBtu	$2.50/MMBtu	$1.10/MMBtu	$0.30/MMBtu	$7.30/MMBtu
Annual operations cost[c]	n.a.	n.a.	n.a.	n.a.	n.a.	n.a.	n.a.

Notes: a Includes interest charges during construction, plant, infrastructure, but not production costs, pipelines and price cost escalations.

b It is entirely possible that 'interest rate' is a project discount rate that would obviate the need for capital/equity ratios, loan repayment, etc.

c In addition to the cost of natural gas (see text for comment).

Bibliography

Adelman, M. A. (1962) *The Supply and Price of Natural Gas* (Oxford: Basil Blackwell).

Adelman, M. A. (1972) *The World Petroleum Market* (Baltimore, Md.: The Johns Hopkins Press for Resources for the Future Inc.).

Anchishkin, A. (1980) *National Economic Planning* (Moscow: Progress; trans. Jane Sayers).

Anderson, Douglas D. (1980) 'State regulation of electric utilities', in James Q. Wilson (ed.) *The Politics of Regulation* (New York: Basic Books, pp. 3–42).

Arnoni, Y. H., Keens, D. and Lewis, J. P. (1981) 'A technical and economic evaluation of utilization of offshore gas reserves', paper presented at 13th Annual Offshore Technology Conference, Houston, Texas, 4–7 May 1981.

Averich, H. and Johnson, L. L. (1962) 'Behaviour of the firm under regulatory constraint', *American Economic Review*, vol. 52, December, pp. 1052–69.

Bates, R. and Fraser, N. (1974) *Investment Decisions in the Nationalized Fuel Industries* (London: Cambridge University Press).

Blair, John M. (1976) *The Control of Oil* (New York: Pantheon).

Bodle, William and Smith, William (1980) 'Reliability of base load delivery', *Sixth International Conference on LNG* (Kyoto), session 4, paper 10.

Bonfiglioli, G. (1980) 'Economics of gas imports to Europe from OAPEC', *Oil and Gas Journal*, 4 August.

Bonfiglioli, G. and Cima, F. (1980) *The Economics of Gas Utilization in Different Fields*, paper to OAPEC Symposium on 'Ideal Utilization of Associated Flared Gas', Algiers, Algeria, 29 June–1 July.

Brecht, C. (1978) 'Gas supplies into Continental Europe in the 80's', *Institute of Gas Engineers Communication 1067* (London: Institute of Gas Engineers).

Breyer, S. G. and MacAvoy, P. W. (1974) *Energy Regulation by the Federal Power Commission* (Washington, DC: Brookings Institution).

Campbell, R. W. (1968) *The Economics of Soviet Oil and Gas* (Washington, DC: The Johns Hopkins University Press for Resources for the Future).

Campbell, R. W. (1976) *Trends in Soviet Oil and Gas Industry* (Washington, DC: Johns Hopkins University Press for Resources for the Future).

Campbell, R. W. (1978) *Soviet Energy Balances* (Santa Monica, RAND R-2257-DOE, December).

Campbell, R. W. (1979) *Basic Data on Soviet Energy Balances* (Santa Monica, Rand N-1332-DOE, December).

Campbell, R. W. (1980) *Soviet Energy Technologies* (Bloomington, Ind.: Indiana University Press).

Centre du Recherche sur le Droit des Marchés et des Investissements Inter-

nationaux (1973) *Les hydrocarbures gazeux et la développement des pays producteurs* (Dijon: Université de Dijon).
Chantler, Philip (1938) *The British Gas Industry: An Economic Study* (Manchester).
La Communauté Europėene et les Entreprises Publiques (1979) *Public Enterprise in the European Community* (Paris: Les Procèdes d'Orel).
Cowan, D. and Hagar, R. (1982) 'Gas surplus puts squeeze on pipelines and producers', *Oil and Gas Journal*, 19 July.
Cowhey, Peter F. (1982) *'The gas policies of the industrialized nations'*; paper to be published in Ragaei El Mallakh (ed.) *Natural Gas: Policy Planning and Technical Issues* (in press).
Dam, K. W. (1970) 'The pricing of North Sea gas in Great Britain', *The Journal of Law and Economics*, vol. 13, no. 1, April, pp. 11–44.
Dam, K. W. (1976) *Oil Resources: Who Gets What How?* (Chicago, Ill.: University of Chicago Press).
Daniels, E. J. and Anderson, P. J. (1977) 'International LNG projects continue to progress as new plans evolve', *Pipeline and Gas Journal*, June, p. 29.
Davis, J. D. (1981a) 'Future prospects for international trade in LNG' (Århus: Consulting Report to INTECS/Crédit Suisse, December).
Davis, J. D. (1981b) *High Cost Oil and Gas Resources* (London: Croom Helm).
Davis, J. D. (1982) 'Making the rules', paper presented to the European Consortium for Political Research, Århus, Denmark, April.
Davis, J. D. (1983) 'The political economy of European natural gas markets', *Cooperation and Conflict*, vol. 18, pp. 2–20.
Davis, J. D., Svendsen, B. A., with Lau, K. L. and Knudsen, T. S. F. (1978) *Dansk naturgas i 80'erne: Problemer og Muligheder* (Århus: Politica).
Davis, Lee N. (1979) *Frozen Fire* (San Francisco, Calif.: FOE Books).
Dienes, L. (1977) 'The Soviet Union: an energy crunch ahead?' *Problems of Communism*, vol. 26, pp. 41–60.
Dienes, L. and Shabad, T. (1979) *The Soviet Energy System: Resource Use and Policies* (New York: John Wiley).
Drayton, G. (1981) *The Market for LPG in the 1980's* (London: Economist Intelligence Unit, Special Report No. 961, March).
Ebel, R. E. (1970) *Communist Trade in Oil and Gas* (New York: Praeger).
Economist Intelligence Unit (1975) *Soviet Natural Gas to 1985* (Quarterly Energy Review, Special No. 24, August).
Elliot, J. F. (1974) *The Soviet Energy Balance: Natural Gas, Other Fossil Fuels and Alternative Energy Sources* (Boston, Mass.: Lexington).
Ellman, M. (1971) *Soviet Planning Today: Proposals for an Optimally Functioning Economic System* (Cambridge: Cambridge University Press).
ELSAM (1982) *Afsvovlingsundersøgelse* (Fredericia, Denmark: September).
'Etat des contrats internationaux de gaz naturel à long terme au ler Mars 1977' (1977), *Le Pétrole et le gaz Arabes*, vol. 9, 199, pp. 21–3.
Frankel, P. H. (1969) *Essentials of Petroleum* (London: Frank Cass).
Galbraith, J. K. (1970) *The New Industrial State* (Harmondsworth, Middx: Penguin).

Gas Committee, UN Economic Commission for Europe (1961–82) *Annual Bulletin of Gas Statistics for Europe* (New York: United Nations).

Gas Committee, UN Economic Commission for Europe (1981) *Annual Bulletin of Gas Statistics 1980* (New York: United Nations).

Gas Gathering Pipeline Systems in the North Sea (1976) (London: Department of Energy).

Gerwig, R. W. (1962) 'Natural gas production: a study of costs of regulation', *The Journal of Law and Economics*, vol. 5, no. 2, October, pp. 69–92.

Goldberg, V. (1976) 'Regulation and administered contracts', *The Bell Journal of Economics*, vol. 7, no. 2, Autumn, pp. 423–45.

Goldman, M. I. (1980) *The Enigma of Soviet Petroleum: Half-Full or Half-Empty?* (London: Allen & Unwin).

Grant, W. and Wilks, S. R. M. (1983) 'British industrial policy: structural change, policy inertia', *Journal of Public Policy*, vol. 3, no. 1, February, pp. 13–29.

Grayson, L. E. (1980) *National Oil Companies* (New York: Wiley).

Harding, R. W., ed. (1963) *Natural Gas Distribution* (University Park, Penn.: The Pennsylvania State University).

Harlow, C. (1977) *Innovation and Productivity under Nationalization: the First 30 Years* (London: Allen & Unwin for PEP).

Hervieu, P. (1969) *Le Gaz natural en Europe Occidentale et les problèmes de sécurité d'approvisionnement* (Paris: Diss).

International Energy Agency (1981a) *Energy Balances of the OECD Countries* (Paris: OECD/IEA).

International Energy Agency (1981b) *Energy Policies and Programmes of IEA Countries: 1980 Review* (Paris: IEA/OECD).

International Energy Agency (1981c) *OECD Energy Statistics. 1975–1979* (Paris: IEA/OECD).

International Energy Agency (1982) *Natural Gas: Prospects to 2000* (Paris: OECD/IEA).

Jensen Associates (1981) 'The demand for Alaskan natural gas' (July), in *Hearings of the Committee on Energy and Natural Resources U.S. Senate: the President's Alaska Natural Gas Transportation Act Wawer Recommendation* (22, 23, 26 October 1981, Appendix E, p. 1082).

Jensen, R. G. (ed.) (1979) *Soviet Energy Policy and the Hydrocarbons* (Washington, DC: AAG Project on Soviet Natural Resources in the World Economy, February).

Jones, F. B. (1963) 'Organization and intra-company coordination', in R. W. Harding (ed.) *Natural Gas Distribution* (University Park, Penn.: The Pennsylvania State University).

Katsenelinboigen, A. (1978) *Studies on Soviet Economic Planning* (White Plains NY: M. E. Sharpe).

Kelly, W. J. (1978) 'Effects of the Soviet price reform of 1967 on energy consumption', *Soviet Studies*, vol. 30, no. 83, pp. 394–402.

Khan, A. (1982), 'Natural gas in the '80's', *Wärme Gas International*, vol. 31, no. 5, pp. 210–15.

Kitch, E. W. (1968), 'Regulation of the field market for natural gas by the

Federal Power Commission', *Journal of Law and Economics*, vol. 11, October, pp. 243–80.

Küster, G. H. (1974) 'Germany', in R. Vernon (ed.) *Big Business and the State* (London: Macmillan).

MacAvoy, P. W. (1962) *Price Formation in Natural Gas Fields. A Study of Competition, Monopsony and Regulation* (New Haven, Conn.: Yale University Press).

MacAvoy, P. W. and Noll, R. (1973) 'Relative prices on regulated transactions of the natural gas pipelines', *Bell Journal of Economics and Management Science*, vol. 4, Spring, pp. 217–34.

MacAvoy, P. W. and Pindyck, R. S. (1975) *The Economics of the Natural Gas Shortage* (Amsterdam: Elsevier).

McDonald, J. (1977) *The Game of Business* (New York: Anchor).

de Man, R. (1982) 'Energy policy in the Netherlands: institutional structure and policy process' (London), paper presented to Dubrovnic Conference on Societal Problems of the Energy Transition, 13–17 September.

Manners, G. (1968) *The Geography of Energy* (London: Hutchinson University Library).

Medici, M. (1974) *The Natural Gas Industry* (London: Newnes–Butterworths).

Meyerhoff, A. A. (1981) *The Oil and Gas Potential of the Soviet Far East* (Beaconsfield, Bucks.: Applied Science Press).

Miller, J. T. (1970) *Foreign Trade in Gas and Electricity in North America: A Legal and Historical Study* (New York: Praeger).

Minchin, L. T. (1966) *The Gas Industry, Today and Tomorrow* (Toronto: n.p.).

Mitchell, E. J. (1979) *Oil Pipelines and Public Policy: Analysis of Proposals for Industrial Reform and Reorganization* (Washington, DC: American Enterprise Institute for Public Policy Research).

Netherlands Ministry of Economic Affairs (1981) *Natural Gas and Oil of the Netherlands: Annual Review 1981* (The Hague, April).

Neuner, E. J. (1960) *The Natural Gas Industry: Monopoly and Competition in Field Markets* (Norman, Okla.: University of Oklahoma Press).

N.V. Nederlandse Gasunie (1979) *1980 Gas Marketing Plan* (Groningen).

Odell, P. R. (1969) *Natural Gas in Western Europe: A Case Study in the Economic Geography of Energy Resources* (Haarlem: De Erven F. Bohn N.V.).

Odell, P. R. (1979) *Oil and World Power: A geographical interpretation* (Harmondsworth, Middx.: Penguin).

Office of Oil and Gas, Energy Information Administration, US Department of Energy (1981) *The Current State of the Natural Gas Market* (Washington DC: USGPO, DOE/EIA-0313, December).

Office of Technology Assessment, Congress of the United States (1980) *Alternative Energy Futures I. The Future of Liquefied Natural Gas Imports* (Washington, DC: USGPO).

Organization for Economic Cooperation and Development, Special Committee for Oil (1969) *Pipelines in the United States and Europe and their Legal and Regulatory Aspects* (Paris: OECD).

Ouwehand, F. (1981) 'The role of natural gas in Western Europe in the next 20 years', paper presented at the European Gas Conference in Oslo, May.

Peebles, M. W. H. (1980) *Evolution of the Natural Gas Industry* (London, Macmillan).

Pegrum, D. F. (1965) *Public Regulation of Business* (Homewood, Ill.: Irwin).

Penrose, E. T. (1968) *The Large International Firm in Developing Countries: The International Petroleum Industry* (London, Allen & Unwin).

Polanyi, G. (1967) *What Price North Sea Gas?* (London: PEP).

Prewitt, L. A. (1942) 'The operation and regulation of crude oil pipelines', *Quarterly Journal of Economics*, vol. 106 (February), pp. 177–211.

Price Commission (1979) *British Gas Corporation – Gas Prices and Allied Charges* (London: HMSO).

Prindle, D. F. (1981) *Petroleum Politics and the Texas Railroad Commission* (Austin, Texas: University of Texas Press).

Prodi, R. (1974) 'Italy', in R. Vernon (ed.) *Big Business and the State* (London: Macmillan).

Pryke, R. (1981) *The Nationalized Industries: Policies and Performance since 1968* (Oxford: Martin Robertson).

ReVelle, C. S., Rosing, K. E., and Rosing-Vogelaar, H. (1977) 'Planning of oil pipeline systems at sea: an application of a branch-bound p-median algorithm', *Series A Working Papers* (Rotterdam: Economic Geography Institute, Erasmus University, August).

Ridgeway, J. (1973) *The Last Play* (New York: Dutton).

Riker, W. M. and Ordeshook, P. C. (1973) *An Introduction to Positive Political Theory* (Englewood Cliffs, NJ: Prentice Hall).

Russell, J. (1976) *Energy as a Factor in Soviet Foreign Policy* (New York: Lexington Books for the Royal Institute for International Affairs).

Sanders, M. E. (1981) *The Regulation of Natural Gas: Policy and Politics 1938–1978* (Philadelphia, Pa.: Temple University Press).

Scholtern, I. (1978) 'Some notes on corporatist integration in the Netherlands', paper presented to European Consortium for Political Research, Grenoble, 6–9 April.

Segal, J. (1980) 'LNG market. Slower growth for the 1980s', *Petroleum Economist*, December, pp. 513–16.

Segal, J. and Niering, F. (1980) 'Special report on world natural gas pricing', *Petroleum Economist*, September, pp. 373–9.

Select Committee on Nationalized Industries, House of Commons (1961), *Report from the Select Committee on Nationalized Industries: The Gas Industry. I. Report and Proceedings* (London: HMSO).

Select Committee on Nationalized Industries, House of Commons (1968) *The Exploitation of North Sea Gas* (London: HMSO).

Select Committee on Nationalized Industries, House of Commons (1975) *Nationalized Industries and the Exploitation of North Sea Oil and Gas. First Report from the Select Committee on Nationalized Industries Report, together with Minutes of the Proceedings of the Committee, Minutes of Evidence and Appendices. Session 1974–1975* (London: HMSO).

Shell International Gas Ltd (1969) *Natural Gas Terms and Measurements* (London).

Smith, H. (1977) *The Russians* (London: Sphere Books).

Société Nationale Elf-Aquitaine, (1977–82) *Annual Reports* (Paris).

Spavins, T. C. (1979) 'The regulation of oil pipelines', in E. J. Mitchell (ed.) *Oil Pipelines and Public Policy* (Washington, DC: American Enterprises Institute, pp. 77–106).

Stern J. P. (1979) *Soviet Natural Gas in the World Economy* (Washington, DC: Association of American Geographers).

Stern, J. P. (1981) *Soviet Natural Gas Development to 1990: The Implications for CMEA and the West* (Lexington, Mass.: Lexington Books).

Stern, J. P. (1982) *East European Energy and East–West Energy Trade* (London: Royal Institute of International Affairs).

Subcommittee on Energy and Power, US House of Representatives (1980) *The Energy Factbook* (Washington, DC: USGPO, November).

Tiratsoo, E. N. (1979) *Natural Gas* (Beaconsfield, Bucks.: Scientific Press).

Tretyakova, A. and Birman, I. (1976) 'Input–output analysis in the USSR' *Soviet Studies*, vol. 28, no. 2, pp. 157–86.

Turner, L. (1978) *Oil Companies in the International System* (London, Allen & Unwin for the Royal Institute of International Affairs).

United Nations (1967–81) *Commodity Trade Statistics* (New York: UN).

United Nations Centre on Transnational Corporations (1980) *Alternative Arrangements for Petroleum Development* (New York: UN).

US Federal Energy Regulatory Commission (1982) *Annual Report 1981* (Washington, DC: USGPO).

US Federal Energy Regulatory Commission (1977a) *Initial Decision Upon Applications to Import LNG from Algeria* (Washington, DC: USGPO, 25 October).

US Federal Energy Regulatory Commission (1977b) *Initial Decision Authorizing Importation of Liquefied Natural Gas from Algeria, Construction and Operations of Facilities and Sales of LNG* (Washington, DC: USGPO, 18 November).

US Federal Power Commission (1972a), *Opinion No. 613* (Washington, DC: USGPO, 9 March).

US Federal Power Commission (1972b) *Opinion No. 622* (Washington, DC: USGPO, 28 June).

Valais, M. and Durand, M. (1975) *L'Industrie du gaz dans le monde* (Paris: Editions Technic, May).

Vernon, R. (ed.) (1974) *Big Business and the State* (London: Macmillan).

Waverman, L. (1973) *Natural Gas and National Policy* (Toronto: University of Toronto Press).

Wei, J., Russell, T. W. F. and Swartzlander, M. W. (1979) *The Structure of the Chemical Processing Industries* (New York: McGraw Hill).

Wellisz, S. H. (1963) 'Regulation of the natural gas pipeline companies: an economic analysis', *Journal of Political Economy*, vol. 71, February, pp. 30–43.

Wilks, S. R. M. (1982) 'UK industrial policy: instruments and the state',

paper presented to the European Consortium for Political Research, Århus, Denmark.
Wilson, J. Q. (ed.) (1980) *The Politics of Regulation* (New York: Basic Books).
Woodcliffe, J. C. (1975) 'North Sea oil and gas: The European Community connection', *Common Market Law Review*, vol. 12, pp. 7–26.

Journals and newspapers
Gas 2000
Gas World
Noroil
Oil and Gas Journal
The Oilman Newsletter
Petroleum Times
Wärme Gas International
World Gas Report

Index

GPSR Compliance

The European Union's (EU) General Product Safety Regulation (GPSR) is a set of rules that requires consumer products to be safe and our obligations to ensure this.

If you have any concerns about our products, you can contact us on ProductSafety@springernature.com

In case Publisher is established outside the EU, the EU authorized representative is:

Springer Nature Customer Service Center GmbH
Europaplatz 3
69115 Heidelberg, Germany

Batch number: 10761599

Printed by Printforce, the Netherlands